Human Population Genetics

A Centennial Tribute to J. B. S. Haldane

Human Population Genetics

A Centennial Tribute to J. B. S. Haldane

Edited by

Partha P. Majumder

Indian Statistical Institute
Calcutta, India

Plenum Press ● New York and London

Library of Congress Cataloging-in-Publication Data

Human population genetics : a centennial tribute to J.B.S. Haldane /
  edited by Partha P. Majumder.
      p.   cm.
    "Proceedings of an international conference on human genetics,
  held December 15-19, 1992, in Calcutta, India"--t.p. verso.
    Includes bibliographical references and index.
    ISBN 0-306-44572-7
    1. Human genetics--Congresses. 2. Human population genetics-
  -Congresses. 3. Medical genetics--Congresses. 4. Haldane, J. B. S.
  (John Burdon Sanderson), 1892-1964--Congresses.   I. Majumder,
  Partha P.
  QH431.M5387 1993
  573.2'1--dc20   93-32262
                                                          CIP

Proceedings of an International Conference on Human Genetics, held December 15–19, 1992,
in Calcutta, India

ISBN 0-306-44572-7

©1993 Plenum Press, New York
A Division of Plenum Publishing Corporation
233 Spring Street, New York, N.Y. 10013

Printed in the United States of America

PREFACE

J.B.S. Haldane, R.A. Fisher and Sewall Wright simultaneously, and largely independently, laid the foundations of population genetics and the mathematical theory of evolution. Haldane was born on November 5, 1892. Although he primarily worked at the University College London (UCL), in 1957 he resigned from the UCL and joined the Indian Statistical Institute, Calcutta (India) as a Research Professor. In celebration of his birth centenary, the Indian Statistical Institute organized an *International Conference on Human Genetics* from 15 to 19 December, 1992. The prime motive in holding this Conference was to bring together a group of scientists — geneticists, anthropologists, clinicians and statisticians — to evaluate the impact of Haldane's contributions to various areas of human genetics, and also to review recent developments in the subject. Session and lecture themes were so chosen that they covered areas theoretical and applied, classical and emerging. Speakers were then identified and invited to deliver lectures on these themes. Manuscripts of all invited presentations and a selected number of contributed presentations were considered for inclusion in this Proceedings Volume. Each manuscript was reviewed by at least one Conference participant, which resulted in revision of several manuscripts and rejection of some. This volume is a collection of the manuscripts which have been 'accepted' after the review-process.

The Conference began with the "J.B.S. Haldane Centenary Lecture" delivered by C.R. Rao. A condensed version of this Centenary Lecture, mainly emphasizing, with examples, the 'scientific method' followed and advocated by Haldane, is the first article of this volume.

The technical papers have been grouped in five sections. Haldane's most profound contributions have been in the area of Population Genetics and Evolution, which is the theme of the first section. Although Haldane, unlike Fisher and Wright, did not propose any evolutionary paradigm, he proposed many important concepts which paved the way for a better understanding of the process of evolution and also for the formulation of new evolutionary theories (e.g., the neutral theory). However, not all the concepts proposed by Haldane are above criticism; the article by Ewens provides some examples. The article by Takahata, on the other hand, is a testimony of the profound impact of some of Haldane's outstanding observations and remarks on the processes of evolution.

In his Croonian Lecture delivered on November 7, 1946 (later published in 1948 in *Proc. Roy. Soc., Lond.*, Ser. B, 135: 142), Haldane defined 'formal genetics of man' as "the study of heredity and variation, based on description and counting of individuals ... enumeration of the results obtained from various matings; and deductions drawn from such enumeration ..." He further stated that "the first step in formal genetics is to establish that certain characters are inherited in accordance with Mendel's laws, and in particular that segregation occurs in Mendelian ratios ... The final aim, perhaps asymptotic, should be the enumeration and location of all the genes found in normal human beings." Haldane contributed greatly to the development of formal genetics. He was the first to derive the maximum likelihood estimate of the segregation ratio for a recessive character. His formulation of ascertainment-bias correction for non-randomly sampled nuclear families continues to be adopted. He was the first to detect X-linkage in man. His mapping function is still widely used. 'Formal genetics of man' is the theme of section 2, in which both segregation and linkage analyses are discussed.

The genetic structure and composition of a population is, to a considerable extent, affected by its mating structure. Haldane mathematically analyzed the consequences of inbreeding and other systems of mating on gene and genotype frequencies in populations. After he migrated to India, he was fascinated by the richness of genetic diversity of human populations in India. He became deeply interested in the study of human population structure and initiated several projects, which were carried out primarily by his students. He even convinced the Government of India to include the enumeration of consanguineous marriages as a part of the Indian decennial census operations. This was undertaken during the 1961 Census of India. Thus, even though Haldane did not make any significant contribution to the study of human population structure, as has been pointed out by Roberts in this volume, his interest in the topic warranted a separate session in the Conference. 'Genetic Structure and Diversity of Human Populations' is, therefore, the theme of section 3.

While classical human genetics considered purely Mendelian traits/disorders, and, therefore, paid little attention to environmental effects, the realization of the need for consideration of the joint effects of genes and environmental factors in the etiologies of disorders gave birth to 'genetic epidemiology'. There have been many significant developments in this area, which have been reviewed in the papers included in section 4. It must, however, be pointed out that some of the papers included in section 2, could also, legitimately, be included in this section.

While physical environment is known to act jointly with genes in determining the etiologies of many diseases, the epidemiological profiles of diseases in defined populations can also be influenced by the 'social environment' (e.g., social customs governing mate selection). Some genetic epidemiologists and anthropologists considered these issues — impacts of physical and social environments on disease profiles — in a panel discussion. 'Genes, Environment and Disease' is the theme of section 5, which opens with a paper by Garruto and Plato, followed by a report of the panel discussion.

J.B.S. Haldane was one of the most socially-conscious scientists of this century. Throughout his life he was acutely concerned about the social implications of genetical findings and theories. This volume, therefore, concludes with an article by James V. Neel entitled 'How would Haldane have viewed the societal implications of today's genetic knowledge?', which was a Special Lecture in the Conference.

I wish to take this opportunity of thanking my colleagues in the Indian Statistical Institute, the participants of the Conference, and various national and international organizations for intellectual and logistical support.

Calcutta, India   Partha P. Majumder
March, 1993

CONTENTS

III. GENETIC STRUCTURE AND DIVERSITY OF HUMAN POPULATIONS

IV. GENETIC EPIDEMIOLOGY

V. GENES, ENVIRONMENT, AND DISEASE

SPECIAL LECTURE

J.B.S. HALDANE CENTENARY LECTURE

J.B.S. HALDANE: A POLYMATH IN THE INDIAN CONTEXT

C.R. Rao

Department of Statistics
The Pennsylvania State University
University Park, PA 16802

A PASSAGE TO INDIA

I have great pleasure in participating in the birth centenary celebrations of Professor J.B.S. Haldane for many reasons. First, Haldane was a great scientist with a very few parallels in the depth of his scholarship, intellectual versatility and prodigious memory. Haldane's fundamental contributions covered a variety of fields — genetics, statistics, biochemistry, and the social applications as well as popularization of science. Second, he was a foreign scientist who had a genuine love for India. He was known to have championed the cause of India in Britain at the time of the British rule. He came to India a few years before his retirement from the University College London against the advice of his colleagues, and took Indian citizenship. He saw in India natural resources and, perhaps, a good deal of human talent which could be utilized to advance natural knowledge. Third, he placed India on the genetic map of the world by starting research on human genetics and motivating and inspiring a number of Indian scholars to do research in India. Before Haldane's arrival in India, there were only two centers of genetical research in India, one in Bombay and the other at the Indian Statistical Institute (ISI), Calcutta. The research at these centers were mainly confined to anthropometric and blood-group surveys. Haldane initiated research in many other directions utilizing local resources and minimum equipment. He advocated pursuing research that was inexpensive, and, if one prefers, non-violent in nature, and showed how such research can yield results of far-reaching importance. Finally, Haldane was a faculty member of the Research and Training School of the ISI, of which I was the Director. I share the credit along with Professor P.C. Mahalanobis of attracting such an intellectual giant as Professor Haldane to the ISI. Perhaps it is appropriate to mention an anecdote connected with Haldane's employment at the ISI. Professor Mahalanobis and I decided that Haldane should have the designation of a Research Professor and that he should be given a high salary consistent with his international reputation and outstanding contributions. The amount we decided was substantially higher than my own salary as the Director of the Research and Training School. When Haldane learnt about this, he protested stating that as an employee

Human Population Genetics, Edited by P.P. Majumder
Plenum Press, New York, 1993

of the Research and Training School, he would not accept a salary higher than that of the Director. I tried to persuade my boss, Professor Mahalanobis, to increase my salary, but he was unable to do so. Haldane's salary was then fixed at the same level as mine.

While the acquisition of Haldane as a regular staff member was a matter of great prestige to the ISI, I was somewhat nervous as I had heard that Haldane was easily irritable and difficult to deal with. To my surprise, I found him to be very cooperative, helpful and deeply absorbed in his work. He was frugal in his life-style, but exhibited generosity in supporting the work of his students. He saw science as an integral part of his life, and made no distinction between his personal and professional lives.

At the time Haldane joined the ISI, we were trying to introduce the Bachelor of Statistics (B. Stat.) and Master of Statistics (M. Stat.) degree courses in the ISI. One of the problems that engaged our attention was the preparation of the curriculum of courses for the students who were admitted to the ISI after completing their high school. Mahalanobis and I sought Haldane's advice to develop inter-disciplinary curricula to emphasize the role of statistics in scientific investigations. Haldane took great interest in our project. He suggested a set of ten practical exercises of an inter-disciplinary nature for the B. Stat. students. I will mention one of them as an example. "After previous training in surveying and determination of elasticity, a tree will be tied by a rope to a neighboring building. The deformation of the tree will be studied with a theodolite; the observations will be repeated under high wind. This experiment may be regarded as an exercise in several subjects — surveying or applied trigonometry, statistics, quantitative biology and meteorology." After the B. Stat. course was started in the ISI, Haldane himself participated in teaching. He taught most of the science courses spending six hours a week.

RESEARCH WORK IN INDIA

Haldane's contributions to genetics and his efforts to popularize science are well known. His own initiation into scientific research seems to have begun at the age of eight when he started assisting his father in scientific experiments. Although he had no formal degree or training in science, he started publishing highly original and fundamental papers in several areas of science. He believed that formal education in a subject where you are forced to learn what is taught and answer typical questions set in class examinations inhibits original thinking in the same area, but can provide ideas for creative work in other areas.

Haldane was a pioneer in using mathematical and statistical methods in genetic research. He was the founder of mathematical genetics along with Sewall Wright and R.A. Fisher. He guided the research work of numerous students, some of whom won the Nobel Prize. I shall describe only the research work done in India under his guidance by his students. Each piece of research work initiated by Haldane is a perfect example of the scientific method: A question is raised after observing an unexplained phenomenon in nature, problems are formulated for investigation, data are generated through direct observation or a designed experiment, inferences are drawn through appropriate statistical analysis of data and new questions are raised for further investigation. Haldane's work in India also demonstrated that research of far reaching importance can be done without expensive equipment, and that nature is an open book of problems accessible to any inquisitive mind.

When Haldane started working at the ISI in 1957, he had one research associate, S.K. Roy. Later he recruited three others — T.A. Davis, K.R. Dronamraju and S.D. Jayakar. Mrs. Haldane was also in Haldane's research team. In 1961, Haldane moved to Bhubaneswar where he established his own Biometric and Genetic Laboratory with support from the Orissa State government. Dronamraju and Jayakar moved with the Haldanes to Bhubaneswar. Two others, A.K. Ray and P. Meera Khan joined the group. Haldane continued to work in Bhubaneswar until his unfortunate death in 1964. I shall give a brief description of the various projects directed by Haldane during the seven year period 1957 - '64.

Distribution of Petal Number of Flowers

Soon after his arrival in India, Haldane, while walking around the garden in ISI one day, noticed that flowers on the same plant had different numbers of petals. This aroused his curiosity and he asked S.K. Roy to study the frequency distribution of the petal number of flowers and provide a possible explanation for its variability. S.K.Roy used to collect everyday the flowers of *Jasminum multiforum* that had fallen to the ground and sort them by the number of petals. The findings were as follows. The petal number varied from 8 to 12 and the frequency distribution was rather skewed. There were outliers, the so-called double flowers with a very large number of petals. An interesting feature of this research conducted over the whole flowering season is the unexpected trends in the mean and variance of the petal number over time. The mean number of petals increased in the beginning, then showed a decreasing trend for a while and then increased again towards the end of the flowering season. The variance in the petal number was small in the beginning but increased towards the end of the flowering season. Haldane explained that "towards the end of the flowering season, the plant gets tired and starts producing petals almost randomly, so that the variation increases." What is interesting about this simple investigation is the number of new questions it raised on the behavior of the plants. What causes the variability in the petal number? Why do the mean and variance of petal number change over time? These questions have not been answered up to now.

Experiments With Crop Mixtures

Haldane was always interested in the interaction between species. He encouraged Roy to conduct experiments on mixed cropping. Roy chose two varieties of rice called BK and CA and used an experimental design to compare the yields per unit area of each of BK and CA when sown on separate plots and on the same plot in some kind of mixture, for instance using each variety in alternate rows. An analysis of the experimental data showed that the yield per unit area for each variety of rice was more in the plots where they were mixed than in the plots where they were sown separately indicating a symbiotic effect. The experiments were repeated under different environmental conditions and a variety of mixing plans. The overall conclusion remained the same. Haldane suggested that the results of these experiments could be exploited to increase food production in India.

Hypertrichosis of the Pinnae

A small percentage of men in a population have a hairy growth on the rim of their ears, a trait known as the *hypertrichosis of the pinnae*. The genetics of this trait was already studied in Europe but its exact mode of inheritance was not known. Haldane noticed that Dronamraju, who joined the ISI to work under him, had this trait. He immediately despatched Dronamraju to Andhra Pradesh to collect data on this particular trait in his family. Fortunately, Dronamraju could contact many of his family members and collect data on a large pedigree. A simple tabulation of these results showed that in every case a father had hairy pinnea, only his adult sons but none of his grand sons through his daughters exhibited the trait. This suggested that the gene causing the hairy pinnae is Y-linked. Later elaborate studies corroborated this finding.

Preference to Colors in Butterflies

While watching the visits of butterflies of different species to flowering plants for nectar, Dronamraju observed an interesting phenomenon. A particular species of butterflies predominantly preferred orange colored flowers, another species pink, and yet another species was indifferent to flower color and visited flowers of different colors at approximately the same frequency. Were these color preferences a learnt characteristic or were these butterflies born with an innate preference for a certain color? Dronamraju and Mrs. Haldane conducted careful experiments to answer these questions. The results of these studies showed that color preference is an innate character. This is an important finding from the point of view of evolutionary biology.

Heterostyly in Flowers

Another investigation by Dronamraju was concerned with the variation in style length of flowers from several bushes of *Bauhinia acuminata* L., that were growing in the campus of the ISI. In the classical type of heterostyly studied by Darwin and others, the flowers with long, short or medium length styles occur on different plants. However, in *Bauhinia*, Dronamraju's measurements over several months showed that both long and short styled flowers occurred throughout the flowering season on the same plant. The proportion of short styled flowers on five plants ranged from 6% - 24%. Since the short styled flowers were female sterile, it was concluded that the variation may possibly represent an intermediate step in the evolution of classical heterostyly studied by Charles Darwin in *Lythrum* and *Primula*.

Studies on Consanguinity

Under the guidance of Haldane, Dronamraju conducted studies on consanguinity in Andhra Pradesh. He found the total inbreeding coefficient to be 0.02, which was much higher than that found in any large open population. To study the effects of inbreeding, data were collected on consanguineous marriages in hospital patients, their parents and their children. Effects of inbreeding on certain diseases, especially pulmonary tuberculosis, were noted.

Right-Left Directional Phenomenon[1]

Soon after his arrival in Calcutta, Haldane recruited T.A. Davis, a coconut research scientist, to work with him. Davis was a devoted, unassuming and energetic scientist with a flair for fieldwork. Among the many projects he undertook at the ISI, the most interesting are his studies on right-left directional phenomenon in nature.

The flowers on the same plant can be of two kinds depending on the way their petals overlap one over the other forming a right or left spiral. We may call them right and left flowers. Davis collected data on different kinds of flowers and found that the ratio of right to left flowers is 1:1, indicating a random phenomenon.

There are creepers of the same species, some of which grow spiralling to the right and some to the left. Davis conducted experiments with cow pea (a creeper) in the ISI garden to see whether he could force them to change their ways. The plants resisted and kept on growing only one way, the way they started from the seedling.

Some coconut trees have right foliar spirality and others left. Davis conducted experiments to show that foliar spirality is not an inherited characteristic. He collected data from all over the world and discovered that the proportion of coconut trees with left foliar spirality is more than half in the northern hemisphere and less than half in the southern hemisphere. This phenomenon is similar to the bathtub vortex which is probably caused by rotation of the earth. A more interesting finding by Davis was that the yield of coconuts is 10% to 12% more on the left trees than on the right trees. This result is of great economic importance as yield of coconuts can be substantially increased by selective planting of trees with left foliar spirality, which can be detected at the seedling stage.

---

[1]for a detailed account of this research work, the reader is referred to the book "Statistics and Truth" by the author

Mathematical Models of Evolution and Biometry

S.D. Jayakar joined the Haldane group in 1960. He had a strong background in mathematics and statistics having received the Master of Statistics degree from the ISI. Haldane considered Jayakar to be "one of the stars which would be seen rising in the Indian scientific firmament". Unfortunately Jayakar died at a very young age.

Jayakar and Haldane worked on temporal variation of natural selection in a two-allele diploid locus model with discrete generations. In this study they introduced a major theme of population genetics the importance of which became widely recognized later in connection with the debate over the possible causes for the surfeit of polymorphism observed in natural populations. Haldane and Jayakar also ascertained that a mutant allele can invade a monomorphic population if the fitness it conveys has, over generations, a geometric mean which exceeds that of the prevailing type. This remains a fundamental result which has applications in other areas of evolutionary biology, besides population genetics, such as in the ecological theories of optimal life strategies under fluctuating environments.

Another important contribution made by Jayakar and Haldane is the enumeration of some human relationships. What do we mean by saying that two individuals A and B are first cousins? Jayakar and Haldane showed, by considering three generations above, that this may mean any one of 48 relationships if the sexes are not specified, and any one of 192 if the sexes are specified. By constructing suitable coefficients of relationship they were able to provide answers to questions like: What correlation may be expected between the phenotypic characters of two biologically related individuals A and B as a result of their relationship? If A and B are of different sexes and have children, how many of these children are expected to differ from the general population?

Jayakar continued his research on genetic polymorphism started in collaboration with Haldane and made significant contributions.

An Anomaly of the Fourth Toe

After Haldane moved to Bhubaneswar, several projects were started in the field of human genetics. At the suggestion of Haldane, A.K. Ray conducted a survey to study foot and toe anomalies in human beings. The survey revealed that the defect known as the short fourth toe (due to an abnormally short metatarsal) is due to an autosomal dominant gene.

I have taken some time to describe to you some of the research projects initiated and directed by Haldane during a short period of seven years to emphasize that Haldane believed that research of far reaching importance in genetics can be done without expensive equipment.

THE HALDANE TRADITION

Haldane was a great scientist, scholar and philosopher. His mission in life was to expand the frontiers of knowledge and spread scientific knowledge among the public. He devoted all his time in doing research and in encouraging and inspiring others to do research. He wrote popular books on science and on all kinds of contemporary topics ranging from religion to politics to educate the public. He considered knowledge of human genetics as important to all individuals. He said:

> "It does not matter if many people know nothing about hybrid maize or progeny testing for milk yield. It does matter a great deal if many people know nothing about human genetics, because it is a topic of interest to all human beings and gaps in their knowledge will be filled by superstition or intellectually dishonest propaganda."

5

Haldane received accolades from many important people. A few examples are:

"A sage of science."

Obituary of Haldane in New York Times (page 1)

"One with a broad understanding of how the world has got the way it is."

Sir Oliver Goonetillake, President of Sri Lanka

"I have never met a man with a better conversation or with a more varied knowledge than J.B.S. Haldane."

Norbert Wiener

Haldane may be regarded in the history of science as a man of many contradictions. He was a renaissance scholar who had no formal training in science, yet made important contributions to several areas of science. He was a bombing officer in World War I who later embraced non-violence and Hinduism. I recall in this connection an incident narrated by his sister. One day she saw a gadfly sitting on Haldane's hand and stinging him. She took a fly-flap to kill it. Haldane said "No, let her take some blood. She needs blood to lay eggs." He was a classicist who was also a highly skilled popularizer. He was a Marxist and a socialist who nevertheless admired the elitist schools and the Indian caste system. I have read somewhere, though I can't remember the source, that he adopted Marxism to cure his stomach ulcers. He was an apologist for vivisection in his youth, but could discuss pain in the insect world in his later years. While such contradictions give the true picture of Haldane, he remained a great scientist throughout these transformations. Finally, Haldane has left us a legacy of ideas and traditions in research characterized by loftiness in objectives, simplicity in tools, meticulousness in observation and rigorousness in analysis. His very presence was a stimulus to the scientific programs of ISI, which received worldwide attention. There is a great need to live up to the ideals and standards of scientific research that were initiated by Haldane in India. The only way we can pay a tribute to his memory is to emulate his example.

I would like to thank Dr. K.R. Dronamraju for sending me his writings on Professor Haldane, which I have used in preparing this talk. I also wish to thank Dr. Partha Majumder for asking me to deliver this Centenary Lecture. I am glad to see the large number of geneticists who have gathered here to discuss the current trends of research in genetics. I wish the conference a great success.

BEANBAG GENETICS AND AFTER

W.J. Ewens

Department of Biology
University of Pennsylvania
Philadelphia, PA 19104

INTRODUCTION

This Session of the Haldane Centenary Conference on Human Genetics is devoted to evolutionary population genetics. This is more than appropriate, recalling Haldane's outstanding contributions to the evolutionary population genetics field. Indeed, many have claimed that his most substantial contribution in all of science, both in genetics and elsewhere, is his theoretical evolutionary work: for example, Crow (1992) in his "Centennial: J.B.S. Haldane", claims that Haldane's best-known work is his mathematical theory of evolution, as given in his sequence of papers in the 1920's on that subject. This lecture is designed to be provocative and I will later dispute this claim, and further be critical of several key aspects of Haldane's evolutionary genetics activities.

The title of my talk, chosen for me by the conference organizers, is "Beanbag Genetics and After". This title suggests various themes. One is an assessment of Haldane's work in evolutionary genetics as a whole, since he was the main protagonist in the beanbag debate. The other theme, suggested by the "and After" part of the title, is an assessment of work done in theoretical population genetics since the time of the beanbag comment in the early 1960's. I will indeed attempt to address both these themes in due course. But it is appropriate to start with an assessment of the specific beanbag controversy: this will in any event lead me into general comments about Haldane's work and also post-1960's developments. I remind you again that, in accordance with what would surely have been Haldane's own preference, I will adopt a provocative approach to these various matters, and in particular to Haldane's part in the beanbag business.

Human Population Genetics, Edited by P.P. Majumder
Plenum Press, New York, 1993

BEANBAG GENETICS

When I was younger, and more inclined to believe without argument what my elders and betters told me, my understanding of the beanbag genetics episode was that Ernst Mayr made some ill-informed remarks about population genetics theory, and was subsequently demolished by Haldane in the role of a white knight riding to the rescue of his slandered subject. As I later read, and thought about, the respective writings of Mayr and Haldane, I found myself more or less entirely on Mayr's side: I came to the view that his comments were apt and appropriate, and centered on an understanding of the way evolution as a genetic process works, while Haldane not only evaded the issue as presented by Mayr by using the debater's trick of inventing a false issue and discussing that, but also did not present good arguments on the false issue he invoked. Let me now expand on, and justify, these comments.

The expression "beanbag genetics" first appears in the opening comments of Chapter 10 of Mayr's (1963) classic "Animal Species and Evolution", and refers explicitly to his closing comments in Chapter 9, in which he discussed the then-new concept of the evolutionary load. It is important to realize this fact, which I have never seen noted before. Any discussion of the real "beanbag genetics" comment, as opposed to the false issue later introduced by Haldane, should properly be made in the actual context raised by Mayr, namely the calculations surrounding the "evolutionary load", or as it is sometimes called the "cost of natural selection".

The two Haldane papers introducing this concept, and carrying out some preliminary calculations, are "The cost of natural selection" (1957) and "More precise expressions for the cost of natural selection" (1961). Let me put my cards on the table at once: I regard these as two of the most unfortunate papers ever written. My concern is with the implications drawn, not only by Haldane but also by various subsequent authors, from the calculations in these two papers.

To justify this position it is sufficient to focus on the first paper. The conclusion of that paper, arising from calculations I shall describe in a moment, is that "... the number of deaths needed to secure the substitution ... of one gene for another at a locus is ... often about 30 times the (population size) ... (so that) the mean time taken for each gene substitution is about 300 generations". The crux of the argument concerns simultaneous substitutions at many loci and it is thus appropriate now to sketch the mathematical calculations leading to the above numbers and the subsequent extension to many loci.

It is convenient, and sufficient, to imagine a sequence of loci substituting a superior allele "A_1" for an inferior allele "A_2" at each locus. The calculations are easier, and the essential features not lost, if there is no dominance in fitness at each locus, so we henceforth make this assumption, and suppose that at a typical locus the fitnesses of the three genotypes are

$$A_1A_1 \qquad\qquad A_1A_2 \qquad\qquad A_2A_2$$
$$1 + s/2 \qquad\qquad 1 \qquad\qquad 1 - s/2 \qquad\qquad (1)$$

When the frequency of A_1 is x, the population mean fitness is $1 + s(x - \frac{1}{2})$, and if we define the genetic load $\ell(x)$ at this locus as the difference, in any generation, between the fitness of the most fit genotype and the current mean fitness, then given the frequency x we have

$$\ell(x) = s(1 - x) . \qquad\qquad (2)$$

The total substitutional load L is defined as the sum of the quantity $\ell(x)$ over all generations require to increase the frequency of A_1 from some value x_1 close to zero to some value x_2 close to unity, and is thus given by

$$L = \int \ell(x) \, dt$$

$$= \int_{x_1}^{x_2} [\ell(x) / \frac{dx}{dt}] \, dx$$

$$= 2 \int_{x_1}^{x_2} x^{-1} \, dx \qquad\qquad (3)$$

(using $dx/dt = \frac{1}{2}sx(1 - x)$), and this integration gives, for all practical purposes, $-2 \log x_1$. The value conventionally taken for x_1 is .0001, (corresponding to a mutation-selection balance frequency .0001 of A_1 before the substitution of A_1 begins), and this yields $L = 18.4$. It is known that the value of L increases somewhat if there is dominance in fitness values, and it is therefore generally accepted, following Haldane, that a "representative value" of 30 may be taken for the substitutional load. This agrees with the value given in the quotation earlier. Haldane then assumed, surely very conservatively, that a species could not cope with an excess reproductive requirement of more than 10%, and thus eventually the limit of 300 generations between the start of consecutive gene substitution processes is arrived at.

As noted earlier, the real essence of the load argument concerns simultaneous substitutions at many loci. Let us therefore suppose that a new substitution process begins, at a new locus, every n generations. The mean time T required for each substitution, with all the various assumptions made above, is

$$T = \int_{.0001}^{.9999} [\tfrac{1}{2}sx(1 - x)]^{-1}dx$$

$$= 36.8/s \quad \text{generations},$$

and thus the number T/n of substitutions in process at any one time is

$$36.8/sn \quad . \tag{4}$$

Now the frequencies of the favored allele "A_1" at these various loci will take, at any one time, a range of values symmetrically around .5 , and we may therefore assume, as a close approximation, that the population mean fitness is unity. The essence of the load argument is to focus attention on the offspring requirement of the most fit genotype, that is to say an individual who is "A_1A_1" at all loci substituting, and the calculation used for this is

$$(1 + s/2)^{36.8/sn} \simeq \exp(18.4/n) \quad .$$

Replacing, as above, the "no dominance" value 18.4 by the "representative value" 30, we get

$$\exp(30/n) \quad . \tag{5}$$

Putting $n = 300$, with Haldane, we recover the value $\exp .1 \simeq 1.1$, leading to the 10% excess reproductive requirement calculated by him.

What do these calculations really mean? It is perhaps best to discuss this question in the context of the neutral theory of evolution, for which these calculations formed the initial main theoretical support. Kimura and Ohta (1971) estimated that in the evolutionary history of mammals, about six gene substitutions began in every generation in any evolutionary line. This value would require us to put $n = 1/6$, and using this value in (5) leads to a substitutional load of $\exp(180)$, or about 10^{78} . This leads to the view of Kimura and Ohta that "to carry out ... substitution at (this) rate, each parent must leave 10^{78} offspring to survive. This was the main reason why random fixation of selectively neutral mutants was first proposed by one of us as the main factor in molecular evolution".

I believe that these arguments are quite misguided. This is a view expressed by many, and I apologize for reviving it here: my excuse is that the error in the argument can be traced back to the thrust of Mayr's beanbag genetics comment, and beanbag genetics is the title of my talk. Mayr argued that evolution concerns actual individuals who live, reproduce, pass on sometimes favorable and sometimes unfavorable genes, and then die, and that the real substitutional load should focus around real living individuals and not around abstract algebraic entities. Let us examine the substitutional load from this point of view.

As noted above, the entire argument centres around the offspring requirement of the most fit genotype, with "A_1A_1" at all loci substituting. What is the probability that an individual of this genotype actually exists? Here we must decide on a value of s, and we use the conventional value $s = .01$. With $n = 1/6$, we find from (4) that at any one time there will be, on average, 22,080 loci substituting. The frequency of the favored allele "A_1" at these loci will take a variety of values between zero and unity, and it is fairly straightforward calculation to show that any given individual is of the optimal genotype with probability $10^{-23,000}$. Clearly no such Napoleonic individual exists or will ever exist, and yet the load arguments used by Kimura and Ohta focus entirely around such an individual. What we should ask ourselves, with Mayr, is this: what are are the genotypes of those individuals we can realistically expect to find in the population? The fact is that most of us are rather ordinary, having a roughly equal assortment of good and bad genes at those loci where the frequencies of the two are approximately equal. None of us is simultaneously Mozart, Newton, da Vinci, Fisher, Haldane and Wright. It is clear by symmetry that the mean fitness of an individual taken at random in the above process is unity, and it is not difficult to see that the standard deviation in fitness is about .25. Thus the great bulk of individuals in the population have fitness within the range .5 to 1.5, and the statistics of extreme values shows that even in a population of size one million, the most fit individual would only have a fitness of about 2. There is no doubt that this is an achievable fitness for a most fit individual - indeed, in human populations, our main problem at the moment is that far too many individuals, and not merely the most fit individuals, are reproducing at this rate.

The essence of this argument is that one must focus on individuals, not beans in bags. Tedious but straightforward calculations show that substitutions at the required rate can occur if this new focus is used, that is if we carry out the entire substitution load argument in terms of the real individuals we can expect in a real population. To this extent, Mayr's beanbag comments are justified and timely.

There are many further arguments supporting this conclusion. The immense fitness values noted earlier arise from the supposition that fitnesses are multiplicative over loci. That is to say, a fitness array (such as (1) above) is constructed for each locus, and it is then assumed that the overall fitness of any individual is found by multiplying appropriate single locus fitness values. This is a reductionist approach to finding fitnesses of individuals: we first look at the separate parts (i.e., separate loci), and attempt to construct the total picture from these. But it is an individual who has a fitness, not genotypes at separate loci, and there is no guarantee that a reductionist approach to fitness construction is appropriate. Indeed I believe it is not, in that it is simply an artifact of the reductionist method (multiplying fitnesses over loci) that leads to the immense fitness values calculated. No genotype can have a fitness 10^{20}, let alone 10^{78}, so the multiplicative assumption must be incorrect. Sewall Wright pointed out as early as 1930 that "... with respect to such a character as fitness, it may safely be assumed that there are always important epistatic interactions" (Wright (1930)). Thus the immense genetic loads calculated to support the neutral theory are not real loads - they are artifacts of incorrect models.

Further complications arise through the fact that genes lie on chromosomes. Leaving aside for a moment the question of recombination, it is the whole chromosome which is passed on to an offspring, carrying possibly many favored genes. Of course recombination complicates this picture: nevertheless, an analysis of gene fixation processes which does not take into account the fact that many of the loci at which substitution is occurring are linked must be incomplete. Recall, for example, that in the numerical example considered above there are 22,080 loci substituting simultaneously: many of these will be extremely closely linked to others.

Many further factors, including the possibility of frequency-dependent selection and important ecological arguments concerning the reasons for selective differences between loci, must also be considered. There is no need to belabor the point: beanbag genetics has failed miserably as a theoretical support for providing limits to the rate at which evolution through natural selection can occur, and incidentally to the neutral theory. Mayr's insistence on considering real individuals, having individual fitnesses, is in my view clearly substantiated.

There are several further points of interest. First, we note that a genetic load refers to a <u>variance</u> in fitness, not a <u>mean</u>: substitutions can occur only if variation in fitnesses exists. Now if a certain allele has substituted for another, it can be argued that it does not matter whether this occurred by chance among selectively equivalent alleles (neutral substitution) or because of selective differences - there must have been the same variation in offspring number in the two cases. Thus, according to this argument, we cannot escape a genetic load by appealing to selective neutrality. This argument becomes extremely complicated when many loci are substituting through the existence of genes on chromosomes, a matter referred to above: many linked substitutions can occur together. We do not pursue this point here, and note again only the obvious fact that an allele will increase in frequency only because its carriers leave more offspring than others, or by chance it is passed on more frequently from heterozygote parents, and that so far as the former process is concerned, given that a certain substitution has occurred, it might not matter whether the variation in offspring numbers is neutral or selective.

A second point concerns the assumption made in load arguments that there exists a pool of low-frequency deleterious alleles (typically at a frequency of about 10^{-4} , as was assumed above and by Haldane), and that, perhaps due to a change in the environment, these become favored and increase in frequency. I have never really liked this concept. Certainly one allele can increase in frequency at the expense of another because of environmental changes - the *Biston betularia* example is a classic case in point. But can such a change be described as evolution, if by this term we mean improvement, for example in improving the structure of the eye? If nothing more is done than keep up with the environmental change, there may be no real improvement in a meaningful sense. We are interested mainly in evolutionary change as it relates to a real improvement in the organism, and for this there might not be a pool of low-frequency alleles - rather, we might have to await the arrival of a clearly beneficial new mutation which, once it occurs, can be expected (leaving aside problems of stochastic effects and random extinction) to increase in frequency essentially immediately. All of this is a standard argument, and I mention it here mainly to emphasize the point that a crucial ingredient in formal load calculations is the initial frequency of the substituting allele, and different values for this might be necessary depending on the appropriateness or otherwise of the above arguments.

A final comment on load "theory" concerns its impact on the empirical side of population genetics. Dobzhansky (1970), apparently influenced by this theory, stated (page 220) that "higher vertebrates and man do not possess enough 'load space' to maintain more than a few balanced polymorphisms" and concluded, as a result, (page 224) that selection favoring heterozygotes "cannot explain the polymorphisms observed in man". I believe that this is a totally incorrect conclusion. The neutral theory of evolution, discussed in more detail below, also relied for its original main theoretical support on load theory - again, in my view, an inappropriate use of theoretical population genetics.

Having discussed the original beanbag theme at length, I turn now to the false issue, or red herring, that Haldane raised in his beanbag paper. While the initial paragraph in "A defense of beanbag genetics" (Haldane, 1964) does indeed refer directly to the original beanbag comments, the first sentence in the second paragraph refers to an entirely different comment made by Mayr, at a different place and time, and the entire remainder of "defense" relates to this second matter. *Thus Haldane never really addressed the main issue of the beanbag argument.* Mayr's new comment was "What, precisely, has been the contribution of (the Fisher, Haldane and Wright) mathematical school to evolutionary theory?" This is indeed a valid question, one which we should continually be asking ourselves, and it is therefore important to discuss Haldane's response to this wider and more important enquiry. To continue with my provocative approach, I will argue that Haldane's defense to this second question, (let alone the original beanbag question), although spirited, is not completely convincing or successful.

Many matters are raised by him, and it is possible here to comment only on what appear to be the salient issues. Haldane first refers to the solutions of the differential, or difference, equations which form the substance of his papers in the 1920's describing the rate of change of gene frequency under natural selection. Here his comments are largely descriptive, and do not form part of his beanbag, or as it should more accurately be called pseudo-beanbag, defense. His real defense starts by taking up a remark of Hogben, that "mutation ... is ... the pacemaker of evolution". Haldane claimed that his calculations around 1930, and those of Wright, refute this claim, asserting that "only algebraic argument can be decisive in such a case". However, algebra is surely only a means of calculation, not the real essence of the argument. Any algebraic analysis must rely on a model of the real world, and on real-world parameter values, which can only come from observation. Furthermore, Haldane argues this point at what I believe to be the incorrect level. It is certainly true that, once a selective process starts, selective forces rather than mutation will determine the rate of change of gene frequencies. But there is a further level at which the argument must be made: a selective process must often await a favourable new mutation to arise in the first place, and it is plausible that often the time spent waiting for this to occur, depending as it must on the rate of mutation to favourable new alleles, is longer than the comparatively quick process of fixation of the favoured allele. It is no doubt for this reason that Wright himself, in referring to his shifting balance theory of evolution, which he felt would imply faster evolution than that arising under Fisher's paradigm, stated that "... (evolution) would, however, be very slow [under his theory], even in terms of geologic time, since it can be shown to be limited by mutation rate". One can therefore reasonably doubt the claim that Wright's calculations do support Haldane's view on the question of what is the the real pacemaker of evolution.

One example of the need to look beyond the algebra relates to the original "beanbag" focus on the evolutionary load. It will be recalled that Haldane's original load calculations assume an initial pool of low frequency deleterious alleles, and the load calculation depends crucially on what this initial frequency is. Now I have raised above the question of whether for "real" evolution, in the sense of "real" improvements in the organism, this is an appropriate concept, and suggested that we should perhaps focus on entirely new mutations when discussing "real" evolution. It is ironic that in calculating the rate of evolution under the neutral theory, whose main original theoretical support derived ultimately from Haldane's "pool of deleterious alleles" concept, Kimura does do precisely this. In assessing the rate at which new neutral mutations arise and spread through a population, he uses the calculation

$$2Nu \times (2N)^{-1} = u .$$

The first factor (2Nu) gives precisely the mean number of totally new mutations per generation - no concept of a pool of alleles here - and the second factor ($(2N)^{-1}$) gives the probability of fixation of any new mutant gene (not a completely accurate value, since it assumes no further mutation at the locus). The resulting value, u , which is the mutation rate, is then used centrally in Kimura's neutral theory arguments and calculations. We already have one irony, namely the use of the pool of mutant genes, and its non-use, in two different parts of the load argument, and now we have another: the mutation rate *has* become the "pacemaker of evolution", in contradiction to one of the claims of Haldane's beanbag defence paper. What indeed is Ernst Mayr to make of all this, and where indeed is there here a "decisive algebraic argument"?

In my view, the remaining matters raised by Haldane in his "defense" paper all concern one or other "local", or "tactical" question, rather than broad strategic issues. I myself would have hoped for a broader defense, on the strategic level, of mathematical population genetics, interesting and valuable though the local tactical questions might be. Some examples of the tactical matters he raises are as follows.

How can we estimate a mutation rate, in particular in humans, without examining millions of individuals and observing the proportion of new mutations? Now Haldane, in the 1920's, (along with Fisher and Wright) established the formula for the mutation-selection balance frequency of the mutant allele, (assumed to be selectively deteterious), and in the "defense" paper he discussed how this could be used to estimate mutation rates, as described in his 1932 book. However inspection of the book reveals only the most cursory calculations. The only case considered is that of a recessive deleterious allele, the mutant homozygote genotype having a selective disadvantage s compared to heterozygotes and the homozygous normal genotype. The frequency of the mutant homozygote genotype is then shown to be approximately u/s , which, when s is quite large (as for a semi-lethal), is of order u . A similar conclusion holds for a sex-linked trait such as haemophilia, where the frequency of haemophiliacs in the population is shown, by a similar calculation, again to be of order u . Haldane then drew the conclusion that u is of order 10^{-5} to 10^{-4} , presumably because this is approximately the frequency of haemophiliacs in the general population. But many difficulties are swept under the carpet. Has a reliable estimate of the frequency of haemophiliacs really been established? Would not a reliable estimate require many times more than u^{-1} individuals, a number similar to that needed for direct observational estimation? What is the (probably very large) standard error of the estimate? What are the (also probably very large) effects of non-stationarity? (The formula u/s assumes that a stationary state has been reached, possibly quite unlikely in humans, as Haldane himself states quite explicitly in his "defense" paper.)

Haldane goes on to state that similar calculations can be made in the "no-dominance" case, but offers no examples in his 1932 book. He also mentions improvements in the estimation procedure, citing two later papers as giving the most accurate estimation methods. But these two papers use no evolutionary population genetics theory whatever, describe only a proposed laboratory experimental scheme involving mice, not humans, and do not really take us into new territory. I detect a far from decisive analysis of the value of theoretical evolutionary population genetics in estimating mutation rates in Haldane's "defense" paper, and indeed in his work generally.

The next, in my view tactical rather than strategic, matter raised by Haldane concerned his classic calculations of the selective differences necessary to explain the rapid increase in the melanic form of *Biston betularia* last century. His calculated values, derived from his differential equations for changes in gene frequency through selection, were very high, being of order 35% or more, and were certainly larger than values of order 1% or less often assumed then (see the load calculations above), and even now, as being typical. There is no doubt that his estimation has been justified by later events. But, from the devil's advocate point of view, what is there here which is novel from the point of view of evolutionary theory? All that is shown is that the selective differences driving evolution can be quite substantial, and this is not under debate: indeed, the one concrete conclusion that can be drawn is that a specific and central assumption made in his load theory calculations, that a species cannot cope with an excess reproductive requirement of more than 10% for the most fit genotype is undermined: selective differences of 50% will certainly require a reproductive excess larger than 10%.

This comment bears directly on the next claim made by Haldane, that his "cost of natural selection" argument provides the main ingredient in determining a limit to the speed of evolution. My negative comments on this are given in detail above - here I merely remark on the contradiction between two consecutive claims made in the "defense" paper.

And so it goes on - there is no need to go through the arguments point by point. $_W$ I believe that the "defense" paper is disappointing, first because it does not address the nub of the beanbag comments, second because it focuses on less important tactical questions, rather than strategic, third that the comments made are sometimes misguided or even mutually contradictory, but most of all, for someone who would wish to see his own subject defended in the most effective way, because it does not give the best possible defense of theoretical population genetics.

It is important, then, to ask what further points could Haldane have raised to defend his, and our, subject. First and foremost, I believe, he should have made the point, stressed repeatedly by Fisher (1930, 1958), that population genetics theory shows that, under a mendelian hereditary scheme, the variation required for natural selection to act is preserved, and not rapidly destroyed (as under the obsolete blending theory of inheritance). This observation derives from the genetical "quantum theory" of the preservation of the gene, and not only disposed of the major scientific objection to the darwinian theory, but gave the impetus to examine evolution as a mendelian phenomenon. I believe that the calculations necessary for this are not trivial, and I believe that they, and their implications, are insufficiently understood today, even in scientific circles, and I wish therefore that Haldane had emphasized their significance. Perhaps he could have also emphasized the point, made by Fisher, that natural selection is a more effective agency than mutation in effecting gene frequency changes, if we accept standard values for selective differences and mutation rates. The differential equations showing this were, after all, primarily associated with Haldane himself.

Next, no mention is made by Haldane of the so-called Fundamental Theorem of Natural Selection. Whatever the correct interpretation of this theorem might be - a matter discussed below - the standard interpretation is a pretty quantification of darwinism in mendelian terms. The theorem also uses the concept of the additive genetic variance, a central concept in almost all areas of quantitative genetics, and shows that it is the additive part of the variance in fitness, rather than the total variance, that has an evolutionary significance. It is difficult to formulate these conclusions other than quantitatively, surely a major argument for the value of theoretical population genetics. Further, the additive genetic variance plays a key part in Fisher's (1918) great paper on the correlation between relatives, the achievements of which cemented the importance of the mendelian theory and led to the examination of its evolutionary implications.

Third, one matter well known to Haldane (and also Fisher and Wright) should, I feel, also have been mentioned in the "defense" paper, since a misunderstanding of it has repeatedly formed the basis of ill-advised criticisms of Darwin's theory. This is the fact that natural selection is an agency, within the mendelian framework, for generating outcomes of very high *a priori* improbability.

The ill-advised argument is that there is a minute probability that the genetic arrangements which account for such intricate structures as the brain, the eye, the heart and so on could have arisen by evolution through natural selection. The most recent resurrection of this old chestnut was made by Hoyle and Wickramasinghe (1981), who argue as a result that living matter must have arrived on earth by extraterrestrial sources, there not being enough time since the formation of the earth for this evolution on this planet. (They do not appear to notice that, under their model of evolution, described below, there has not been enough time since the big bang, either: thus their extraterrestrial solution also fails.) A correct model, or description, of evolution is of course very complex, and any quantitative model is bound to contain great simplifications. Nevertheless it is possible to illustrate the essential error of Hoyle and Wickramasinghe by a simple quantitative model.

Suppose we wish to attain some desired sequence of 27 letters, such as

THEGREATWALLOFCHINAFELLDOWN

by some random procedure. The Hoyle and Wickramasinghe paradigm of the evolutionary process would be analogous to selecting a random sequence of

27 letters and comparing it to the desired sequence. If it is correct, we have "evolved" to the desired state. In the much more likely event that it is incorrect, we generate a new random sequence of 27 letters, and continue in this way until the desired sequence is reached. The mean number of trials necessary to reach the desired sequence is 26^{27}, and even if we conducted a billion trials a second, we would still require on average 5×10^{21} years to reach the correct sequence - certainly longer than the age of the earth, but also far longer than the time since the big bang.

But this is an incorrect paradigm. The separate letters are the "optimal" alleles, or perhaps DNA nucleotides, at the various loci (sites). Now an allele optimal in one sequence is often optimal in another. Thus a more plausible model of the real nature of evolution by natural selection is as follows. We select a sequence of 27 letters at random. Any letter matching the "correct" letter in the corresponding position is then forever retained. A new random choice is then made over the incorrect letters. In this new random choice, perhaps further letters will correspond to the "correct" letter. These are also retained forever. A third choice, over the remaining incorrect letters, is then made, and so on until the desired sequence is reached. The mean number of trials is now only about a hundred. In other words, evolution under this paradigm is much faster than under the Hoyle and Wickramasinghe model. Of course this new model is still only a very crude representation of evolution as a mendelian process (in particular because evolution is not aiming at a particular target, so that the evolutionary procedure might eventually arrive at the equally effective THELARGECHINESEWALLCOLLAPSED) - but it is far more accurate than the Hoyle and Wickramasinghe model, and is sufficient to expose the invalidity of their arguments.

This matter has been discussed at some length by Denton (1986), who points out, quite correctly, that the actual process of biological evolution is far more complex than either of the two simplistic models outlined above. While this is so, Denton still appears to believe that the Hoyle-Wickramasinghe argument has some validity and that our theory of evolution is in jeopardy - the very title of his book ("Evolution: a Theory in Crisis") makes this clear. However, I believe that the Hoyle-Wickramasinghe argument is totally incorrect and valueless, and that the alternative argument outlined above, while it no doubt leaves out major aspects of the real nature of evolution, points us in the correct direction and indicates that we are in no difficulties in our neo-darwinian position.

Fourth, some mention could have been made of the results of the stochastic theory. Here the conclusions reached could not easily have been obtained by non-mathematical methods, in particular that even a small selective advantage will usually be sufficient to dominate chance factors and lead to the fixation of favourable alleles.

Finally, I wish in effect that Haldane had made, as the leading argument of his "defense" paper, the point that population genetics theory has succeeded in its major aim, namely to validate the darwinian theory as a mendelian process. That is to say, given mutation rates as we know them, selective differences which are plausible, the time scales that are available and, most of all, using (unlike Hoyle and Wickramasinghe) a reasonably correct model of evolution as a genetic process, we can account for the biological world as we see it as the result of an evolutionary process proceeding in accordance with presently-observed phenomena. Nor do I think it is a coincidence that this validation took place quantitatively, or that quantitative geneticists were perhaps the first to accept darwinism. It could have been conceded that many details remained to be filled in - indeed, the real point of Mayr's beanbag argument being one. I have elsewhere (Ewens, 1979) called the era of Fisher, Haldane and Wright the "golden age", indicating my intense admiration of their achievements, but I do believe that these achievements were not best defended in the beanbag episode.

AND AFTER

In order to get away from the misleading phrase "beanbag genetics", I will call the period after 1955 not the post-beanbag, but the post-golden age (PGA), era. What happened in theoretical population genetics in this era?

I believe that PGA research can be divided, broadly, into two main areas. The first is a fleshing-out of the Fisher, Haldane and Wright "prospective" theory, i.e. the theory which takes the mendelian mechanism, and the present genetic situation, as given, and looks to evolution in the future. The second is a comparatively new "retrospective" theory, in which the present genetic situation is again taken as the starting point, but in which we enquire into the evolutionary process which got us to this position. I will now outline briefly several aspects of PGA research in both of these areas.

The golden-age validation of darwinism was a prospective activity, looking to the future. Mathematically, the tools used were differential (or difference) equations in gene (or gametic) frequencies, stationary points of these equations and their stability properties, all applying where random frequency fluctuations can be ignored (the deterministic theory), together with a corresponding stochastic theory, applying when random fluctuations must be taken into account, involving in effect Markov chains, branching processes and the partial differential equations of diffusion theory.

The deterministic theory for one gene locus, especially under random mating, was thoroughly analysed in the early days, and little PGA work was necessary, or carried out, in this area. My own interest in this theory centres around the Fundamental Theorem of Natural Selection (FTNS), referred to above. It was shown, more or less independently, by several authors in the one-locus k-allele case, that under random mating the mean fitness increases. (This was never explicitly shown by Fisher.) This is usually taken as the content, in the one-locus case, of the FTNS. However I believe that this is a quite erroneous interpretation of the theorem as it was conceived by Fisher. Fisher stressed repeatedly in his statement of the theorem that, in respect of most interesting phenotypic characters, mating would not be at random, and thus would not be at random in the genes determining those phenotypes. Now it is obvious that if random mating does not obtain, mean fitness can decrease, indeed in some cases indefinitely, and this was also well known to Fisher. He cannot, then, have had the "standard" interpretation of the theorem, as outlined above, in mind. I have expanded on this theme elsewhere (Ewens, 1989) and will return to it, in the multi-locus case, in a moment.

The most spectacular advances in the deterministic theory in the PGA era concern multi-locus processes, that is, processes where the fitness of an individual is treated, more realistically, as depending on the genes at many loci. It was, first of all, shown by Kimura (1956) (implicitly) and by Lewontin and Kojima (1960) (explicitly) that the correct vehicle for examining evolution, in the multi-locus case, is the set of gametic frequencies (rather than gene frequencies) defined by the loci involved. This viewpoint then introduces the concept of linkage disequilibrium, whose evolutionary significance is great. It is curious that Wright essentially never used gametic frequencies in his analysis, using instead gene frequencies (possibly under the argument, still hotly contested, that linkage disequilibrium can be ignored to a first approximation). The curiosity arises because the main debate between Fisher and Wright was not over drift versus selection (despite the accepted wisdom), but rather over the question of whether evolution can be described, for all important practical purposes, by using gene frequencies only (Fisher) or whether a more complex multi-locus analysis is necessary (Wright). Thus Wright did not even use, in any systematic way, the vehicle necessary for an analysis of his favoured model of evolution.

It is indeed a matter of considerable debate, even today, as to which of the two approaches is to be preferred. Ohta and Kimura (1975), for example, support the Fisherian view and claim that linkage disequilibrium is comparatively rare in nature and that the Fisherian approach, which certainly allows us to carry the theory a considerable distance, is acceptable. Lewontin (1974), on the other hand, has claimed that any adequate analysis must take linkage disequilibrium into account, so that a Wrightian multi-locus analysis is required.

Several conclusions arise in a multi-locus analysis which do not hold in single-locus analyses. Even under random mating the mean fitness can decrease over substantial time periods. This occurs because of recombination: without recombination, the equations for gametic frequency changes reduce (as is otherwise obvious) to the form of a (large number of "alleles") single-locus system, for which, under random mating, mean fitness cannot decrease. A second unexpected conclusion of the multi-locus theory is that, under natural selection only, gametic frequencies can cycle indefinitely around a path of fixed length. This phenomenon contradicts notions about evolution under selection in single-locus systems, in which this behaviour cannot occur. Since mean fitness is not held constant during the cycle, this confirms the view that mean fitness can decrease in multi-locus systems.

These observations lead to further remarks about the mean fitness. The first is that, in my view, despite potential decreases in mean fitness, the FTNS continues to hold in a multi-locus setting (Ewens, 1989). This is because, in the interpretation which I favour, the theorem does not concern changes in mean fitness but rather what I have called the "partial" change in fitness, which is the change due, according to Fisher, to changes in gene frequencies.

There are two further remarks, both negative, about the mean fitness. It may be asked whether mean fitness and the FTNS (to which it is so closely linked in the conventional wisdom) are really central and important in evolutionary thinking outside the mathematical theory. Thus the only reference by Dobzhansky (1970) to the FTNS is in a seemingly irrelevant addendum (page 204) in a section devoted to the concept of heritability, nor do the classic neo-Darwinian works of Mayr, Simpson and Julian Huxley appear to use either mean fitness, or the FTNS, in any central way. Next, the question of whether the mean fitness of a population can be used as a measure of the capacity of that population to out-compete another is highly controversial. Much debate, too long to summarize here, has surrounded this issue. Fisher certainly appeared, on occasion, to be using mean fitness in this way in his argument for the evolutionary decrease of recombination rates, in noting that lower recombination rates tend to increase (equilibrium) mean fitness values and then appearing to use group-selection arguments in claiming that recombination would then diminish over time. A multi-locus approach to this problem, to be described shortly, does not appeal to mean fitness or group selection, and appears to have substantial advantage over the Fisherian argument.

Before describing PGA work in multi-locus systems, it is appropriate to comment on Wright's shifting balance theory of evolution. As noted above, Wright focussed on multi-locus systems and the shifting balance theory was designed to promote the build-up of harmonious multi-locus collections of genes. I myself believe that his theory does not work, in that the combination of migration frequency, mutation rate, selective advantage and recombination fraction required for his argument to succeed can essentially never exist. Many problems with the theory, for example the break-up by recombination of favourable combinations built up in one sub-population when exported to another, (a matter explicitly raised by Haldane), have never been satisfactorily addressed. I am myself quite pessimistic about any future validation of this theory, either theoretically or empirically.

What advances in multi-locus theory have occurred in the PGA era? Perhaps the most interesting is in a rather full theory and application of modifier theory. We consider first the evolution of recombination rate, referred to above. The modern approach imagines a modifying locus M, for which the recombination fraction between the primary loci, A and B, depends on the genotype at the modifier locus. The joint evolution of the MAB system is then considered, and it appears that evolution at the M locus will occur in such a way as to decrease the recombination fraction between A and B loci. Thus recombination decreases under natural selection (even though there are no primary fitness differentials at the M locus), and no appeal is needed to group selection.

It should of course be noted that there are countervailing arguments for an increase in recombination rate, namely that higher recombination favours the joint incorporation into the genome of favourable alleles at different loci. Quantitative arguments (Crow and Kimura (1965) and Maynard Smith (1968)) have been brought to bear on this matter, but again group selection arguments seem to be needed.

Fisher, as early as 1928, also used modifier arguments in his analysis of the evolution of dominance. This analysis, perhaps more than any other, fueled the Fisher-Wright antagonism. PGA analyses have shown that several calculations made by Fisher are incorrect, and a curious calm seems to have settled onto this debate, with perhaps the majority accepting Wright's views on the matter.

Several further modifier theory analyses have been made in the PGA literature. Various important general principles of the theory were obtained by Karlin and McGregor (1974), who show in particular that any argument using mean fitness and group selection can be paralleled by a corresponding argument using modifiers.

The second major area of the fleshing-out of the prospective theory is that of stochastic processes. Random changes in gene frequencies become important when population sizes are small, and the quantification of the effects of these changes, for example in calculating survival probabilities, stationary distributions of gene frequencies, and so on, formed a central part of the classical theory. This has been continued in the PGA era, with emphasis on mean fixation times, simplifications of the procedures (for example in using standard results of Markov chains), processes relating to many loci and, perhaps most important, processes associated with molecular genetics.

We start with the simplest example, namely the description of the density function f (= $f(x;t)$) of the frequency x of some allele at time t. In a two-allele system without selection or mutation, this density function satisfies, in cases considered by Fisher and Wright, the partial differential equation

$$\partial f/\partial t = \frac{1}{2} \partial^2 \{x(1 - x)f\}/\partial x^2. \tag{6}$$

The "solution" of this equation, used often by Fisher and Wright, was given as

$$f = \exp(-t). \tag{7}$$

Now it is certainly true that f, as given by (7), does satisfy (6). But so do many other functions, and what is needed is a solution satisfying the initial condition $x = p$ when $t = 0$. Kimura (1955) pointed this out and gave the correct solution of (6), which is in the form of an infinite eigenfunction expansion for which (7) provides only the first term. Much early unreliable theorizing on the effects of drift in small populations - as late as Dobzhansky (1970), fifteen years after Kimura's correct solution was put forward - resulted from using (7) rather than the correct expression.

Equation (6) is the prototype diffusion equation of population genetics. The use of diffusion theory blossomed in the PGA era, with the effects of selection and mutation being taken into account, multi-locus problems resolved, and conditional processes (where for example the condition is made that a given allele eventually fixes) being introduced. A noteworthy development was the introduction and systematic use of the backward Kolmogorov equation (in contrast to forward equations such as (6)), which gave easy access to calculation of fixation probabilities and mean fixation times. These developments have led to a vastly improved theory and a far more accurate use of the theory for practical purposes. Quite difficult mathematical problems, for example the classification of boundary types, (Feller, 1954), were solved during this time.

Of particular interest was the development of the stochastic theory appropriate for the "infinitely many alleles" model of Kimura and Crow (1964). Here new alleles arise by mutation, each new mutant being of a novel allelic type. This model was inspired by the (then new) molecular genetics and the picture of the gene as a sequence of nucleotides. In the "neutral" case, all alleles are assumed to be selectively equivalent. Clearly no allele, in this model, has a permanent existence, being destined to be lost from the population sooner or later. All questions of interest thus relate to the number, and patterns of frequency, of those alleles present in the population at any one time. A very elegant theory surrounding this model, in particular in the properties of samples of genes, has grown up over the last twenty years, much of which I have summarized elsewhere (Ewens, 1990).

One interesting property of the frequency pattern of the alleles present in any one generation is as follows. Suppose a sample of genes is taken and the frequencies of the alleles present are observed: what frequencies would tend to suggest selective differences between these alleles, and what frequencies would suggest selective equivalence? In carrying out a non-mathematical discussion on this matter, Wright (1978) stated that approximately equal frequencies of the alleles observed would suggest selective equivalence of those alleles, and thus rejected the hypothesis of selective equivalence in the case he was examining, where the frequencies were markedly unequal. However the sampling theory of selectively neutral alleles (Ewens, 1972) shows that we would indeed expect to see unequal frequencies in the selectively equivalent case. The common-sense explanation of this, perhaps initially surprising, result is that different mutations arise at different times and thus those present at any one time can expect to occur in significantly unequal frequencies.

Other areas, apart from the stochastic theory, were developed in the PGA era. Perhaps the most important of these is that of the concept of inclusive fitness. The origins of this concept certainly lie in the golden age - one recalls specifically Haldane's remark about how many brothers, cousins, etc he would be prepared to die for. But the fully-fledged development of the concept, and its associated theory, came in the PGA era, in particular with the outstanding work of Hamilton (1964 and subsequent papers). These developments have had a major impact on all areas of thinking in population genetics and evolutionary thinking, and furthermore have linked research in population genetics with anthropology and sociology, in particular in the analysis of altruistic behaviour.

One interesting aspect of population genetics theory concerns the concept of optimality. Many optimality concepts exist in the physical sciences - nature often seems to act optimally. Is this also true in evolution? Here we consider the recurrence relations describing changes in gene frequency under natural selection. The simplest non-trivial example concerns a one-locus, k-allele (k > 2) system under random mating. Here natural selection will induce, in one generation, certain changes $(\delta_1, \delta_2, ..., \delta_k)$ in these frequencies, with $\sum \delta_i^2 = c$, say. We ask whether, of all changes $(d_1, d_2, ..., d_k)$ in allelic frequencies satisfying $\sum d_i^2 = c$, the natural selection values correspond to the largest increase in mean fitness. The answer to this question is "no", and in this sense genetic evolution under natural selection does not, immediately, appear to satisfy an optimality principle. However if we ask an amended question, in which we replace the two conditions above respectively by $\sum \delta_i^2/p_i = c^*$, $\sum d_i^2/p_i = c^*$, where $(p_1, ..., p_k)$ are the current allelic frequencies, the answer becomes "yes", and this comment introduces a new metric in the space of allelic frequencies, one that is natural in connection with the important, and perhaps unappreciated, Fisherian concept of the average effect of an allele.

I turn finally, in describing PGA activities, to the retrospective theory. In one sense, retrospective questions were already taken up by Fisher, Haldane and Wright: that is to say, given that evolution has occurred, they asked what properties can we expect in current genetic systems. The evolution of dominance question is a good example of such a retrospective question, as are the evolution of mutation and recombination rates.

PGA retrospective questions are of a rather different nature, and are given their impetus by what we know of the genetic properties of contemporary populations. Three in particular are worth singling out. The first is the reconstruction of phylogenetic trees, the two main examples being the phylogenetic tree of species and that of the different races in man. The methodology here is rather distinct from that of theoretical population genetics and relates largely to finding effective algorithms for tree reconstruction, although in an outstanding series of papers Felsenstein has amalgamated population genetics theory with this algorithmic procedure.   No   further   description   of   this   activity   will therefore be given here: an excellent summary is given by Swofford and Olsen (1990).

The   second   retrospective   question   centres   around   the   neutral hypothesis of Kimura (1968).   Under this hypothesis it is claimed that the great bulk of the genetic variation which we observe was not caused by natural selection, but rather by purely random changes in gene frequency acting   on   selectively   equivalent   alleles.   The   sampling   theory   of selectively neutral alleles, referred to above, was introduced to assess the likely allelic composition of a sample of genes when in fact selective neutrality is the case.   The neutral theory has been controversial, is still unresolved, and is probably unresolvable.   Further, the theory might be plausible at one genetic level but not another, or for one part of the genome and not other parts.   For contrasting views on the theory, see Kimura (1983) and Gillespie (1991).

The   final   retrospective   matter   is   in   a   sense   largely   a   technical device, but it has played so important a part in population genetics theory that it deserves special mention.   This is the theory of the coalescent, introduced by Kingman (1982) and used decisively by a series of authors since   its   introduction.   Here   the   starting   point   is   the   fact   that   the properties of a sample of genes are often best examined by considering their co-ancestry, which eventually coalesces in one common ancestor.   An outstanding description of the properties of the coalescent, and of its use in   various   applications,   is   given   by   Hudson   (1991),   and   an   equally outstanding description of the mathematical properties of the coalescent is given by Tavaré (1992).   In these two reviews a beautiful amalgamation of high-level theory and important applications of the theory is reached.

There is therefore a massive volume of PGA work in population genetics theory.   Naturally this builds on the work of the pioneering masters.   But very substantial developments of their work have been made, entirely new areas opened up, and with the acquisition of even more genetic data on which to work, one can look forward to a successful continuation of what has been one of the major areas of research in all of twentieth century science.

REFERENCES

Crow, J.F., 1991, Centennial: J.B.S. Haldane, *Genetics* 130 : 1.
Crow,   J.F.   and   Kimura,   M.,   1965,   Evolution   in   sexual   and   asexual populations. *Amer. Nat.* 99 : 439.

Denton, M., 1986, Evolution: A Theory in Crisis,   Adler and Adler, Bethesda, MD.

Dobzhansky, T., 1970, "Genetics of the Evolutionary Process,"   Columbia University Press, New York.

Ewens, W.J., 1972,   The sampling theory of selectively neutral alleles, *Theoret. Pop. Biol.* 3 : 87.

Ewens, W.J., 1979, "Mathematical Population Genetics," Springer-Verlag, Berlin, Heidelberg, New York.

Ewens, W.J., 1989, An interpretation and proof of the fundamental theorem of natural selection. *Theoret. Pop. Biol.* 36 : 167.

Ewens, W.J., 1990, Population genetics theory - the past and the future. *in*: "Mathematical and Statistical Developments of Evolutionary Theory," S. Lessard, ed., Kluwer, Dordrecht.

Feller, W., 1954, Diffusion processes in one dimension. *Trans. Amer. Math. Soc.* 77 : 1.

Fisher, R.A., 1918, The correlation between relatives on the supposition of Mendelian inheritance. *Trans. Roy. Soc. Edinburgh* 52 : 399.

Fisher, R.A., 1928, The possible modification of the response of the wild type to recurrent mutation. *Amer. Nat.* 62 : 115.

Fisher, R.A., 1930, "The Genetical Theory of Natural Selection," Clarendon Press, Oxford.

Fisher, R.A., 1958, "The Genetical Theory of Natural Selection,"   Dover, New York.

Gillespie, J.H., 1991, "The Causes of Molecular Evolution,"   Oxford University Press, Oxford.

Haldane, J.B.S., 1932, "The Causes of Evolution," Longmans Green, London.

Haldane, J.B.S., 1957, The cost of natural selection, *J. Genet.* 55 : 511.

Haldane, J.B.S., 1961, More precise expressions for the cost of natural selection. *J. Genet.* 57 : 351.

Haldane J.B.S., 1964, A defense of beanbag genetics. *Persp. Bio. Med.* 343.

Hamilton, W.D., 1964, The genetical evolution of social behaviour. *J. Theoret. Biol.* 7 : 1.

Hoyle, F. and Wickramasinghe, N.C., 1981, "Evolution from Space," Dent, London.

Hudson, R.R., 1991, Gene genealogies and the coalescent process, *in*: "Oxford Surveys in Evolutionary Theory," Volume 7, D. Futuyma and J. Antonovics, eds., Oxford University Press, Oxford.

Karlin, S. and McGregor, J., 1974, Towards a theory of the evolution of modifier genes. *Theoret. Pop. Biol.* 5 : 59.

Kimura, M., 1955, Solution of a process of random genetic drift with a continuous model. *Proc. Nat. Acad. Sci.* 41 : 144.

Kimura M., 1956, A model of a genetic system which leads to closer linkage under natural selection, *Evolution* 10 : 278.

Kimura, M., 1968, Evolutionary rate at the molecular level, *Nature,* 217 : 624.

Kimura, M., 1983, "The Neutral Theory of Molecular Evolution," Cambridge University Press, Cambridge.

Kimura, M., and J.F. Crow, 1964, The number of alleles that can be maintained in a finite population, *Genetics* 49 : 725.

Kimura, M. and Ohta, T., 1971, "Theoretical Aspects of Population Genetics," Princeton University Press, Princeton.

Kingman, J.F.C., 1982, The coalescent, *Stoch. Proc. Appl.* 13 : 235.

Lewontin, R.C., 1974, "The Genetic Basis of Evolutionary Change," Columbia University Press, New York.

Lewontin, R.C., and Kojima, K., 1960, The evolutionary dynamics of complex polymorphisms, *Evolution* 14 : 458.

Maynard Smith, J., 1968, Evolution in sexual and asexual populations. *Amer. Nat.* 102 : 469.

Mayr, E., 1963, "Animal Species and Evolution," Harvard University Press, Cambridge MA.

Ohta, T. and Kimura, M., 1975, The effect of selected linked locus on heterozygosity of neutral alleles (the hitch-hiking effect). *Gen. Res. Camb.* 25 : 313.

Swofford, D.L. and Olsen, G.J., 1990, Phylogeny reconstruction. *in:* "Molecular Systematics" D.M. Hillis and C. Moritz, eds., Sinauer, Sunderland MA.

Tavaré, S., to appear, An introduction to the coalescent, *in:* "Mathematics and Molecular Biology" E.S. Lander, ed.

Wright, S., 1930, The Genetical Theory of Natural Selection - a Review. *J. Hered.* 21 : 349.

Wright, S., 1978, "Evolution and the Genetics of Populations. Vol. 4. Variability within and among natural populations," University of Chicago Press, Chicago.

POPULATION GENETICS OF QUANTITATIVE CHARACTERS

P. Narain

Indian Agricultural Research Institute
Pusa, New Delhi 110 012
INDIA

INTRODUCTION

Even before the science of genetics was born in 1900, evolution and inheritance of quantitative characters were topics of intensive research in the hands of stalwarts like Galton, Pearson, etc. The law of ancestral heredity propounded by Galton (1889) involved relating father's height with those of sons resulting in invention of the regression method. Pearson (1904) studied correlation between relatives in large populations and concluded that it was incompatible with Mendelian hypothesis of genetic factors. Such attempts on evolutionary issues were basically statistical in nature and developed into what came to be later known as biometrical school. Subsequently, it was the monumental work of Fisher (1918) which proved that the biometrical approach is compatible with Mendelism in providing a sound and theoretical basis for the inheritance of quantitative characters. In doing so, Fisher for the first time introduced genetics in an otherwise statistical approach. There are thus two ways to approach the problem of the inheritance of quantitative characters. Even in the theory and practice of plant and animal breeding also, particularly in regard to artificial selection for the improvement of economic characteristics, the biometric and genetic approaches could be distinguished. The formula for the expected response to selection can be established either from a purely statistical approach - in fact relying on regression method of Galton (1889) - or from a genetic approach invoking genetic based concepts of breeding value and gene frequency.

On the other hand, while attempts were made by evolutionary biologists to develop the theory in quantitative terms using principles of population genetics, these were mostly confined to discrete variation caused by single-gene mutations with major effects which could be followed at the phenotypic level. They did not use the tools of quantitative genetics to study the progress in evolution by natural selection of quantitative characters. It was during the last decade or so that considerable interest was generated on the

Human Population Genetics, Edited by P.P. Majumder
Plenum Press, New York, 1993

evolutionary dynamics of quantitative characters. Several workers, quite a few being mathematicians and statisticians, put forth different models for evolution of polygenic traits and often adopted the quantitative genetic principles already used in plant and animal breeding. However, as is noticed in the selection theory of domesticated plants and animals, there is variation in the approach adopted from a purely statistical one leading to what is being termed as 'phenotypic evolution' to the one based on genetic principles.

In this paper, I focus attention on the equivalence of the two approaches and present an overview of the recent developments - mostly theoretical - in the population genetics of quantitative characters in the context of evolution by natural selection. The issue on the maintenance of genetic variation in natural population which has been extensively studied since 1920 and involves interesting methodological aspects has also been discussed.

MODELLING OF EVOLUTIONARY QUANTITATIVE GENETICS

As already stated the evolution of quantitative traits can be modelled in either of the two ways. One way would be to define and study the underlying models at the level of phenotype avoiding any reference to gene frequency. Lande (1975, 1976, 1980, 1988) adopted this approach on the assumption of univariate or multivariate normal distribution of the phenotypes. The other way would be to define and study a genetic model anticipating a complete genetic analysis of the trait more or less in the same tradition as that of Fisher (1918). In this case one has to start from the simplest situation of a single locus with two alleles and build over it the more complex systems of infinitely many alleles at the locus, several loci, linkage and epistatic effects, non-random mating, finite population, etc. Wright (1931), Haldane (1932), Robertson (1956), Latter (1960, 1970), Bulmer (1972), Narain and Chakraborty (1983, 1987), Turelli (1984, 1988), Narain (1988, 1990, 1992), Slatkin (1987), Barton and Turelli (1987), Turelli and Barton (1990) and several others adopted this approach. The essential step in this case is to set up the genotypic-phenotypic relationship, particularly for natural selection, a general method for which is given by Kimura and Crow (1978).

PHENOTYPIC EVOLUTION APPROACH

Lande (1976, 1979, 1988) popularised the Gaussian phenotypic analyses of quantitative trait on the basic premise that genetic variances and covariances are known *a priori*. Stimulated by Simpson's (1944) concept of phenotypic adaptive zones which itself was prompted by Wright's (1931, 1932, 1940) adaptive landscape for gene frequency change, he gave an expression for the change per generation in the mean phenotypic value of a quantitative trait and showed how it is affected by the force of selection acting directly on it as well as by the forces of indirect selection on the other genetically correlated traits. In doing so he bypassed all considerations of gene frequency change in more or less the same spirit as Galton (1889) and Pearson (1904). Ronghgarden (1979) has introduced the concept of *segregation kernel* - the offspring phenotype distribution - for this purpose but with respect to a single character. Here we take a general case of k correlated characters so as to reveal the hidden pleiotropic effects of the loci involved.

Consider a set of k quantitative phenotypic variables denoted by the column vector $z = (z_1, z_2, \ldots, z_k)^T$ where the superscript T indicates the transpose of a matrix. As is common in quantitative genetics, each z_i is assumed to be the sum of a variable representing the additive genetic effects or breeding value A_i and an independent environmental effects variable including dominance, epistasis, environment and genotype x environment interaction effects etc. R_i. A_i's and z_i's are assumed to follow a multivariate normal distribution in the population before selection each generation. With these assumptions, the natural selection can be studied in its most general form by assigning expected fitness $W(z)$ to individuals with phenotype vector z. Let $p_t(z)$ be the phenotype distribution in the tth generation before selection. The mean fitness $\overline{W}_t$ in a population in this generation can be expressed as

$$\overline{W}_t = \int w(z) p_t(z) dz \qquad (1)$$

If $p_{w,t}(z)$ is the phenotypic distribution in the tth generation after selection, we have

$$p_{w,t}(z) = W(z) p_t(z) / \overline{W}_t \qquad (2)$$

If we assume random mating and infinitely large population with discrete non-overlapping generation, the phenotypic distribution in the $(t+1)^{th}$ generation, denoted by $p_{t+1}(z)$ can be expressed as

$$p_{t+1}(z) = (\overline{W}_t)^{-2} \int\!\int L[z-\theta] W(z_1) p_t(z_1) W(z_2) p_t(z_2) dz_1 dz_2 \qquad (3)$$

where L(.) denotes the k-variate segregation kernel, being function of k variables, with a mean vector $\ominus$ of dimension kx1

$$\theta = \frac{1}{2} H(z_1 + z_2) + (I-H) E_t(z) \qquad (4)$$

and variance-covariance matrix D_L of order kxk; z_1 and z_2 are the two vector-valued parents in the selected population in the t-th generation, H is the k-variate counterpart of heritability as given in Narain (1990), and $E_t(z)$ is the mean vector of the population in the t-th generation. The expression (3) is a k-variate version of the population genetics of quantitative characters with natural selection in its most general form.

The selection pressure on quantitative characters can be

either of stabilizing type or of directional type. In the former case, the individuals close to the mean of the population are favoured at the expense of those deviating far away from the mean. The fitness function $w(x)$ could be of Gaussian type as we will discuss later. Here mean coincides always with the optimum in each generation and therefore, the mean phenotypic value in the next generation does not change but the variance is reduced. On the other hand, with directional selection, those individuals get selected which are away from the mean on one of the two sides of it. The mean phenotypic value in the next generation moves in the direction of selection. A variant of this type of selection is truncation selection where $w(x)$ is a step function, all individuals above a fixed threshold being involved in the breeding at the expense of others. This is the type of selection mostly used in plant and animal breeding. In what follows, we discuss directional selection. If we multiply both sides of (3) by z and integrate over z, we get, after simplification

$$E_{t+1}(z) = E_t(z) + H\{E_{w,t}(z) - E_t(z)\} \qquad (5)$$

where $E_{w,t}(z)$ is the mean vector after selection but before reproduction in the t-th generation. The change per generation due to directional selection in the mean phenotypic vector, denoted by $\Delta E(z)$, is therefore $H[E_{w,t}(z) - E_t(z)]$. This result can be expressed in an alternative form by making use of Price's (1972) selection mathematics extended to vector-valued variables and is given by

$$\Delta E(z) = H \; \text{Cov} \; (w,z) \qquad (6)$$

where $\text{Cov}(w,z)$ is a vector of covariances between w and z_i, $i=1,2,\ldots,k$. Now the mean fitness of the population in the t-th generation $\bar{W}_t$ as given by (1) can change as a result of change in any of the components of the vector $E_t(z)$. Let us therefore define the gradient operator

$$\nabla = [\frac{\partial}{\partial E_t(z_1)} , \frac{\partial}{\partial E_t(z_2)} , \cdots , \frac{\partial}{\partial E_t(z_k)}] \quad \text{and apply it to}$$

the scalar $\bar{W}_t$. This operation converts a scalar into a vector. Assuming frequency-independent case, (1) becomes

$$\nabla \bar{W}_t = \int \bar{W}(z) \frac{\partial p_t(z)}{\partial E_t(z)} \; dz \qquad (7)$$

If we specify $p_t(z)$ as multivariate normal distribution, we get

$$\nabla \bar{W}_t = \Sigma_P^{-1} \bar{W}_t \text{Cov}(w,z) \qquad (8)$$

Hence, dropping the suffix for generation number, (6) becomes

$$\Delta E(z) = H\left(\Sigma_P/\ \overline{W}\ \right) \nabla\ \overline{W}\ =\ \Sigma_A \nabla\left(Log\ \overline{W}\ \right) \tag{9}$$

This is the basic evolutionary dynamic equation for the per generation change in the average phenotypic vector under directional selection when we assume constant fitnesses, infinite population and multivariate Gaussian distribution for the phenotypic values. For a particular trait, this shows that the evolution of the character depends not only on selection acting directly on it, but also on the indirect selection on all genetically correlated characters, the so called hidden effects due to pleiotropy of the loci involved. If we take k=1, we recover from (9), the classic equation for response to directional selection.

GENE FREQUENCY MODEL AT A LOCUS WITH PLEIOTROPIC EFFECTS

In this approach, the genetic model is first defined and studied at the simplest level of a single locus with two alleles. For quantitative variation, we normally use a polygenic system in which the character is affected by many genes. But each gene typically affects many quantitative characters - the so called pleiotropic effects or side effects - which leads to correlation between characters both at the genotypic as well as phenotypic levels. When we consider selection at the phenotypic level, we assume that the selection acts on the relevant loci only through the character under study. But the complex pathways between gene product and phenotype are such that the segregation of alleles contributing variation in any one character also produces variation in several other characters which are not under consideration but are also under selection. This requires that instead of a single character, we have to consider a set of correlated characters and postulate that selection acts on the relevant loci through the set of characters instead of a single character.

The consequences of pleiotropism or manifold effects of genes in quantitative inheritance was probably first studied by Mode and Robinson (1959) who extended the concept as well as the partition-ing of genetic variance for a character to genetic covariance between characters. Recently, this was used by Narain (1985) to give a generalised treatment of progeny testing and sire evaluation for a given character like milk production. In this paper, a model was developed wherein the breeding value of a sire for milk production is strictly dependent on the phenotypic values of the daughters for milk production as well as for several other characters correlated with milk production. The progeny test based on daughters' milk production alone, as commonly in vogue, is at best an approximation.

As earlier, we consider the set of correlated characters indexed by the vector z which are controlled by the same locus with two alleles A_1-A_2. The population, considered as infinitely large, consists of three genotypes A_1A_1, A_1A_2 and

A_2A_2. Due to pleiotropic effects, this simple genetic model produces variation and co-variation amongst the set of characters z. The conditional phenotypic distribution $p(z|A_iA_j)$ for genotype A_iA_j is taken as multivariate normal with z_{ij} as the mean phenotypic vector for the genotype A_iA_j and Σ_e as the dispersion matric of variances and covariances amongst the k characters at the environmental level only. Using the selection function $w(z)$, the relative selective values of the genotype, w_{ij} is obtained as

$$w_{ij} = \int p(z|A_iA_j)w(z)dz \tag{10}$$

which gives for $(ij)=(11,12,22)$ the average fitnesses of the three genotypes. In fact, phenotypic selection function $w(z)$ is converted into genotype selection w_{ij} by this equation.

With p_t and $q_t=1-p_t$ as the frequencies of genes A_1 and A_2 respectively in the t-th generation and on the assumption of random mating, the gene frequency of A_1 in next generation i.e. (t+1)th generation would be

$$p_{t+1} = p_t(p_tw_{11}+q_tw_{12})/\ \bar{w} \tag{11}$$

where $\bar{w} = p_t^2w_{11}+2p_tq_tw_{12}+q_t^2w_{22}$.

The average value of z in the (t+1)th generation is then given by

$$E(z_{t+1}) = p_{t+1}^2z_{11}+2p_{t+1}q_{t+1}z_{12}+q_{t+1}^2z_{22} \tag{12}$$

These equations thus form a complete model for the evolution of a set of characters z on the basis of gene frequency approach. The essential difference between this approach and modelling phenotypic evolution is that here the problem is translated at the genetic level and the iteration is carried out at the level of gene frequency (instead of directly at the phenotypic level) and then the genetic description is translated back into phenotypic terms.

RELATION BETWEEN MULTIVARIATE PHENOTYPIC SELECTION
AND SINGLE LOCUS GENOTYPIC SELECTION

We have separately discussed the purely phenotypic model as well as the purely genetic model. The task is now to connect the two approaches. For this purpose we follow and extend the method of Kimura and Crow (1978) for establishing a general relation between the selection made at the overall phenotypic level on multi-trait basis and the consequent selection induced at the genotypic level at individual loci.
Let the mean vector and dispersion matrix of z be represented by M and Σ_p. As before, we consider the locus A_1-A_2 with z_{ij} as the vector of average phenotypic values of

A_iA_j individuals. Expressed as deviations from the mean vector M, we express them as $a_{ij}=(z_{ij}-M)$. Let z_{OP} be the vector of optimum phenotypic values for the k traits which gives maximum fitness. Without loss of generality, we take z_{OP} as the null vector, the optimum phenotype being at the origin. Then w_{ij}, the relative fitness of the genotype with vector-value z_{ij} is

$$w_{ij} = \int w(z)p(z-a_{ij})dz \qquad (13)$$

This is because the k-variate distribution of z in the sub-population of genotype A_iA_j with vector-value z_{ij} is shifted by the vector-value a_{ij} and the k-variate density function in the sub-population before selection is $p(z-a_{ij})$. Now we apply Taylor series expansion to $p(z-a_{ij})$ in k-dimension around the elements of $a_{ij}=0$. This gives

$$p(z-a_{ij}) = p(z)-a_{ij}^T(\partial/\partial z)p(z)+\frac{1}{2}a_{ij}^T \left(\frac{\partial^2}{\partial z^2}\right)p(z)a_{ij} \qquad (14)$$

where $(\partial/\partial z)$ is a kx1 column vector and $\left(\frac{\partial^2}{\partial z^2}\right)$ is kxk matrix with elements $\partial^2/\partial z_i\partial z_j$ for $i,j=1,2,\ldots,k$. Then, neglecting third order terms, in the elements of a_{ij}, we get

$$w_{ij} = \beta -a_{ij}^T b+\frac{1}{2}a_{ij}^T B\ a_{ij} \qquad (15)$$

where

$$\beta = \int w(z)p(z)dz \qquad (15a)$$

$$b = \int w(z)(\partial/\partial z)p(z)dz \qquad (15b)$$

$$B = \int w(z)(\partial^2/\partial z^2)p(z)pz \qquad (15c)$$

Let the effect of substituting A_2 for A_1 on the set of k-characters be vector d, with elements $d_1,d_2,\ldots,d_k$. Then the three genotypes A_1A_1, A_1A_2 and A_2A_2 with genotypic frequencies p^2, 2pq, q^2 will have vector-valued average genotypic values as :
$a_{11} = -2qd$, $a_{12} = (1-2q)d$, and $a_{22} = 2(1-q)d$. This when substi-tuted in (15) gives w_{ij} in terms of q, d^Tb and d^TBd. The mean fitness of the whole population is then $\beta+pq(d^TBd)$. The per generation change in gene frequency of A_2 due to natural selection then becomes

$$\Delta q = q(1-q)[-(d^Tb/\beta)+(\frac{1}{2}-q)(d^TBd/\beta)] \qquad (16)$$

where $\bar{w}$ is approximated by β, neglecting terms of higher order than the coefficients of genic effects. Assuming no dominance, we know Δq is also approximated by $sq(1-q)$ where s is coefficient of selective advantage of A_2 over A_1 in the single locus two allele case. Thus we have

$$s = -(d^T b/\beta) + (\tfrac{1}{2} - q)(d^T B d/\beta) \qquad (17)$$

When k=1, this reduces to (10.9) of Narain (1990). This is the generalization of the single character situation to multivariate case. This expression, given probably for the first time in the literature, is applicable to a set of k correlated traits affected by the same locus on the basis of pleiotropy. It is seen that the selective advantage of A_2 over A_1 is firstly dependent on its gene frequency, q and secondly on quantities $(d^T b)/\beta$ and $(d^T B d)/\beta$. The meaning of these latter quantities will be clear shortly when we apply them to specific cases.

Application to Truncation Selection

In the case of truncation selection practised in plant and animal breeding, the cut-off points for the k traits can be set up independently. Narain (1979, 1990) discusses this aspect under the topic of selection on the basis of independent culling levels. If we truncate the distribution of z rectangularly, keeping the same cut-off point c for all traits, by:

$$S_k(c) = \{z : z_1 \geq c, z_2 \geq c, \ldots . z_k \geq c\}$$

we have

$$P_k(c; \Sigma_P) = \int_c^\infty p(z)\,dz = \beta \qquad (18)$$

as the proportion of the population selected since $w(z) = 1$ for those individuals included in P_k and 0 otherwise. Incidentally, this result is true even if the distribution of z is not assumed as multivariate normal. For b and B, we get,

$$b = -\beta \Sigma_P^{-1} \Delta_S M \qquad (19)$$

$$B = \beta \Sigma_P^{-1}[(\Delta_S \Sigma_P) + (\Delta_S M)(\Delta_S M)^T]\Sigma_P^{-1} \qquad (20)$$

where $\Delta_S M = (E_w(z) - M)$, $\Delta_S \Sigma_P = (\Sigma_{P_w} - \Sigma_P)$,

$E_w(z)$ and Σ_{P_w} being respectively the mean vector and dispersion matrix of the truncated distribution. Now with these expressions for b and B, the selection coefficient s given by (17) becomes

$$s = d^T \Sigma_P^{-1}\Delta_S M + (\tfrac{1}{2} - q)d^T \Sigma_P^{-1}[\Delta_S \Sigma_P + (\Delta_S M)(\Delta_S M)^T]\Sigma_P^{-1}d \qquad (21)$$

Thus the selective advantage of A_2 over A_1 gets expressed in terms of d and $\Delta_S M$ and $\Delta_S \Sigma_P$, the changes in the mean vector and dispersion matrix respectively due to selection. This generalises (10.27) of Narain (1990) and reduces to it for k=1. Assuming additive gene action (no dominance, no linkage, no epistasis over all loci), the vector of the total genetic change due to truncation selection is given by

$$\Delta\bar{z} \;=\; \sum_a 2q_a(1-q_a)\,\mathbf{d}_a\mathbf{d}_a^T\Sigma_P^{-1}\Delta_S\mathbf{M}$$

$$+ \;\sum_a 2q_a(1-q_a)\,(\tfrac{1}{2}-q_a)\,\mathbf{d}_a\mathbf{d}_a^T\Sigma_P^{-1}[\Delta_S\Sigma_P+(\Delta_S\mathbf{M})(\Delta_S\mathbf{M})^T]\Sigma_P^{-1}\mathbf{d}_a \quad (22)$$

To the first degree approximation, therefore

$$\Delta\bar{z} \;=\; \Sigma_A\Sigma_P^{-1}\Delta_S\mathbf{M} \;=\; H\Delta_S\mathbf{M} \tag{23}$$

where the matrix of additive genetic variance and covariances for the set of k-characters, under pleiotropic effects, $\Sigma_A = 2\sum_a q_a(1-q_a)\mathbf{d}_a\mathbf{d}_a^T$. This result is what we get from (5) also where we resorted to modelling at the phenotypic level. The equivalence of the two approaches is thus established. At the same time, this relation gives the responses in the set of k correlated characters and is the multivariate generalisation of the relation (10.37) given in Narain (1990). For any given character we find that the response to individual selection is not only the direct response in it but also includes the correlated responses in the rest of genetically correlated characters. In spite of the equivalence shown above, it is important to note that the gene frequency approach gives more information than the phenotypic modelling approach due to the generalised nature of the coefficient of selection in terms of b and B. If we do not approximate to the first degree, the total response given by (22) reveals additional features. For a single character, this leads to

$$\Delta\bar{z} \;=\; h^2\Delta_S\mathbf{M} + \frac{\mu_{3A}}{2\,(\sigma_P^2)^2}\left(\Delta_S\sigma_P^2+(\Delta_S\mathbf{M})^2\right) \tag{24}$$

where σ_A^2 and μ_{3A} are respectively the second and third moments of the breeding values when dominance effects are not existent (vide p.293 in Narain, 1990). This is the same expression which Turelli (1988) gets if we neglect the term in $(\Delta_S\mathbf{M})^2$ on the condition that $(\Delta_S\mathbf{M})^2 \ll \Delta_S\sigma_P^2$. However, although he adopts the allele frequency dynamics to get the expression, his approach is via recursions derived by Barton and Turelli (1987) and Turelli and Barton (1990). When we consider two characters, we have to introduce as many as six moments of breeding values. The expressions for the total response in the two characters are similar to those of Turelli (1988) who obtained them by a different approach. Thus we see that the strategy adopted in this paper for determining the total response is a general one. We have also determined the per generation changes in higher order moments adopting this approach. The results are discussed elsewhere.

Stabilizing Selection

In stabilizing selection, as we have noted earlier, the extreme deviants from the norm are weeded out by higher fitnesses of the individuals near the optimum value giving the maximum fitness. The most plausible model of this type of selection, for the set of k correlated characters is the Gaussian fitness function given by

$$w(z) = w_{max} \exp\left[\left(-\tfrac{1}{2}\right)(z-z_{opt})^T \Sigma_w^{-1}(z-z_{opt})\right] \tag{25}$$

where z_{opt} is the vector of optimum phenotypic values with maximum fitness so that when $z = z_{opt}$, the fitness $w(z)$ is w_{max} which can, without loss of generality, be taken as unity; Σ_w is the positive definite dispersion matrix giving widths or ranges of the vector-valued phenotypic values around the optimum. With $p(z)$ as multivariate normal $N(M, \Sigma_p)$, the expressions for β, b and B are obtained as

$$\beta = Q|\Sigma_p|^{-1/2}|\Sigma_p^*|^{1/2} \tag{26}$$

where $\Sigma_p^* = \left(\Sigma_w^{-1}+\Sigma_p^{-1}\right)^-$ and

$$Q = \exp\left[\left(-\tfrac{1}{2}\right)\{z_{opt}^T \Sigma_w^{-1}\left(I-\Sigma_p^*\Sigma_w^{-1}\right)z_{opt} + M^T\Sigma_p^{-1}\left(I-\Sigma_p^*\Sigma_p^{-1}\right)M\right.$$

$$\left. -2M^T\Sigma_p^{-1}\Sigma_p^*\Sigma_w^{-1}z_{opt}\}\right] \tag{26a}$$

$$\beta = Q|\Sigma_p|^{-1/2}|\Sigma_p^*|^{1/2}\Sigma_s\left(M-z_{opt}\right) \tag{27}$$

with $\Sigma_s = \left(\Sigma_w+\Sigma_p\right)^{-1}$ and

$$B = Q|\Sigma_p|^{-1/2}|\Sigma_p^*|^{1/2}\Sigma_s\left[\left(M-z_{opt}\right)\left(M-z_{opt}\right)^T-\Sigma_s^{-1}\right]\Sigma_s \tag{28}$$

The selection coefficient s then becomes

$$s = -d^T\Sigma_s m + \left(\tfrac{1}{2}-q\right)d^T\Sigma_s[m\,m^T-\Sigma_s^{-1}]\Sigma_s d \tag{29}$$

where $m = (M-z_{opt})$.

If the deviations of the means from the optima are much larger than the effects of allele substitution for all the k characters, s can be approximated by

$$s = -d^T\Sigma_s m \tag{30}$$

Natural selection, of the directional type, thus shifts the means towards the optima. On the other hand, if $M = z_{opt}$ i.e. $m = 0$, i.e. means are at the optima, we have

$$s = -\left(\tfrac{1}{2}-q\right)d^T\Sigma_s d \tag{31}$$

which is strictly selection of stabilizing type but frequency dependent as s is positive or negative according as $q > \tfrac{1}{2}$ or $< \tfrac{1}{2}$ respectively. The alleles behave as if negatively overdominant and therefore leads to unstable equilibrium. With no dominance, no linkage, and no epistasis etc., the vector of total genetic change due to selection can be obtained by summing over loci, the contribution of single locus for a single character being $2da\Delta q_a$. This gives

$$\Delta \bar{z} = -\left(\sum_a 2q_a(1-q_a)\mathbf{d}_a\mathbf{d}_a^T\right)(\Sigma_s \mathbf{m})$$
$$+ \sum_a q_a(1-q_a)(1-2q_a)\mathbf{d}_a\mathbf{d}_a^T\Sigma_s(\mathbf{m}\ \mathbf{m}^T-\Sigma_s^{-1})\Sigma_s\mathbf{d}_a \qquad (32)$$

To the first degree approximation, therefore

$$\Delta \bar{z} = -\Sigma_A \Sigma_S \mathbf{m} = -\mathbf{H}\ \Sigma_P \Sigma_S \mathbf{m} \qquad (33)$$

This result generalises Bulmer's (1971) result on the response to natural selection under Gaussian fitness function and reduces to his result for k=1.

STEP-WISE DISCRETE MUTATION AND STABILIZING SELECTION

As already stated, to explain the nature of genetic variability found in natural populations of various organisms for most polygenic traits, mathematical analyses based on different models have been attempted by different workers since 1930s, the problem being initiated by Fisher (1930) and Haldane (1932). But mechanism for the maintenance of this variability is still far from being well understood. In almost all studies published on this topic before 1982, the model of mutation adopted has been either one of diallelic loci with equal mutation rates between the two alleles or of the one with infinitely many alleles at the locus where alleles are distinguished according to the distribution of their additive effects and their frequencies do not enter into the analysis (Crow and Kimura, 1964). A more realistic model of mutation involving discrete change of state was introduced by Chakraborty and Nei (1982) to examine the extent of genetic variation for the quantitative trait but under the forces of mutation and random drift only. The model happens to be a discretized version of continuum-of-alleles models of Kimura (1965). Narain and Chakraborty (1983, 1987), Slatkin (1987) and Narain (1988, 1990) considered such a model in relation to stabilizing selection. In particular, the transient behaviour of the approach to equilibrium when there are infinite number of alleles at the locus and population evolves from monomorphism (at an optimum phenotype) was studied. The equilibrium distribution of genotypic values is found to be non-normal. Further, Narain (1992) considered in general the case of a finite number of alleles at the locus and showed how the results published so far fit into this general scheme. In fact, he showed that for mutation-selection equilibria in large natural populations, the number of alleles at the locus could be a crucial factor. For details, Narain (1992) may be referred to.

MULTILOCUS POPULATION GENETICS THEORY

For a complete dynamics of the genetic situation

underlying quantitative traits which involves recombination, mutation, mating scheme and selection, the composition of the population at any time t has to be considered in terms of a multilocus probability density function $p_t(x_1, x_2, \ldots, x_n)$, where x_i refers to the gene frequency at the ith locus, $i=1,2,\ldots,n$, there being n loci. Even without pleiotropy, it is quite difficult to describe the dynamics of p_t. Most of the treatment in the literature involve simplifying assumption of the Gaussian distribution for p_t so that a direct analysis of the dynamics of p_t is not necessary. The dynamics of mutation, meiosis, mating and selection can then be modelled by deriving recursions for the means, variances and covariances of effects within and between loci. If, in addition, the linkage disequilibrium is ignored, the dynamics can be studied separately at each locus, p_t being then the product of p_{it}'s at individual loci, n in number.

Recently, a general analysis of the multilocus population genetics theory underlying quantitative characters has been presented by Turelli and Barton (1990). Selection, recombination and drift have been included in the most general form on the assumption of the additivity of genetic effects (no dominance and no epistascy) but without any assumptions regarding the distribution of breeding values or number of loci or number of alleles. In the previous analyses, Barton and Turelli (1987) and Turelli (1988) derived standard equations for the evolution of allele frequencies and then approximated the changes in mean, variance and higher order moments of the selected character, assuming that enough loci contribute variation to ensure that breeding values are nearly Gaussian and that linkage disequilibrium is negligible. Their treatment was complementary to Bulmer's (1971) who ignored allele frequency evolution and considered only changes in linkage disequilibria. In the latest analysis of Turelli and Barton (1990), both the dynamics of allele frequencies and linkage disequilibria are included by studying the multilocus selection recursions in terms of genotypic frequencies rather than gene frequencies.

The investigation reveals how non-Gaussian distribution of breeding values alters the dynamics of mean and variance of the breeding values. The results are approximated for the weak selection case to illustrate the effects of linkage disequilibria on the dynamics of the mean of a character determined by a very large number of segregating loci. It is found that the selection generally drives the distribution of breeding values away from a Gaussian distribution by creating multilocus linkage disequilibria. Long-term dynamics of mean depart substantially from the prediction of the standard selection recursions but the discrepancy may often be negligible for short-term selection. To the eternal problem of the maintenance of genetic variance under a balance between recurrent mutation and stabilizing selection, Turelli and Barton (1990) show that, for realistic parameter values, linkage disequilibrium has little effect on the amount of equilibrium additive genetic variance - a result also obtained by Narain and Chakraborty (1983, 1987).

RANDOM GENETIC DRIFT

The effect of random genetic drift, the chance loss of alleles due to random sampling of gametes in reproduction, in evolution has been a controversial subject right since Fisher's (1930) time when he underplayed its role. Although, Haldane (1927)did investigate the chance fixation of mutants, it was Wright (1931) who emphasised the importance of random genetic drift and wrote several papers on the subject culminating in his *shifting balance theory* of evolution which regards it as a trial and error process in terms of a multidimensional adaptive surface. However, the main credit for recognising the part played by random drift in evolution goes to Kimura who using diffusion models (1964) founded the neutral theory of molecular evolution (1968) which led to fierce controversy as it was misunderstood as non-Darwinian.

The incorporation of random drift in any treatment of the mechanism for the maintenance of polygenic variability is however beset with complexities not easily resolvable. The early workers to deal this problem are Latter (1970) and Bulmer (1972). The contribution of random drift in phenotypic evolution was studied by Lande (1976) based on Simpson's (1953) concept of adaptive zones. Within such zones stabilizing selection operates and between which there is a threshold of low adaptive value. Once the threshold has been crossed, rapid evolution into the new adaptive zone ensues. He developed a phenotypic model of the stochastic evolution of the average phenotype in a finite population under stabilizing selection based on diffusion equations. It was found that if stabilizing selection is weak and an adaptive threshold is not very far away, random genetic drift between adaptive zones may be an important mechanism of evolution in populations of effective size in the hundreds or thousands. This approach of the phenotypic evolution of a single character was generalised for several correlated characters by Lande (1988) using multivariate diffusion equation. The divergence of a polygenic system subject to stabilizing selection, mutation and drift was investigated by Barton (1989). When the alleles responsible for variation are rare, many classes of equilibria may be stable. The rate at which drift causes shifts between equilibria was studied by integrating the gene frequency distribution $(\bar{w})^{2N}\Pi(pq)^{4Nv-1}$. Over a wide range of population sizes, drift keeps the population near an equilibrium which minimizes the genetic variance and the deviation from the selective optimum. Shifts between equilibria in this class occur at an appreciable rate if the product of population size and selection on each locus is small ($Nsa^2 < 10$). Hill and Keightley (1988) discussed the interrelations of mutation, population size, artificial and natural selection. They found that compared to directional selection where the variance maintained increases without bounds, that for stabilizing selection reaches a limit. The value of this limit is relatively large if we consider models with high mutation rate (but with small effects) and weak selection (Kimura, 1965) but much smaller if we assume low mutation rate (but with large effects) and strong selection (Turelli, 1984).

STABILIZING SELECTION-BALANCE MODELS

The most popular model to study the maintenance of polygenic variability in natural populations which has been extensively studied is that of mutation-selection balance already discussed in the above sections. However, there are other competing hypotheses on which much less attention has been paid so far. One such hypothesis is that of the *balancing selection* at the phenotypic level. Kimura (1983) however proved that such a case leads to the imbalance of the polymorphic state of the genes involved since negatively overdominant genes are for more likely to be fixed by random genetic drift than the unconditionally deleterious alleles having the same magnitude of selection coefficients. Recently, Gillespie (1984), Gillespie and Turelli (1989), Barton (1990) and Zhivotovsky and Feldman (1992) have considered stabilizing selection-balance models wherein mutation is not required for the maintenance of polymorphism.

Barton (1990) showed that the variation could be maintained as the pleiotropic side effect of the balanced polymorphism at n loci by overdominance. Such a model could account for high heritability despite stabilizing selection directly on the character. In fact, overdominance produces variation in a manner analogous to mutation to restore the elimination of variability by stabilizing selection. Robertson's (1956) treatment, in such a case, applicable to the biallelic case is discussed by Barton (1990) by introducing random variation in the allelic effects across loci with a symmetric distribution. Gillespie and Turelli (1989) introduced a simple model of additive polygenic inheritance in which the additive contributions of alleles vary with the environment. The average phenotype of a given genotype is therefore a random variable which depends on the macro-environment and leads to a genotype-dependent component of the environmental effects. Under certain simplifying symmetry assumptions, this model implies that the variation of phenotypes over environments decreases as the number of heterozygous loci increases. This means developmental homeostasis of Lerner (1954). Introducing Gaussian fitness function for the phenotype and taking expectation of the relative fitness of a given genotype over the environmental distribution lead to the main result that the mean fitness of the genotype is an increasing function of the number of heterozygous loci. This is what leads to balancing selection, polymorphism and potentially high level of additive genetic variance even though all allelic effects remain additive within each specific environment. Zhivotovsky and Feldman (1992) presented a more elaborate model of stabilizing selection-balance by generalising the genotype-dependent environmental component of Gillespie and Turelli (1989) to include unequal allele frequencies, unequal proportions of complementary gametes and a reduction of the genetic (and phenotypic) variance by linkage disequilibrium. The motivation of the modelling is partly due to polygenic balance model of Mather (1942), which is invoked to explain the greater fitness under stabilizing selection of individuals with complementary pairs of chromosomes (with genotype balance of the type +-+-/-+-+...), relative to those in which configurations like ----... or ++++... predominate. Their results are in agreement, qualitatively, to those of Gillespie and Turelli (1989) who do not keep track of linkage equilibrium. Due to negative linkage disequilibrium, the heritability is expected to be less than half.

REFERENCES

Barton, N.H., 1989, Divergence of a polygenic system subject to stabilizing selection, mutation and drift, *Genet. Res.* 54: 59.

Barton, N.H., 1990, Pleiotropic models of quantitative variation. *Genetics* 124: 773.

Barton, N.H. and Turelli, M., 1987, Adaptive landscapes, genetic distance and the evolution of quantitative characters, *Genet. Res. (Camb.)* 49: 157.

Bulmer, M.G., 1971, The effect of selection on genetic variability, *Amer. Nat.* 105: 201.

Bulmer, M.G., 1972, The genetic variability of polygenic characters under optimizing selection, mutation and drift, *Genet. Res. (Camb.)* 19: 17.

Chakraborty, R. and Nei, M., 1982, Genetic differentiation of quantitative characters between populations or species. I. Mutation and random genetic drift, *Genet. Res. (Camb.)* 39: 303.

Crow, J.F. and Kimura, M., 1964, The theory of genetic loads, *Proc. XI Internat. Cong. of Genet.* 2: 495.

Fisher, R.A., 1918, The correlation between relatives under the supposition of Mendelian inheritance, *Trans. R. Soc. Edinburgh* 52: 399.

Fisher, R.A., 1930, "The Genetical Theory of Natural Selection," Clarendon Press, Oxford.

Galton, F., 1889, "Natural Inheritance," MacMillan, London.

Gillespie, J.H., 1984, Pleiotropic overdominance and the maintenance of genetic variation in polygenic characters, *Genetics* 107: 321.

Gillespie, J.H. and Turelli, M., 1989, Genotype-environment interactions and the maintenance of polygenic variation, *Genetics* 121: 129.

Haldane, J.B.S., 1927, The mathematical theory of natural and artificial selection. Part V, *Proc. Cambridge Philos. Soc.* 23: 838.

Haldane, J.B.S., 1932, "The Causes of Evolution," Longman Green, London.

Hill, W.G. and Keightley, P.D., 1988, Interrelations of mutation, population size, artificial and natural selection, *in:* "Proceedings of the Second International Conference on Quantitative Genetics," B.S. Weir, E.J. Eisen, M.M. Goodman and G. Namkoong, eds., Sinauer, Sunderland, MA. p. 57.

Kimura, M., 1964, Diffusion models in population genetics, *J. Appl. Prob.* 1: 177.

Kimura, M., 1965, A stochastic model concerning the maintenance of genetic variability in quantitative characters, *Proc. Natl. Acad. Sci., USA,* 54: 731.

Kimura, M., 1968, Evolutionary rate at the molecular level, *Nature* 217: 626.

Kimura, M. and Crow, J.F., 1978, Effect of overall phenotypic selection on genetic change at individual loci, *Proc. Natl. Acad. Sci., USA,* 54: 731.

Lande, R., 1975, The maintenance of genetic variability by mutation in a polygenic character with linked loci, *Genet. Res. (Camb.)* 26: 221.

Lande, R., 1976, Natural selection and random genetic drift in phenotypic evolution, *Evolution* 30: 314.

Lande, R., 1979, Quantitative genetic analysis of multivariate evolution applied to brain, body size allometry, *Evolution* 34: 402.

Lande, R., 1980, The genetic covariance between characters maintained by pleiotropic mutations, *Genetics* 94: 203.

Lande, R., 1988, Quantitative genetics and evolutionary theory, *in:* "Proceedings of the Second International Conference on Quantitative Genetics," B.S. Weir, E.J. Eisen, M.M. Goodman and G. Namkoong, eds., Sinauer, Sunderland, MA. p. 71.

Latter, B.D.H., 1960, Natural selection for an intermediate optimum, *Aust. J. Biol. Sci.* 13: 30.

Latter, B.D.H., 1970, Selection in finite populations with multiple alleles. II. Centripetal selection, mutation and isoallelic variation, *Genetics* 66: 165.

Lerner, M., 1954, "Genetic Homeostasis," John Wiley, New York.

Mather, K., 1942, The balance of polygenic combinations, *J. Genet.* 43: 309.

Mode, C.J. and Robinson, H.F., 1959, Pleiotropism and the genetic variance and covariance, *Biometrics,* 15: 518.

Narain, P., 1979, Certain generalisations in response to genetic selection, *in:* "Contributions to Statistics," S.C. Das Memorial Vol., Utkal University, Bhubaneswar, p. 71.

Narain, P., 1985, Progeny testing with auxiliary traits, *Biometrics* 41: 895.

Narain, P., 1988, The evolutionary dynamics of quantitative characters, *Ind. J. Pure Appl. Math.* 19: 125.

Narain, P., 1990, "Statistical Genetics," Wiley Eastern Ltd., New Delhi.

Narain, P., 1990, Genetic models of evolution for polygenic traits, *in:* "Proceedings R.C. Bose Symposium on Probability, Statistics and Design of Experiments," R.R. Bahadur, ed., Wiley Eastern Ltd., New Delhi. p. 527.

Narain, P., 1992, Genetic variability under step-wise discrete mutation and stabilizing selection, *J. Ind. Soc. Agri. Stat.* 42: 171.

Narain, P. and Chakraborty, R., 1983, Genetic differentiation of quantitative characters between populations or species. II. Optimal selection in infinite populations, Unpublished.

Narain, P. and Chakraborty, R., 1987, Genetic differentiation of quantitative characters between populations or species. II. Optimal selection in infinite populations, *Heredity* 59: 199.

Pearson, K., 1904, Mathematical contributions to the theory of evolution. XI. On the influence of natural selection on the variability and correlation of organs, *Phil. Trans. R. Soc., London*, Ser A 200: 1.

Price, G.R., 1972, Extension of covariance selection mathematics, *Ann. Hum. Genet.* 35: 485.

Robertson, A., 1956, The effect of selection against extreme deviants based on deviations or on homozygosis, *J. Genet.* 54: 236.

Roughgarden, J., 1979, "Theory of Population Genetics and Evolutionary Ecology: An Introduction," MacMillan, New York.

Simpson, G.G., 1944, "Tempo and Mode in Evolution," Columbia University Press, New York.ork.

Slatkin, M., 1987, Heritable variation and heterozygosity under balance between mutation and stabilizing selection, *Genet. Res. (Camb.)* 50: 53.

Turelli, M., 1984, Heritable Genetic Variation via Mutation-Selection Balance: Lerch's Zeta meets the Abdominal Bristle, *Theo. Pop. Biol.* 25: 138.

Turelli, M., 1985, Effects of pleiotropy on prediction concerning mutation-selection balance for polygenic traits, *Genetics* 111: 165.

Turelli, M., 1988, Population genetic models for polygenic variation and selection, *in:* "Proceedings of the Second International Conference on Quantitative Genetics," B.S. Weir, E.J. Eisen, M.M. Goodman and G. Namkoong, eds., Sinauer, Sunderland, MA. p.601.

Turelli, M. and Barton N.H., 1990, Dynamics of polygenic characters under selection, *Theo. Pop. Biol.* 38: 1.

Wright, S., 1931, Evolution in Mendelian populations, *Genetics* 16: 97.

Wright, S., 1932, The roles of mutation, inbreeding, cross-breeding and selection in evolution, *Proc. Sixth Internat. Cong. Genet.* 1: 356.

Wright, S., 1940, Breeding structure of populations in relation to speciation, *Amer. Nat.* 74: 232.

Zhivotovsky, L.A. and Feldman, M.W., 1992 On models of quantitative genetic variability: A stabilizing selection-balance model, *Genetics* 130: 947.

HALDANE'S CONTRIBUTIONS TO THE UNDERSTANDING OF THE EVOLUTION OF VERTEBRATE IMMUNE SYSTEM

Naoyuki Takahata

National Institute of Genetics
The Graduate University for Advanced Studies
Mishima 411, Japan

FROM HALDANE TO MAJOR HISTOCOMPATIBILITY COMPLEX (*MHC*)

Researches on tumor transplantation and blood groups simultaneously began at the turn of this century. However, they had nothing in common for the following three decades. The man who eventually brought them together was J.B.S. Haldane (see Klein, 1986; p. 7). In 1933, Haldane made a tour of the United States during which he talked to C.C. Little, then at the Jackson Laboratory in Bar Harbor, about his work on tumor transplantation. In a paper published in *Nature*, Haldane (1933) wrote:

> The genetics of reaction to transplantable tumours have been very fully worked out by Little and his colleagues. The laws disclosed are precisely similar to those which govern the transplantation of normal tissue or the transfusion of blood or of leukæmic corpuscles.

Haldane was impressed by the observation that the tumor resistance factors are akin to blood group antigens and the failure of the transplants to grow is similar to the destruction of incompatible blood cells following transfusion. To study new antigens, he encouraged one of his students, P.A. Gorer, and provided him with three inbred strains of mice, I, II and III, which were brought from the Jackson Laboratory. Three years later, Gorer discovered a new antigen; histocompatibility antigen II now designated as H-2. Histocompatibility antigens have major effects on tumor transplants or skin grafts, and subsequent studies have revealed that in most vertebrates the genes form a complex located on a single chromosome. These genes are now referred to as the major histocompatibility complex (*Mhc*).

The *Mhc* genes consist of class I and II, and the allelic product is a glycoprotein, a polypeptide with one or more polysaccharide side chains. It is expressed on cell surfaces; the bulk of it represents the extracellular region, a small part spans the plasma membrane, and another short segment protrudes into the cytoplasm. One of the earliest studies of the structure of Mhc molecule was conducted by S.G. Nathenson in the 1960s (see Nathenson, Uehara and Ewenstein, 1981; for a review), but only recently has the folding pattern of class I molecules been determined by X-ray crystallography (Bjorkman et al., 1987a, b). There exists a distinct region which forms a groove. The groove of a class I molecule contains 57 amino acid residues to which intracellularly processed peptides are held by noncovalent interaction, and these residues are collectively called the peptide-binding region (PBR), the antigen recognition site (ARS), etc. In the functional *Mhc* genes the rate of nonsynonymous (amino acid replacement) substitutions in the PBR is higher than that of synonymous substitutions, whereas the opposite is true for the rest of the gene. The high nonsynonymous substitution rate in the PBR is incompatible with neutrality (Kimura, 1968), and it is this region to which positive selection is conspicuously directed (Hughes and Nei, 1988, 1989). In view of these recent findings, it is rather surprising to come across the following statement made as early as in 1949 by Haldane:

> Now every species of mammal and bird so far investigated has shown a quite surprising biochemical diversity revealed by serological tests. The antigens concerned seem to be proteins to which polysaccharide groups are attached. We do not know their functions in the organism, though some of them seem to be part of the structure of cell membranes.

Haldane might have imagined the new antigens as something similar to those of classical blood groups. Nevertheless, it is not clear from his writing how he could infer the structure of the antigens so precisely. And while he said he did not know their functions, he speculated about them:

> I wish to suggest that they (antigens) may play a part in disease resistance, a particular race of bacteria or virus being adapted to individuals of a certain range of biochemical constitution, while those of other constitutions are relatively resistant.

This speculation too is remarkable. For one thing, the idea of pathogen adaptation or molecular mimicry (see the following section) was suggested and for the other, the function was rightly visualized such that the antigens are an agent to confer susceptibility or resistance of individuals to particular pathogens. A recent definition of *Mhc* genes in terms of function is given as a group of genes coding for molecules that provide the context for the recognition of foreign antigens by T lymphocytes (Klein, 1986; p. 21). The primary function of Mhc molecules is thus to present foreign antigens to cytotoxic or helper T cells and trigger the host immune response. Although an Mhc molecule can bind a variety of peptides, there is indeed specificity and different Mhc molecules have different repertoires for association with peptides (Carreno et al., 1990; Falk et al., 1991; Guo et al., 1992). Hence, an individual can be either resistant or susceptible to particular pathogens, depending on whether Mhc molecules possessed by the individual can or cannot form a complex with pathogen peptides.

Of course, the whole story concerning the vertebrate immune response is not this simple (Klein, 1990; for review). One complication occurs during the self tolerance induction (Fig. 1). For a thymocyte to develop into a functional T cell, an appropriate interaction between T cell receptors (TCRs) and Mhc molecules is necessary (Blackman et al., 1990). Experiencing positive selection, T cell clones undergo negative selection in response to the interaction with the Mhc and self peptide complex on the plasma membrane (Ramsdell and Fowlkes, 1990; Boehmer and Kisielow, 1990). Because of intense competition among self peptides, T cells choose, as self epitopes, only those self peptides that have a high binding affinity to Mhc molecules. Ohno (1992) argued, in the case of class I *Mhc*, that "self" is at most 1000 self peptides that can strongly bind to Mhc molecules, deleting autoreactive T cell clones in the thymus. The specificity is higher in class II molecules than in class I molecules. Rudensky et al. (1991) suggested that the "self" group of class II *Mhc* may be even less complex than 100 different peptides, since more than 200 (or 1%) of the Mhc molecules on the cell surface are required to activate naive T cells. The repertoire of mature T cell clones is thus *Mhc*-restricted, resulting in the so called blind spot (Klein, 1986; p. 578-581). There may therefore be a situation in which individuals lack T cells that recognize non-self epitopes even when Mhc molecules present them. The existence of the blind spot in the T cell repertoire is a potential disadvantage in the host immune response.

MINORITY ADVANTAGE

The maintenance mechanism of antigen diversity suggested by Haldane (1949) was minority advantage. He argued that, in every species, at least one of the factors which kills it or lowers its fertility must increase in efficiency as the species becomes denser.

> Probably a very small biochemical change will give a host species a substantial degree of resistance to a highly adapted microorganism. It means that it is an advantage to the individual to possess a rare biochemical phenotype. For just because of its rarity it will be resistant to diseases which attack the majority of its fellows. And it means that it is an advantage to a species to be biochemically diverse, and even to be mutable as regards genes concerned in disease resistance.
>
> ....
>
> We have here, then, a mechanism which favours polymorphism, because it gives a selective value to a genotype so long as it is rare.

These statements quoted from "Disease and evolution" (Haldane, 1949) appear to have been a stimulus to subsequent studies of maintenance mechanisms of genetic variation in natural populations, in particular for coevolution of host and parasite (see, for example, Clarke, 1976; Hamilton, 1980; May, 1983; Seger, 1988; Takahata and Nei, 1990).

To Haldane, individuals with predominant alleles are susceptible to pathogens while those with rare alleles are resistant because they are not attacked by pathogens. Why not attacked is simply because of its rarity. Haldane might have meant that there is a critical population density of individuals of each genotype below which infection among these individuals of the same genotype is hindered. He does not explicitly mention whether the hindrance is due to a result of the coupled evolutionary histories of host and parasite, as assumed in most of the subsequently developed models, or due to the prerequisite population density of a particular host genotype for the transmission of virulent parasites to occur. As regards *Mhc* polymorphism, the former possibility has received much attention, while the latter very little.

One familiar hypothesis is molecular mimicry (Damian, 1964, 1987) which postulates that microorganisms try to escape the host immune recognition by mimicking host antigens. Molecular mimicry must occur against host *Mhc*; otherwise it cannot be a driving force of *Mhc* polymorphism. The idea of molecular mimicry is related to minority advantage if rare alleles have less opportunity to be mimicked. Similarly, the hypothesis by Snell (1968) and Bodmer (1972) assumes that host individuals carrying new antigens are at an advantage because pathogens have not had time to adapt to infecting these individuals. It is now well documented that host-cell proteins and infectious agents share T cell epitopes (Fig. 2; Klein, 1990; p. 477-481). This may be due either to molecular mimicry or to pathogen adaptation, although Ohno (1991) interpreted it as a chance event. Whatever the cause is, if the self and non-self epitopes are identical, pathogens evade the host immune response. If the self and non-self epitopes are similar but not identical, the host can respond immunologically to the microbial epitopes. The response may then break the immunological tolerance and lead to the development of autoimmune disease. In either case, the fitness (viability) of host individuals decreases. However, there seems no experimental evidence that molecular mimicry or pathogen adaptation is directed only toward common *Mhc* alleles. A recent finding of association between a particular *Mhc* allele/haplotype and malaria (Hill et al., 1991) indicated that the *Mhc*

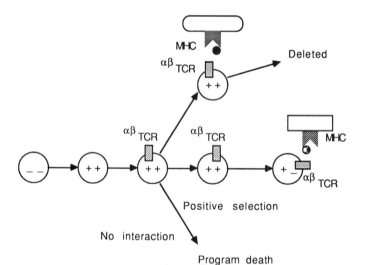

Figure 1. A hypothetical tolerance induction pathway (modified from Ramsdell and Fowlkes, 1990). A filled circle represents a polypeptide recognized as self, while a partially filled one represents one not so recognized. The order of positive and negative selection is controversial (Takahama, Shores and Singer, 1992). In the text, it is assumed that deletion of autoreactive T cell clones takes place after moulding (positive selection).

H-2	K^d	1	2 Y	3	4	5	6	7	8	9 I(L)	
H-2	K^b	1	2	3	4	5 F(Y)	6	7	8 L		
H-2	D^b	1	2	3	4	5 N	6	7	8	9 M	
HLA	A2	1	2 L	3	4	5	6	7	8	9 V	
HLA	Aw68	1	2 V(T)	3	4	5	6	7	8	9 R	10-11
HLA	B27	1	2 R	3	4	5	6	7	8	9 K(R)	

Figure 2. Dominant Anchor Residues (DARs) in self-peptide binding-motifs of class I molecules. DARs are applied if a peptide position (numbered) is occupied by one or a few residues with closely related side chains. Self-peptide binding-motifs are often shared by non-self epitopes (after Falk et al., 1991; Jardetzky et al., 1991; Guo et al., 1992). Here, only DARs are shown by single amino acid codes (bold-face letters). K^d and K^b are allelic forms at the mouse H-2 K locus, and A2 and Aw68 are alleles at the HLA-A locus, so that "self" appears to be completely different between these alleles. The five most frequently used amino acid residues in 2681 sequenced human genes are L, S, G, A, and Q ($\geq$ 7%) whereas the six most infrequently used are F, Y, H, C, M, and N ($\leq$ 4%). Intermediate are V (6.3%), P (5.9%), K (5.7%), T (5.6%), and R (5.3%): The codon usage pattern is fairly similar among mammals (Wada et al., 1992).

The putative class II DR1-binding motif is at variance in length and the key residues in the motif are referred to as the index (I). The first key position (I) is occupied by a positively charged residue, I+5 by a hydrogen-bond donor, and I+9 by a hydrophobic residue (Chicz et al. 1992). Main self-peptides associated with DR1 are those from class I molecules (HLA-A2), transferrin receptor, (Na^+ + K^+) ATPase and bovine fetuin-derived peptides, all of which are recyclying membrane proteins and endocytosed endogenous proteins that travel the pathway of incoming endocytotic vesicles. The precise motif of DR1 is not deduced because of the irregular length of bound peptides and an ambiguity in the position of the first amino acid residue.

allele/haplotype confers resistance, rather than susceptibility. This means that inasmuch as malaria prevails, there is no advantage of the *Mhc* allele/haplotype being rare. What is needed to support the idea of minority advantage seems to be evidence for susceptibility of common *Mhc* alleles to pathogens. It was also shown theoretically that indirect forces mediated through interaction between host and parasite are very model-dependent (Seger, 1988) and that the instability of such dynamical systems results in a relatively small number of segregating alleles in a finite population (Takahata and Nei, 1990). Molecular mimicry or pathogen adaptation may actually occur, but it alone does not seem to be a sufficient mechanism to explain the extraordinary polymorphism at *Mhc* loci.

MHC POLYMORPHISM

The polymorphism at a functional *Mhc* locus is unusual in many respects (see, for example, Klein, Gutknecht and Fischer, 1990) despite the fact that the gene is not particularly large; about 360 and 240 amino acid residues in a class I and II gene, respectively. Importantly, the mutation rate at a primate *Mhc* locus (*DRB1*) estimated from the synonymous substitution rate may be as low as $1.2 \pm 0.4 \times 10^{-9}$ per site per year (Satta et al., 1991). These imply that *Mhc* loci are not at all highly mutable, contrary to Haldane's expectation as well as an early view held among many immunologists around the time when the true extent of *Mhc* polymorphism was first realized (Klein, 1986; p. 747-748).

Table 1 Polymorphism at human *Mhc* (*HLA*) loci (modified from Satta 1992)

	class I loci			class II locus
	A (19 [a])	B (26 [a])	C (6 [a])	DRB1 (19 [a])
n_a	19	37	9	47
D_{mean} [b]	38.2	38.8	28.5	35.9
	(4.2%)	(3.9%)	(3.9%)	(8.0%)
D_{max} [b]	66.9	63.2	47.0	68.6
	(8.2%)	(7.6%)	(6.6%)	(14.6%)
γ	4.50	5.49	1.98	6.06
h	–	0.12	–	0.30

[a] Number of DNA sequences analyzed; The number of codons compared is 361 for class I genes, and 189 for a class II gene.
[b] The number of nucleotide substitutions per gene in all pairwise comparisons (D_{mean}) and that between most distantly related alleles (D_{max}). In parentheses, the synonymous substitutions per site in the corresponding comparisons are given. If the synonymous substitution rate is 1.2×10^{-9} per site per year (Satta et al., 1991), 1% synonymous substitutions per site amounts to 4 million-year separation of two alleles.
n_a, number of known alleles; γ, the ratio of the nonsynonymous substitution rate per site in the PBR to the synonymous substitution rate; h, heterozygosity per PBR nonsynonymous site (after Klein et al., 1993).

Table 1 summarizes the unusual polymorphism revealed by DNA sequences of human *Mhc* (*HLA*) alleles. Most noteworthy is the long persistence time of allelic lineages [the so called *trans*-species mode of polymorphism by Klein (1980)], often being over tens of millions of years (Fig. 3). This is underscored by the observation that the separation time of two alleles at a non-*HLA* locus rarely exceeds one million years, values consistent with the neutrality (Kimura, 1968) in a population of effective size of 10^4 (Nei, 1987; p. 402). Because of the long persistence, two *Mhc* alleles differ by a large number of nucleotide changes not only at the PBR

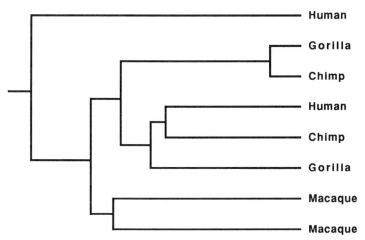

Figure 3. *Trans*-species mode of primate *Mhc* polymorphism (*DRB1*). Because of the long persistence time of allelic lineages, alleles within a species often diverged long before speciation. As a result, different species share the same allelic lineages and the genealogy of orthologous alleles does not reflect the phylogenetic relationships of species (modified from Klein et al., 1993).

nonsynonymous sites on which positive selection operates but also at the linked synonymous sites. The mean pairwise nucleotide differences per site are about 4% at class I *HLA* loci and 8% at the class II *DRB1* locus (Table 1), whereas the mean value over 49 non-*HLA* loci is very small, ranging from 0.03% to 0.11% (Li and Sadler, 1991). Furthermore, the ratio of the PBR nonsynonymous substitution rate to the synonymous one (γ) is larger than 1; largest at the *DRB1* locus (6.1) and smallest at the *C* locus (2.0) (Takahata, Satta and Klein, 1992a). These features of *Mhc* polymorphism are clearly incompatible with neutrality, so that some form of natural selection must have been involved (Hughes and Nei, 1988, 1989; Klein et al., 1990; Klein and Takahata, 1990; Nei and Hughes, 1991; Takahata, Satta and Klein, 1992b; Klein et al., 1993).

WHY OVERDOMINANCE?

In self and non-self discrimination, some self peptides 8-11 amino acids long play an essential role in association with class I Mhc molecules (Falk et al., 1991; Jardetzky et al., 1991; Hunt et al., 1992; Guo et al., 1992), and some other self peptides of 13-24 amino acids play a same role in association with class II Mhc molecules (Rudensky et al., 1991; Chicz et al., 1992). To be T epitopes, there is a requirement for the amino acid residues at critical peptide positions. When the requirement is stringent, or infrequently used amino acids are preferred at critical positions as of mouse class I genes (Ohno, 1992), there may not be a large number of self peptides which are presented in sufficient density and have a high binding affinity to self Mhc molecules (Schild et al., 1990). A similar situation may occur for class II Mhc molecules. Chicz et al. (1992) showed that naturally processed peptides eluted from *HLA-DR1* are limited and predominantly derived from Mhc-related molecules. The tolerance may then be rendered to all of them so that the difference between self and non-self becomes certain. Because of this certainty, there is less danger in responding to non-self epitopes which also have a high binding affinity to Mhc molecules. However, being very selective, T cell epitopes may not be found at all on given antigens.

The requirement for the amino acid residues at critical positions of self and non-self epitopes may be permissive, so that commonly used amino acids are preferred as in human class I genes (Ohno, 1992). There are possibly too many self peptides as candidates of self epitopes and the tolerance may not be directed toward all of self epitopes. Under this circumstance, the distinction between self and non-self becomes uncertain. If the immune response is directed only to peptides which necessarily have a high binding affinity to Mhc, such response is dangerous because some untolerized self epitopes happen to be recognized as non-self. Ohno (1992) argued that T cells therefore ignore the issue of self and non-self, and take action only to predominantly expressed non-self peptides which have a low binding affinity to Mhc. How T cells avoid the issue and what on earth then is the self tolerance are not clear. But provided that this can happen, the situation concerning the blind spot is very different from the one expected under the stringent case. Since immature T cells do not undergo negative selection with low affinity self peptides, there should be no blind spot for non-self epitopes which have a low binding affinity to Mhc. The lack of the blind spot has an obvious advantage in the immune response. It is tempting to speculate that the *Mhc* might have evolved from being stringent to permissive. (If self peptides are required in positive selection of T cell clones, there must always be a blind spot for non-self epitopes with a low binding affinity to Mhc molecules. The fact that the immune response can be directed to low affinity non-self peptides whenever expressed in sufficient density would imply that self peptides may not be required in positive selection.)

Now consider a heterozygote and its respective homozygotes at an *Mhc* locus. Because two allelic products in the heterozygote are equally expressed or codominant in phenotype (Hood et al., 1984; p. 180), the size of the blind spot should be no smaller than that of either homozygote. Therefore, when exposed to the same pathogens, the heterozygote is possibly more disadvantageous than the homozygotes at the T cell level, leading to underdominance selection. When this type of selection is prevalent, no stable polymorphism is maintained at an *Mhc* locus. (This argument may apply only to stringent Mhc.) Equally important is selection occurring at the Mhc level. Since the heterozygote can produce two different types of Mhc molecules, the repertoire of non-self epitopes should be broader in the heterozygote than that in either homozygote. Hence, there is an obvious advantage to being heterozygous in the presentation of various non-self epitopes (Doherty and Zinkernagel, 1975).

For a more quantitative argument, suppose that there are two *Mhc* alleles, A_1 and A_2, segregating in a population and that there is a pathogen which infects host individuals regardless of their genotypes and frequencies in the population. Since the presentation of non-self peptides is a dominant character, there are four situations in which the *Mhc* genotypes react differently. The first situation is that neither of the *Mhc* allelic products can present any processed non-self peptide, or all the three genotypes are susceptible to the pathogen. If this happens, natural selection does not favor any genotypes, so that there are essentially three situations to be focused on (Table 2a). Also, there are four situations with respect to the presence or absence of T cell clones that can recognize the complex of Mhc and processed non-self peptides. If none of the three genotypes possesses appropriate T cell clones, they are equally disadvantageous: There are again essentially three situations to be considered at the T cell level (Table 2b).

Under linkage equilibrium, the genetic background of three *Mhc* genotypes is the same among all individuals. Therefore the T cell repertoire or the blind spot should depend only on the differences in *Mhc* alleles. When infected by a particular pathogen, host individuals process the over-expressed foreign peptides. Suppose that there are a variety of such peptides. If there are too many non-self epitopes, all *Mhc* genotypes would present some of them and be equally fit at this first stage of immune response. Also, the differences in T cell repertoire would be immaterial if the number of non-self epitopes is large; there always exist some T cell clones that recognize some of non-self epitopes. Hence, pathogens of this type, if present, are unlikely the cause of natural selection.

In order for the genotype-dependent immune response to occur, the number of non-self epitopes presented by a pathogen must be relatively small. Indeed, this seems to be the rule; there are only two epitopes in the influenza hemagglutinin (Braciale et al., 1989; Falk et al., 1991). If there exist only a few non-self epitopes derived from the pathogen, then it is conceivable that the presentation to T cells depends on *Mhc* alleles. This would be particularly so for stringent class I as well as class II *Mhc*. For instance, the first position of two dominant anchor residues of H-2 K^b is different from that of another allele H-2 K^d and their non-self epitopes appear to be entirely different from each other (see Table 1 and Table 3 in Falk et al., 1991; Fig. 2). Similarly, even in the permissive case, dominant anchor residues may be entirely different between alleles as suggested by *HLA-A2* and *HLA-Aw68* (Fig. 2). However, if one non-self epitope is presented by a heterozygote, there is necessarily at least one type of homozygote that can present the same epitope. The genotype-dependent resistance/susceptibility can occur against a single non-self epitope, but it alone never confers a heterozygote advantage.

In order for the immune response to be something like heterozygote advantage, there must be multiple non-self epitopes or pathogens to which a population of individuals is exposed. For convenience, assume that there is only one non-self epitope from each pathogen, but that the population is exposed to many pathogens. We define p_1, p_2 and p_3 as the fractions of pathogens whose epitopes can bind to A_1, A_2 or both allelic products, respectively ($p_1 + p_2 + p_3 = 1$). We further define q_1 as the fraction of pathogens against which there exist T cell clones that recognize the complex with A_1. The q_2 and q_3 are defined in similar ways if the complex with A_2 or with both A_1 and A_2 in heterozygotes is recognized by T cell clones ($q_1 + q_2 + q_3 = 1$). The necessary and sufficient condition for heterozygote advantage or overdominance selection becomes

$$p_1 > \frac{q_2}{q_2 + q_3} \quad \text{and} \quad p_2 > \frac{q_1}{q_1 + q_3}$$

(Table 2). Clearly, it is necessary that most of the time there exist responding T cell clones independently of genotypes or no virtual differences in the blind spot determined by the *Mhc* allelic products ($q_3 > q_1$ and q_2). It is also necessary that A_1 and A_2 allelic products can present different sets of non-self epitopes equally well. In other words, of importance is the small extent of overlap (p_1 and $p_2 \gg p_3$) between the peptide repertoire of A_1 and that of A_2. When *Mhc* alleles differ at a number of amino acids in the PBR, the peptide-binding specificities may become nonoverlapping, as demonstrated by H-2 K^b

and K^d, HLA-A2 and $Aw68$, or HLA-A2 and -B37 at different loci (Carreno et al., 1990). If on the other hand the peptide binding specificities of A_1 and A_2 are largely overlapping ($p_3 \gg p_1$ and p_2), the selection pressure becomes weak or it may be offset by the possible existence of blind spots.

In short, the conditions for heterozygote advantage or overdominance are that Mhc alleles are sufficiently diverged at the PBR in peptide presentation, but that the T cell repertoires restricted by these alleles do not differ significantly among the genotypes. As mentioned earlier, the latter condition may be easily met for permissive class I molecules. The model of overdominance selection for Mhc polymorphism has thus its biological basis consistent with the function of Mhc and it can also explain essential features of the polymorphism (Takahata and Nei, 1990; Takahata et al., 1992b; Klein et al., 1993).

Table 2a Mhc recognition patterns of pathogens and the fractions of pathogens to which a host population is exposed

fractions	A_1A_1	A_1A_2	A_2A_2
p_1	+	+	−
p_2	−	+	+
p_3	+	+	+

Table 2b T cell repertoires for the Mhc complex with non-self epitopes derived from various pathogens and their relative fractions

fractions	A_1A_1	A_1A_2	A_2A_2
q_1	+	−	−
q_2	−	−	+
q_3	+	+	+

The immune response occurs against pathogens that can be presented by Mhc (+ sign in Table 2a) and if there exist T cells which recognize the complex (+ sign in Table 2b). There are nine different ways in the response. The relative frequency when A_1A_1 and A_2A_2 genotypes are resistant to all the pathogens is $(p_1 + p_3)(q_1 + q_3)$ and $(p_2 + p_3)(q_2 + q_3)$, respectively, while heterozygotes can be resistant with probability q_3.

AN ALTERNATIVE MODEL SUGGESTED BY HALDANE

Nevertheless, it is important and interesting to examine other possibilities. One such is minority advantage as discussed by Haldane (1949). His argument was rather vague by present standards, and there are a number of ways to model his idea of minority advantage. One way is based on Mhc-restricted T cell repertoires.

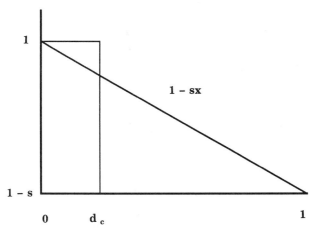

Figure 4. Truncation model of minority advantage. Viability (fitness component; ordinate) vs. allele frequency x (abscissa).

Klein (1991) asked which parasites, among many (more than 100 in humans), select *Mhc* variants and in what manner, and suggested that the main reason for the *Mhc* polymorphism *is* the clonal evolution of the parasites in a geographically subdivided population. Along this line of thought, it is essential in modeling Haldane's idea to consider the fact that Mhc must cope with not one but many pathogens and that each pathogen is likely to have multiple host species (like *Trypanosomes*). There is no one-to-one correspondence between an *Mhc* allele and a pathogen, or even anything similar to it. Under this situation, we cannot expect that any change in frequency of an *Mhc* allele greatly affects the constitution of parasites or vice versa. Because of a large number of different pathogens, assume that an *Mhc* allele is intrinsically susceptible to *some* of the pathogens and this susceptibility is due to the T cell repertoire restricted by the Mhc molecules and self peptides. The latter assumption may simply result from unfortunate chance sharing of self and non-self epitopes owing to their short lengths as well as a small number of dominant anchor residues (Fig. 2). It is not assumed either that the susceptibility is a result of some type of group selection (Wills, 1991) nor that the susceptibility is a molecular device to prevent the whole population from extinction. To model minority advantage, assume that these evading pathogens can be prevalent among host individuals carrying a particular *Mhc* allele only when the allele frequency in a population exceeds a certain critical value. (J. F. Crow informs me that C. Wills and D. R. Green have independently arrived at a very similar model.) This might be expected if the parasite can spread from one individual to another only when the host density is high enough.

The model is summarized as follows: (1) There is a threshold frequency (d_c) of an *Mhc* allele, (2) above this frequency, those individuals with this *Mhc* allele are infected by some pathogens, reducing the relative fitness to $1 - s$, (3) s is constant, or the constitution of the pathogens is stable, not affected by changes in *Mhc* allele frequencies, and (4) the susceptibility is a dominant character because it is related to the blind spot (Fig. 4). Using this selection scheme and the infinite allele model of mutation (the per-generation rate being u), the allelic genealogy (the family relationships among *Mhc* alleles) was generated in a finite population of size N [see Takahata and Nei (1990) for the simulation method]. Table 3 shows some simulation results. This type of truncation selection can be very efficient, under certain conditions, in maintaining a large number of alleles. Mean heterozygosity (H) can be very large and the rate of codon substitutions (γ) is moderately enhanced. More importantly, a fairly large number of allelic lineages may persist in a population for a quite long time (Fig. 5). To examine the allelic genealogy thus generated, we may compute the ratio of the separation time between the two most distantly related alleles (T_c) to that in all pairwise comparisons (T_d). If the allelic genealogy is similar to that expected under random bifurcation of alleles, the ratio becomes $2(1 - H)$, as in a model of overdominance selection (Takahata, 1990). The ratio in the simulation is often close to this expectation. All these features are consistent with those actually observed at functional *Mhc* loci (Table 1; see also Takahata et al., 1992b; Satta, 1992).

For these features to be manifest, however, a condition seems necessary: The value of Nu must be much larger than 0.01. In fact, simulation for the case of Nu < 0.03 (Table 3) shows that the allelic genealogy becomes similar to that expected under neutrality. This may be explained by noting that once the frequency of a new allele exceeds d_c, it becomes neutral so that it does not have a high probability of replacing one or two existing alleles at high frequencies. The situation is different under frequent introduction of new alleles into a population. When a number of alleles are segregating, their frequencies must be low in which case a new allele can replace one of them with appreciable probability. For this situation to be realized, a relatively high mutation frequency (u), a large population number (N), or both are required. However, since u is estimated as 2.4×10^{-6} per class I PBR per generation and 7.2×10^{-7} per class II PBR per generation (Takahata et al., 1992a; Klein et al., 1993), N must be at least as large as 10^4. Conversely, under the present model, a fairly large number of breeding individuals in human populations is suggested, as in an model of overdominance selection (Takahata et al., 1992b).

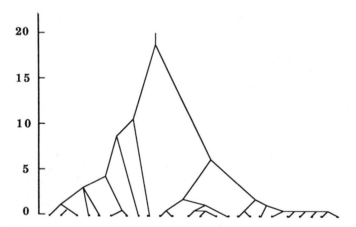

Figure 5. Allelic genealogy under a truncation model of minority advantage. Ns = 90, Nu = 0.1 and d_c = 0.05. The unit of time (ordinate) is N generations. There are 24 alleles (tips) in this simulation. γ = 5.7 and H = 0.94.

Table 3 Simulation results in a model with threshold frequency d_c

	H	n_a	T_c	T_d	γ
			Nu = 0.1		
Ns (d_c)					
0.0 (0.05)	0.27 ± 0.27	2.8 ± 1.0	0.8 ± 1.0	0.7 ± 1.0	0.9 ± 0.1
1.0 (0.05)	0.33 ± 0.20	3.3 ± 1.4	2.2 ± 1.9	1.6 ± 1.4	1.0 ± 0.2
50. (0.05)	0.54 ± 0.14	4.7 ± 1.1	6.3 ± 3.3	3.7 ± 1.4	2.4 ± 0.2
90. (0.10)	0.91 ± 0.01	15.7 ± 2.2	22.2 ± 18.9	11.9 ± 8.7	7.2 ± 0.6
90. (0.05)	0.94 ± 0.01	21.0 ± 2.0	31.2 ± 16.6	18.4 ± 2.9	7.2 ± 0.7
90. (0.025)	0.95 ± 0.01	25.4 ± 2.6	48.4 ± 25.5	23.9 ± 10.4	5.9 ± 0.7
			Ns = 100, d_c = 0.05		
Nu					
0.01	0.13 ± 0.21	1.6 ± 0.8	1.1 ± 1.9	1.0 ± 1.9	2.8 ± 0.2
0.02	0.19 ± 0.21	1.8 ± 0.8	1.3 ± 1.6	1.2 ± 1.5	3.1 ± 0.6
0.03	0.29 ± 0.19	2.9 ± 1.6	1.8 ± 1.9	1.4 ± 1.4	3.2 ± 0.8
0.04	0.40 ± 0.31	5.2 ± 6.1	6.9 ± 8.3	4.8 ± 6.7	3.7 ± 0.9
0.045	0.58 ± 0.32	7.4 ± 7.3	9.4 ± 17.0	5.6 ± 8.9	3.9 ± 0.8
0.05	0.89 ± 0.15	19.5 ± 6.3	41.6 ± 24.6	27.0 ± 14.5	4.8 ± 0.9
0.10	0.95 ± 0.00	24.8 ± 1.3	45.0 ± 26.8	27.0 ± 12.7	4.4 ± 0.6
0.50	0.96 ± 0.00	31.0 ± 2.6	17.9 ± 10.1	9.9 ± 4.9	3.0 ± 0.3
1.00	0.96 ± 0.00	36.7 ± 1.6	14.2 ± 6.4	7.1 ± 1.9	2.5 ± 0.2

Ten replications. H, heterozygosity; n_a, number of alleles in a sample of size 2N; T_c, coalescence time for different alleles; T_d, pairwise coalescence time for different alleles. T_c and T_d are measured in units of N generations. γ, rate of codon substitutions relative to u. The numbers after the ± sign are the standard deviations. See Takahata and Nei (1990) for details of the simulation method.

The fitness reduction of individuals caused by the susceptibility to pathogens may not be truncation-type. It has been shown that if the fitness is proportional to $1 - s_i x_i$ where s_i and x_i are, respectively, the selection coefficient and frequency of the i-th *Mhc* allele under study, the dynamics of the allele frequency change is exactly the same as that under overdominance selection (Takahata and Nei, 1990; Denniston and Crow, 1990). Thus, there can exist many models of frequency-dependent selection, some of which certainly explain the *Mhc* polymorphism as well as or better than overdominance. In general, however, the expression "frequency-dependent selection" does not specify the cause and mechanism. While the cause for *Mhc* polymorphism is most likely pathogens, data for the precise mechanism are still scanty and make hypothesis testing difficult. For example, the threshold model predicts that the number of allele (n_a) is largely determined by the reciprocal of d_c (see n_a for Nu ≥ 0.05 and large values of Ns in Table 3). Therefore, if we set $d_c = 0.01$ and assume a large value of Ns, it is not difficult to generate the polymorphism compatible with the observed *Mhc* polymorphism. There are no data which allow rejection of an *ad hoc* hypothesis that, only when there exist more than 200 copies of an *Mhc* allele in a population, some pathogens can become prevalent among the individuals carrying this allele. At present, it seems that a model of overdominance selection is most parsimonious and provides a legitimate explanation, but other possibilities cannot be ruled out.

SPECIATION

Crow (1990a) remarked that Haldane's papers abound in speculations, and naturally many of them have turned out to be wrong, but much more remarkable are the ones that have turned out to be on target and path-breaking. Crow (1990b, 1992a, b) provides more detailed comments about the breadth of Haldane's knowledge and interest. Here, I have chosen only two papers from his 23 books, 400 scientific papers, and 345 articles in the *Daily Worker*, and focused on his insightful speculations about Mhc antigens. The most fascinating speculation I have ever seen is this (Haldane, 1949):

> Once a pair of races is geographically separated they will be exposed to different pathogens. Such races will tend to diverge antigenetically, and some of this divergence may lower the fertility of crosses.

There has been no proof or disproof of this theme, but the truth may be that no one has seriously considered this problem (see Klein, 1991, for effects of population subdivision on *Mhc* polymorphism). Elucidation of the molecular basis of speciation has become a target in the field of evolutionary biology, yet it is a difficult task even in Drosophila species.

Of course the self defense system in vertebrates is different from that in insects and other invertebrates (Klein, 1986; p. 716-732), and some reproductive isolation mechanisms in vertebrates may be associated with Mhc antigens and/or the T cell repertoire. The reason why *Mhc* does not have a large number of functional copy genes despite the apparent advantage in presenting foreign peptides must lie in that too many diverged *Mhc* genes result in too large a blind spot. By the same token, it is possible, and likely that the blind spot becomes too large if an individual has very different alleles at all the functional class I and class II *Mhc* loci. Hybrids between antigenically diverged individuals will then have a molecular basis for reduced fitness in relation to T cells. A blind spot would increase susceptibility to disease, and most infectious diseases are a greater danger to infants and young children than to adults. If this turns out to be the case, the fitness reduction in hybrids should be related to viability, not to fertility.

It remains to see whether Haldane once again was right in his speculation on speciation mechanisms. Cichlid fishes under explosive evolution in the East African Lakes Victoria, Malawi and Tanganika (Sturmbauer and Meyer; 1992) might be an excellent material to examine this possibility.

ACKNOWLEDGMENTS

I thank Drs. Yoko Satta, Keiko Udaka, Jan Klein and James F. Crow for their many useful comments on an early version of this paper. Thanks are also due to Drs. Masatoshi Nei and Partha P. Majumder for providing me with an opportunity of writing this paper. This paper is supported in part by a grant from the Third World Academy of Sciences to the Organizing Committee of Haldane Centenary International Conference on Human Genetics, Indian Statistical Institute, Calcutta, and in part by grants from the Ministry of Education, Science and Culture, Japan.

REFERENCES

Braciale, T.J., Sweetser, M.T., Morrison, L.A., Kittlesen, D.J., and Braciale, V.L., 1989, Class I major histocompatibility complex-restricted cytolytic T lymphocytes recognize a limited number of sites on the influenza hemagglutinin, *Proc. Natl. Acad. Sci. USA* 86:277.

Bjorkman, P.J., Saper, M.A., Samraoui, B., Bennet, W.S., Strominger J.L., and Wiley, D.C., 1987a, Structure of the human class I histocompatibility antigen, *HLA-A2, Nature* 329:506.

Bjorkman, P.J., Saper, M.A., Samraoui, B., Bennet, W.S., Strominger, J.L., and Wiley, D.C., 1987b, The foreign antigen binding site and T cell recognition regions of class I histocompatibility antigens, *Nature* 329:512.

Blackman, M., Kappler, J., and Marrack, P., 1990, The role of the T cell receptor in positive and negative selection of developing T cells, *Science* 248:1335.

Boehmer, H.V., and Kisielow, P., 1990, Self-nonself discrimination by T cells, Science 248:1369-1373.

Bodmer, W.F., 1972, Evolutionary significance of the *HLA* system, *Nature* 237:139.

Carreno, B.M., Anderson, R.W., Coligan, J.E., and Biddison, W.E., 1990, *HLA-B37* and *HLA-A2.1* molecules bind largely nonoverlapping sets of peptides, *Proc. Natl. Acad. Sci. USA* 87:3420.

Chicz, R.M., Urban, R.G., Lane, W.S., Gorga, J.C., Stern, L.J., Vignali, D.A.A., and Strominger, J.L., 1992, Predominant naturally processed peptides bound to *HLA-DR1* are derived from MHC-related molecules and are heterogeneous in size, *Nature* 358:764.

Clarke, B., 1976, The ecological genetics of host-parasite relationships, *in*: "Genetic Aspects of Host-Parasite Relationships," A.E.R. Taylor and R.M. Muller, eds., Blackwell, Oxford.

Crow, J.F., 1990a, Foreword, *in*: "Selected Genetics Papers of J.B.S. Haldane," K.R. Dronamraju, ed., Garland, New York & London.

Crow, J.F., 1990b, Mapping functions, *Genetics* 125:669.

Crow, J.F., 1992a, Erwin Schrödinger and the hornless cattle problem, *Genetics* 130:237.

Crow, J.F., 1992b, Centennial: J.B.S. Haldane, 1982-1964, *Genetics* 130:1.

Damian, R.T., 1964, Molecular mimicry: antigen sharing by parasite and host and its consequences, *Am. Nat.* 98:129.

Damian, R.T., 1987, Molecular mimicry revisited, *Parasitol. Today* 3:263.

Denniston, C., and Crow, J.F., 1990, Alternative fitness models with the same allele frequency dynamics, *Genetics* 125:201.

Doherty, P.C., and Zinkernagel, R.M., 1975, Enhanced immunological surveillance in mice heterozygous at the *H-2* gene complex, *Nature* 256:50.

Falk, K., Rötzschke, O., Stevanović, S., Jung, G., and Rammensee, H.-G., 1991, Allele-specific motifs revealed by sequencing of self-peptides eluted from MHC molecules, *Nature* 351:290.

Guo, H.-C., Jardetzky, T.S., Garrett, T.P.J., Lane, W.S., Strominger, J.L., and Willy, D.C., 1992, Different length peptides bind to HLA-Aw68 similarly at their ends but bulge out in the middle, *Nature* 360:364.

Haldane, J.B.S., 1933, The genetics of cancer, *Nature* 132:265.

Haldane, J.B.S., 1949, Disease and evolution, *La ricerca scientifica, supplemento* 19: 2.

Hamilton, W.D., 1980, Sex versus non-sex versus parasite, *Oikos* 35:282.

Hill, S.A., Allsopp, C.E.M., Kwiatkowski, D., Anstey, N.M., Twumasi, P., Rowe, P.A., Bennett, S., Brewster, D., McMichael, A.J., and Greewood, B.M., 1991, Common West African HLA antigens are associated with protection from severe malaria, *Nature* 352:595.

Hood, L.E., Weissman, I.L., Wood, W.B., and Wilson, J.H., 1984, "Immunology," Benjamin/Cummings, California.

Hughes, A.L., and Nei, M., 1988, Pattern of nucleotide substitution at major histocompatibility complex class I loci reveals overdominant selection, *Nature* 335:167.

Hughes, A.L., and Nei, M., 1989, Nucleotide substitution at major histocompatibility complex class II loci: Evidence for overdominant selection, *Proc. Natl. Acad.Sci. USA* 86:958.

Hunt, D.F., Henderson, R.A., Shabanowitz, J., Sakaguchi, K., Michel, H., Sevilir, N., Cox, A.L., Appelia, E., and Engelhard, V.H., 1992, Characterization of peptides bound to the class I MHC molecule *HLA-A2.1* by mass spectrometry, *Science* 255:1261.

Jardetzky, T.S., Lane, W.S., Robinson, R.A., Madden, D.R., and Wiley, D.C., 1991, Identification of self peptides bound to purified HLA-B27, *Nature* 353:326.

Kimura, M., 1968, Evolutionary rate at the molecular level, *Nature* 217:624.

Klein, J., 1980, Generation of diversity at *MHC* loci: Implications for T cell receptor repertoires, *in*: "Immunology 80," M. Fougereau and J. Dausset, eds., Academic, London.

Klein, J., 1986, "Natural History of the Major Histocompatibility Complex," John Wiley & Sons, New York.

Klein, J., 1990, "Immunology," Blackwell, Oxford.

Klein, J., 1991, Of *HLA*, tryps, and selection: An essay on coevolution of *MHC* and parasites, *Hum. Immunol.* 30:247.

Klein, J., Gutknecht, J., and Fischer, N., 1990, The major histocompatibility complex and human evolution, *Trends Genet.* 6:7.

Klein, J., and Takahata, N., 1990, The major histocompatibility complex and the quest for origins, *Immunol. Rev.* 113:5.

Klein, J., Satta, Y., O'hUigin, C., and Takahata, N., 1993, The molecular descent of the major histocompatibility complex, *Ann. Rev. Immunol.* (in press)

Li, W.-H., and Sadler, L.A., 1991, Low nucleotide diversity in man, *Genetics* 129:513.

May, R., 1983, Parasitic infections as regulators of animal populations, *American Scientist* 71:36.

Nathenson, S.G., Uehara, H., and Ewenstein, B.W., 1981, Primary structural analysis of the transplantation antigens of the murine H-2 major histocompatibility complex, *Ann. Rev. Biochem.* 50:1025.

Nei, M., 1987, "Molecular Evolutionary Genetics," Columbia University Press, New York.

Nei, M., and Hughes, A.L., 1991, Polymorphism and evolution of the major histocompatibility complex loci in mammals, *in*: "Evolution at the Molecular Level," R.K. Selander, A.G. Clark and T.S. Whittam, eds., Sinauer Associates INC. Sunderland, Massachusetts.

Ohno, S., 1991, To be or not to be a responder in T-cell responses: ubiquitous oligopeptides in all proteins, *Immunogenetics* 34:215.

Ohno, S., 1992, "Self" to cytotoxic T cells has to be 1000 or less high affinity nonapeptides per MHC antigen, *Immunogenetics* 36:22.

Ramsdell, F., and Fowlkes, B.J., 1990, Clonal deletion versus clonal anergy: The role of the thymus in inducing self tolerance, *Science* 248:1342.

Rudensky, A. Yu., Rath, S., Preston-Hurlburt, P., Murphy, D.B., and Janeway Jr., C.A., 1991, On the complexity of self, *Nature* 353:660.

Satta, Y., 1992, Balancing selection at *HLA* loci, *in*: "Molecular Paleopopulation Biology," N. Takahata, ed., Japan. Sci. Soc. Press, Tokyo.

Satta, Y., Takahata, N., Schönbach, C., Gutknecht, J., and Klein, J., 1991, Calibrating evolutionary rates at the major histocompatibility complex loci, *in*: "Molecular Evolution of the Major Histocompatibility Complex," J. Klein and D. Klein, eds., Springer-Verlag, Heidelberg.

Schild, H., Rötzschke, O., Kalbacher, H., and Rammensee, H.-G., 1990, Limit of T cell tolerance to self proteins by peptide presentation, *Science* 247:1587.

Seger, S.W., 1988, Dynamics of some simple host-parasite models with more than two genotypes in each species, *Phil. Trans. R. Soc. Lond. B* 319:541.

Snell, G.D., 1968, The *H-2* locus of the mouse: observations and speculations concerning its comparative genetics and its polymorphism, *Folia Biol.* 14:335.

Sturmbauer, C., and Meyer, A., 1992, Genetic divergence, speciation and morphological stasis in a lineage of African cichlid fishes, *Nature* 358:578.

Takahama, Y., Shores, E.W., and Singer, A., 1992, Negative selection of precursor thymocytes before their differentiation into CD4+CD8+ cells, *Science* 258:653.

Takahata, N., 1990, A simple genealogical structure of strongly balanced allelic lines and trans-species evolution of polymorphism, *Proc. Natl. Acad. Sci. USA* 87:2419.

Takahata, N., and Nei, M., 1990, Allelic genealogy under overdominant and frequency-dependent selection and polymorphism of major histocompatibility complex loci, *Genetics* 124:967.

Takahata, N., Satta, Y., and Klein, J., 1992a, Trans-species polymorphism of the major histocompatibility complex (*Mhc*) loci, *in*: "Progress in Immunology, Vol. 8," (in press)

Takahata, N., Satta, Y., and Klein, J., 1992b, Polymorphism and balancing selection at major histocompatibility complex loci, *Genetics* 130:925.

Wada, K., Wada, Y., Ishibashi, F., Gojobori, T., Ikemura, T., 1992, Codon usage tabulated from the GenBank genetic sequence data, *Nucl. Acids Res. Suppl.* 20:2111.

Wills, C., 1991, Maintenance of multiallelic polymorphism at the *MHC* region, *Immunol. Rev.* 124:165.

GENETIC VARIABILITY OF MAJOR HISTOCOMPATIBILITY COMPLEX IN HUMAN POPULATIONS

Takashi Gojobori[1] and Tadashi Imanishi[2]

[1]DNA Research Center, National Institute of Genetics
Mishima 411, Japan

[2]Department of Anthropology, Faculty of Science
the University of Tokyo, Tokyo 113, Japan

INTRODUCTION

Human major histocompatibility complex (MHC) molecules are called human leukocyte antigens (HLA), which were discovered in 1958 (Dausset, 1958). A variety of knowledge about HLA structures, functions, and genes have been obtained particularly at the molecular level. The HLA gene complex in human is 4Mbp long and is located on chromosome 6. HLA genes are classified essentially into two groups, class I and class II, depending on their structures and functions (Klein, 1986). Sometimes, complement loci which are also located within the MHC gene region are called as class III. Genes in each class exist as a multigene family in the MHC gene region.

It is well known that several HLA genes exhibit an extremely high degree of polymorphism. In particular, the HLA-A, HLA-B, and HLA-C genes in class I and the HLA-DR and HLA-DQ genes in class II are highly polymorphic in various ethnic groups. Thus, these genes are very useful for tracing an evolutionary history of human populations.

In this paper, we describe the outline of our phylogenetic analysis of 80 ethnic groups by use of HLA typing data that were collected in the 11th International MHC Workshop at Yokohama (Inoko et al., 1992; Gojobori et al., 1992a; 1992b).

METHODS

HLA typing data used

Serological and DNA typing data for HLA genes for various ethnic groups have been collected in the Eleventh International Major Histocompatibility Workshop (Gojobori et al., 1992a; 1992b; Inoko et al., 1992). From these data, we have chosen the data only for unrelated and normal individuals because the Workshop data contains those for familial individuals and patients for particular diseases. We also selected a total of 80 ethnic groups in each of which the number of sampled individuals is more than 50 and the quality of data is believed to be good. Q-scores were used as criteria of quality (Imanishi et al., 1992a). In practice, individuals whose Q-scores are larger than 0.55 were used for our phylogenetic analysis. The names of 80 ethnic groups are listed in Table 1.

Table 1   The list of ethnic groups analyzed in the present study

Ethnic name	Ethnic name
African	**Asiatic Indian**
African in England (Blacks)	Bhargavas
Bushman (San)	Indian
Hottentot (Khoi)	Iyers
North African (Blacks)	Tribal
North American (Blacks)	
Senegalese	**North Asian**
South African (Blacks)	Buriat
South African (Blacks; Capetown)	Hui
South American (Blacks)	Inner Mongolian
West African (Blacks)	Japanese (Wajin)
Zairean	Kazakstan
Zimbabwean	Korean
	Manchu
European	Mongolian
Albanian	Northern Han
Armenian	Orochon
Australian (Whites)	Tibetan
Austrian	Uigur
Basque	Yakut
Belgian	
Brazilian (Whites)	**Southeast Asian**
British	Buyi
Canadian	Miao
Cornish	Singapore Chinese
Czech	Southern Han
Dane	Thai Chinese
French	Thailander
German	Taiwan Aborigine
Greek	Vietnamese
Gypsy (Spanish)	
Hungarian	**Amerind**
Italian	Bolivian
Mexican	Inuit (Eskimo)
Norwegian	Native Brazilian (Amerinds)
Polish	North American (Amerinds)
Portuguese	Tlingit
Rumanian	
Sardinian	**Oceanian**
Slovak	Australian Aborigine
South African (Whites)	Highlanders (Papua New Guinea)
Spanish	Javanese
Swedish	Maori
Ukrainian	Timorese

Uralic
Uruguayan
U.S.A. (Whites)
Yugoslavian

Estimation of allele frequencies in each ethnic group

For serological HLA types, we used the standard nomenclature of HLA alleles. Split specificities that could not be definitely separated were included into broad specificities. A similar procedure was taken for DNA typing data. Using these typing data, we estimated the frequencies of HLA alleles in each population of 80 ethnic groups by the maximum likelihood method (Imanishi et al. 1992a; 1992b). For estimating allele frequencies from HLA typing data, the maximum likelihood method is known to be statistically more satisfactory than any other methods. This is because the HLA typing data often contains blank alleles whose specific allele types are unknown. Thus, allele frequencies must be determined taking into account these blank alleles.

In estimating frequencies of alleles, we also made the assumption that all HLA loci have codominant alleles except one recessive allele that corresponds to blank and that the samples represent a population at a Hardy-Weinberg equilibrium. In the computations of allele frequencies, initial values were set to the values obtained by a standard square-root method (Bernstein, 1930) for estimating allele frequencies. Then, recurrence equations were used for calculating allele frequencies at each locus in each ethnic group examined. This procedure is repeated until the frequencies reach equilibrium values. These equilibrium values are used as the maximum likelihood estimates of allele frequencies.

As a result, the frequency data for 19 HLA-A and 42 HLA-B alleles were obtained from the computation. We excluded the data for the HLA-C locus from our present analysis, because the frequency of C-blank alleles was quite high in many ethnic groups and the estimated frequencies of other alleles were not very reliable. We also excluded the data for class II loci because for these loci the typing data are limited to a small number of ethnic groups.

Calculation of genetic distances between different ethnic groups

Using the allele frequencies, we estimated genetic distances for all possible pairs of ethnic groups. We used Nei's standard distances [D, (Nei, 1972)] as genetic distances between different ethnic groups compared. The definition of D has been given as follows:

$$D = Jxy / (JxJy)^{1/2},$$

where $Jxy = \text{Sum}_i (x_i y_i)$, $Jx = \text{Sum}_i (x_i^2)$, and $Jy = \text{Sum}_i (y_i^2)$. Here, x_i and y_i are frequencies of the i-th allele in two ethnic groups x and y, respectively.

HLA loci are tightly linked to each other on chromosome 6. It is also likely that the HLA loci may not be selectively neutral and positive selection such as overdominance selection may be operating on these genes (Hughes and Nei, 1988; 1989). Thus, D may, in principle, not be suitable for using it as a genetic distance of HLA genes. However, it can be used as a first approximation, because the degree of genetic diversity between ethnic groups can be reasonably measured by this distance. In this case, we can keep in mind that D may not be proportional to the number of codon substitutions taken place during the divergence between two ethnic groups compared.

Construction of phylogenetic trees

To construct phylogenetic trees, we used the UPGMA (Sokal and Sneath, 1963; Nei,

1975) and the Neighbor-Joining (NJ) method (Saitou and Nei, 1987). The former method is very popular for anthropological studies, but it requires an assumption that changes of the genetic distance must be constant over time. On the other hand, the NJ method does not require any assumption on the rate constancy. Instead, the NJ method usually gives only an unrooted tree unless an outgroup is properly given. Thus, we used both methods for constructing phylogenetic trees of 80 ethnic groups.

RESULTS AND DISCUSSION

Phylogenetic tree

Phylogenetic trees constructed were shown in Figures 1 and 2. Figure 1 shows a phylogenetic tree which was constructed by the UPGMA. In this tree, the allele frequencies of HLA-A and HLA-B were used. We showed the trees only for the serological typing data. We did not show the trees for the DNA typing data. The name of each ethnic group represents mostly the region of the world where it inhabits. As for the name of each ethnic group, we followed the original description that was given to the 11th MHC International Workshop by the data submitters in the participating laboratories (Gojobori, 1992b). Thus, the names of some ethnic groups are sometimes very vague and may not be well-defined from anthropological point of view. However, we are not going to discuss detailed relationships between specific ethnic groups. We are rather interested in identifying major clusters of ethnic groups in the phylogenetic trees and in knowing how human populations are genetically divergent in the HLA gene region. For these purposes, the original names of ethnic groups given by the data submitters are enoughly useful for the present study as long as they can roughly represent anthropological context and the geographical regions where they live. Therefore, we used the original names of ethnic groups as they were.

Figure 2 shows a phylogenetic tree which was constructed by the NJ method. The naming of each ethnic group in this tree is identical to that in the UPGMA tree.

Five major groups in the UPGMA tree

In the UPGMA tree, a total of 80 ethnic groups are roughly classified into five distinct groups except Spanish Gypsy and Sardinians (Figure 1). Group 1 contains Hottentot (Khoi), Bushman (San), South African (Blacks), Zimbabwean, South African (Blacks sampled particularly from Capetown), Zairean, South American (Blacks), North African (Blacks), North American (Blacks), Senegalese, and West African (Blacks). Group 2 has Taiwan aborigine, Maori, Australian aborigine, Highlander at Papua New Guinea, Javanese, Timorese, Yakut, Orochon, and Inuit. Group 3 contains Native Brazilian, North American (Amerinds), Tlingit, and Bolivian. Group 4 is composed of Thai-Chinese, Vietnamese, Miao, Thailander, Buyi, Southern Han, Singapore Chinese, Uralic, Inner Mongolian, Kazakstan, Buriat, Mongolian, Manchu, Northern Han, Hui, Tibetan, Japanese (Wajin), and Korean. Group 5 is formed by Uigur, Uruguayan, Basque, Australian (Whites), Belgian, South African (whites), Portuguese, Spanish, Hungarian, Austrian, French, Canadian, German, an undefined white population in the United States, Czech, Dane, Slovak, Polish, Ukrainian, African Blacks in England, British, Cornish, Norwegian, Swedish, Bhargavas, Indian, Iyers, Tribal, Armenian, Mexican, Brazilian (Whites), Greek, Yugoslavian, Albanian, Italian, and Rumanian. Sardinians and Spanish Gypsy are somewhat located outside Groups 4 and 5 in the UPGMA tree.

As mentioned before, the UPGMA requires an assumption that changes of the genetic distance must be constant over time. In reality, however, this assumption may not be valid because an ethnic group with the smaller population size is known to have the higher rate of changes in genetic distances if the alleles are not strictly neutral. Thus, the UPGMA tree may

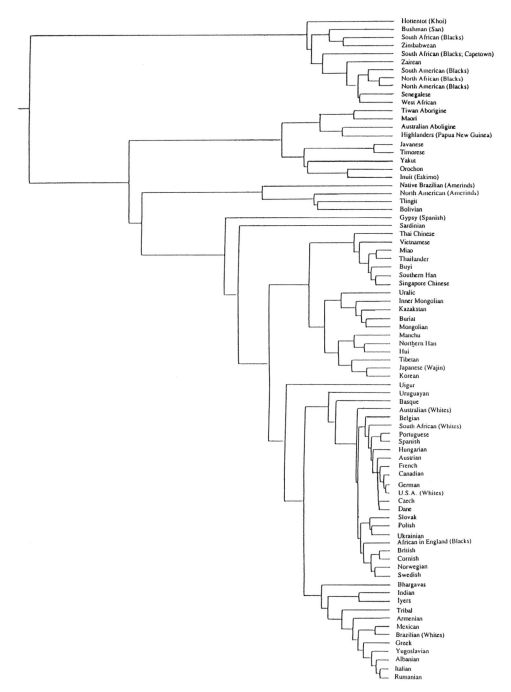

Figure 1. The phylogenetic tree constructed by the UPGMA using allele frequencies of HLA-A and B genes, which shows evolutionary relationships among 80 ethnic groups.

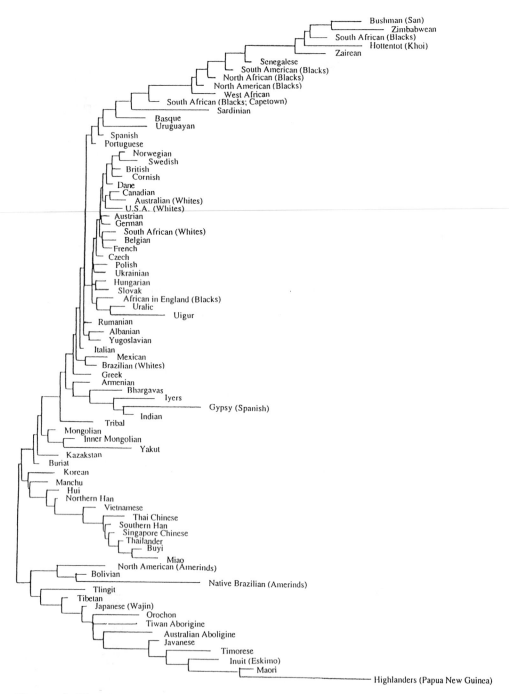

Figure 2. The phylogenetic tree constructed by the Neighor-Joining method using allele frequencies of HLA-A and B genes,which shows evolutionary relationships among 80 ethnic groups.

give a biased classification of ethnic groups in which grouping of some ethnic populations is incompatible with the present anthropological understanding. It can also explain topological differences between the UPGMA and NJ trees, as will be discussed later.

Reproducibility of the tree topology is another problem to be considered. The trees in Figure 1 as well as Figure 2 may be distorted by statistical fluctuations due to sampling errors in estimating allele frequencies, variance of genetic distances, statistical errors in constructing the tree, and so on. Obscure relationships of Sardinians and Spanish Gypsy to other ethnic groups may be also explained by some sort of statistical fluctuations.

Roughly speaking, however, groups 1 to 5 seem to correspond to African populations, Oceanian populations together with some Northern Asians, Amerind populations, Asian populations, and European populations with Asiatic Indians, respectively.

African origin of human populations

According to the UPGMA tree, the 'African' group is separated first from the other four groups. Thus, the UPGMA tree supports an idea that the origin of human populations may be located in the African continent. In fact, Cavalli-Sforza et al.'s (1988) tree also indicated that African populations diverged first from other human populations. However, it is quite possible that distant relationships among ethnic groups are less reliable than closer relationships in the tree, because estimation of the genetic distances becomes less reliable when the genetic relationships are more distant. A further analysis is required to arrive at more valid relationships among distantly-related ethnic groups.

Differences in topologies between the UPGMA and NJ trees

The tree in Figure 2 was constructed by the NJ method based on the same set of genetic distances as that in Figure 1. The root of the tree was tentatively determined at the midpoint between the two most distant ethnic groups.

In the NJ tree, Sardinian is located just at the border between African and European populations. Moreover, Spanish Gypsy is the closest to Indians. This observation is consistent with the present anthropological knowledge that Gypsy is of Indian origin. In these points, the NJ tree appears more reliable than the UPGMA tree, though the former gives only the unrooted tree.

It is of particular interest to point out that Japanese and Tibetans lie on the border between the 'Asian' group and the 'Oceanian" group. This is mainly because Japanese and Tibetans share a high frequency of the HLA-A24 allele with the 'Aborigine' populations.

Although both UPGMA and NJ trees in the present study did not show a fully reasonable picture of the evolutionary process of human populations, they certainly gave a rough idea about how various ethnic groups over the world evolutionarily diverged from their ancestral population. In particular, the trees given in the present study can be valuable in the anthropological study, when we notice that these trees contain the largest number of ethnic groups ever examined using the HLA genes.

Acknowledgments

We would like to thank 242 laboratories of the world that participated in the 11th International Major Histocompatibility Workshop at Yokohama for providing the typing data of HLA genes. Finally, we are very grateful for Dr. Partha P. Majumder's kindness and patience. Without him, the present paper could not have appeared in this wonderful volume.

REFERENCES

Bernstein, F.Z., 1930, Fortgesetzte Untersuchungen aus der Theorie der Blutgruppen, *Ind. Abst. Vereb. Lehre* **5 6**:233.

Cavalli-Sforza, L.L., Piazza, A., Menozzi, P., and Mountain, J., 1988, Reconstruction of human evolution: Bringing together genetic, Archaeological, and linguistic data, *Proc. Natl. Acad. Sci.* **8 5**:6002-6006.

Dausset, J., 1958, Iso-leuco-anticorps, *Acta Haemotal.* **2 0**:156-166.

Gojobori, T, Imanishi, T., Ina, Y., Saitou, N., Tateno, Y., Sato, K., Takata, H., Kimura, A., Nishimura, Y., Akaza, T., Miyoshi, H., Mura, T., Tokunaga, K., Tanaka, H., Konoeda, Y., Wakisaka, A., Juji, T., Sasazuki, T., Tsuji, K., and Inoko, H., 1992a, Construction of the MHC database, *in*: "HLA 1991," Vol.1, Tsuji, K. et al. eds, pp.66-67, Oxford University Press, Oxford.

Gojobori, T, Imanishi, T, and Ina, Y., 1992b, Statistics in the numbers of panels stored in the MHC database *in*: "HLA 1991," Vol.1, Tsuji, K. et al., eds., pp.67-76, Oxford University Press, Oxford.

Hughes, A.L., and Nei, M., 1988, Pattern of nucleotide substitution at major histocompatibility complex class I loci reveals overdominant selection, *Nature* **3 3 5**:167-170.

Hughes, A.L., and Nei, M., 1989, Nucleotide substitution at major histocompatibility complex class II loci: Evidence for overdominance selection, *Proc. Natl. Acad. Sci. USA* **8 6**:958-962.

Imanishi, T., Akaza T., Kimura A., Tokunaga K., and Gojobori T., 1992a, Estimation of allele and haplotype frequencies for HLA and complement loci, *in*: "HLA 1991," Vol.1, Tsuji, K. et al., eds, pp76-79, Oxford University Press, Oxford.

Imanishi, T., Wakisaka, A., and Gojobori, T., 1992b, Genetic relationships among various human populations indicated by MHC polymorphisms, *in*: "HLA 1991," Vol. 1, Tsuji, K. et al., eds, pp.627-632, Oxford University Press, Oxford.

Inoko, H., Sato K., Takano H., Imanishi T., Ina Y., Saitou N., Tateno Y., Kimura A., Nishimura Y., Akaza T., Miyoshi H., Mura T., Tokunaga K., Tanaka H., Konoeda Y., Wakisaka A., Juji T., Sasazuki T., Tsuji K., and Gojobori T., 1992, Data flow from collection of raw data to construction of the MHC database, *in*: "HLA 1991," Vol.1, Tsuji K. et.at., eds., p.65, Oxford University Press, Oxford.

Klein, J., 1986, "Natural History of the Major Histocompatibility Complex," John Wiley & Sons, New York.

Nei, M., 1972, Genetic distance between populations, *Amer. Natur.* **1 0 6**:283-292.

Nei, M., 1975, "Molecular Population Genetics and Evolution," North-Holland, Amsterdam.

Saitou, N., and Nei, M., 1987, Neighbor-joining method: A new method for reconstructing phylogenetic trees, *Mol. Biol. Evol.* **4**:406-425.

Sokal, R.R., and Sneath, P.H.A., 1963, "Principles of Numerical Taxonomy," Freeman, W.H., San Francisco.

Mt DNA, A MATERNAL PERSPECTIVE, AND THE ORIGIN OF MODERN HUMANS

Rebecca L. Cann

Department of Genetics and Molecular Biology
John A. Burns School of Medicine
University of Hawaii at Manoa
Honolulu, HI, 96822

INTRODUCTION

For the last ten years, molecular approaches to studying human diversity have altered the way biologists interpret the evolutionary history of anatomically modern humans. Using a maternally inherited system of genes found in the cell's mitochondria, researchers have suggested that all humans alive today can trace their maternal genealogies to an African woman who lived roughly 200,000 years ago. Haldane might have appreciated this emphasis on mutation and the importance of simple models for inference, but I also wonder what he might have thought about a "molecular clock" for evolutionary studies derived from counting mutations. Our discovery with mtDNA was unexpected and raises a number of questions about biological speciation and anthropological models for the origin of modern people. The ecological perspective that comes from working with endangered mammalian species, especially those having certain life-history characteristics similar to humans, also suggests some questions to be asked about social behaviors and environmental conditions that ensured the survival of our African ancestors.

Evolutionary studies of humans should seek to make sense of information derived from research on behavior, anatomy, fossils, and molecules. This approach can potentially provide an integrated view of human history, in contrast to the narrow focus which results from concentration on a single functional complex, genetic or morphological adaptation, or subdiscipline in science. Also, it is a truism that when technology advances, facts change. New "facts" suggest both new models for research and raise deeper questions.

By the early 1970's, genetic studies were an easily accepted part of biological anthropology. Human racial classifications and their importance for anthropologists had been replaced by more sophisticated research efforts aimed at understanding the limits of human morphological plasticity. Human diversity, manifested by different forms of proteins present in blood samples, hinted that the scenarios describing the evolution of our subspecies, <u>Homo sapiens sapiens,</u> were inaccurate. Microchange, clines, and gradualism were the dominant intellectual paradigm, but both the time scale and the degree of geographic isolation thought necessary to evolve distinctive human morphologies were open to question. The descriptions from textbooks of that era communicated the complexities of human phylogeny by analogies. These described different geographic lineages as a series of intertwined, interdigitating vines.

Anthropologists stressed that geographic isolates of humans were ephemeral, interbreeding and not separate branches of a tree representing distinct gene pools with sometimes different genetic or cultural trajectories. Only one species of humans were ever supposed to exist in one place at one time. New fossils from East Africa would change that view, but high levels of gene flow were universally implied by these models. However, low resolution genetic studies of religious, ethnic, or geographically isolated peoples sometimes identified large numbers of differences in the allelic frequencies of proteins, usually attributed to genetic drift. Anthropologists could therefore choose from a multitude of conflicting explanations to account for observed differences in shape, stability, and abundance of various important biological molecules within modern human populations.

We geneticists use the word polymorphism to describe alternative forms of genetic characters (alleles) present in a population at levels greater than those which would be expected from just newly occurring mutations, i.e.,usually an allele present at greater than a one per-cent frequency in the gene pool. Shared polymorphisms in circulating serum proteins, thought to be functionally equivalent and therefore, nearly neutral with respect to natural selection, provided evidence that some populations had shared a more recent common ancestor than others. Some molecular geneticists questioned the simplistic concepts of natural selection and gene flow and the significance of the biological differences in modern human populations. They sought more satisfying mechanistic explanations for these biological patterns[1]. Their approach was very different from previous generations of geneticists who shared a fascination with eugenics. This time, their explanations focused on estimated ages of populations, effective population size, mutation rates, the level of diversity in natural populations, and the relative strength of natural selection for determining gene frequencies.

Human genetic polymorphism studies also experienced a minor revolution in phylogenetic analysis, similar to that affecting the interpretation of fossil evidence [2]. How one experimentally grouped observed genetic similarities to calculate a heterozygosity statistic could influence the conclusions of a study. Cladistic analysis of DNA sequence variability held the promise of allowing an investigator to extract more information from a data base than was previously possible using similarity methods which broadly grouped gene frequency data. The wide-scale availability of DNA-based technologies using restriction enzymes also made it possible to begin human diversity studies at a more direct level of analysis than the indirect studies of proteins. Changes in direct sequencing of DNA from enzymatically amplified materials have now taxed our ability to analyze laboratory results mathematically. How close are our approximations, our reconstructions? Researchers rightfully expected that a clear picture of recent human history might eventually emerge from the chaos and conflict. I now think that we have arrived at this point.

Considerable heat now characterizes the study of modern human phylogeny, and that heat has not been fueled by discovery of new fossils alone. Rather, it stems from the necessary condition that behaviorists, morphologists and geneticists must reconstruct the history of the same species. DNA is being extracted from hair, blood, muscle, urine, and saliva of modern donors [3]. DNA is being extracted from the bones and skin of ancient donors. Ancient bones and their behaviorally associated morphologies have been placed on lineages leading to modern people, but it is the contention of many geneticists is that this practice shows questionable logic. Although DNA data confirms continuity of discrete maternal lineages over a thousand years in Japanese ancient donors [4] it also shows replacement of those lineages in populations when long-range comparisons (over thousands of generations) are made. If we ask what is the biological fate of earlier populations, some of us now answer that their genes were diluted in frequency from those of modern populations by the massive reproductive success of immigrants who replaced them.

Full understanding of this process had to wait until geneticists had sufficiently concentrated on a non-traditional genetic system with sufficient experimental resolution to reveal breaks in maternal lineages. Some DNA sequences present in the cell's nucleus are either highly conserved or selected for extreme variation, and antedate the origin of modern species. Individual genes, especially the biparentally inherited ones, will not always provide unambiguous confirmation of population replacement. Neither is the data from a single set of genes, such as the mitochondrial ones, absolute proof of a pattern. Enough evidence has accumulated, however, to convince many geneticists that molecular biology will show replacement. By this, I mean that the populations certain fossils represent, found in a given geographic locality may not be the closest direct ancestors of modern people who reside in that same place today. At the same time, morphologists are reassessing their interpretations of the significance attached to previously studied fossils, and also correcting dates for these fossils. Linguists have joined in the process, responding to the speculation that language-based adaptations could have driven the replacement process.

BIOTECHNOLOGY CHANGED THE "FACTS"

For the past five years, a number of molecular geneticists have been carrying forward the program of building a new understanding of human diversity based on mitochondrial DNA polymorphisms. Molecular biologists had published DNA sequence information derived from the mitochondrial genes of a single woman. This study allowed researchers to begin intensive restriction endonuclease analysis of modern populations [5]. The information discovered by these studies was tantalizing.

A preview of what was soon to come with new genetic data was presented in a publication on human hemoglobin gene polymorphisms [6]. Prior to this time, if genes entered into any conversation on human evolution, they were usually filed quickly away as p's, q's, on chromosomes in the nucleus, and overall, rather boring to keep track of if you were a non-specialist. Genetic recombination made analysis of linked sequences in the hemoglobin gene family and, therefore most known molecular genetic analyses, very complicated. Time to last shared common ancestor of a given DNA sequence was not something too many geneticists worried about themselves, although coalescence and its associated theorists would soon change that.

Quick and easy experimental approaches to the DNA sequences underlying genetic polymorphisms were few in the early 1980s, due to the complex linear organization of most genes. Of the 100,000 estimated genes in humans, few were actually known functionally, and even fewer could be studied by traditional biochemical methods. The genes of the nucleus are large, often interrupted by non-coding regions, and from a maternal perspective, inherited in a difficult manner to model over time, because of genetic recombination. Heroic efforts to chemically sequence a small (less than 4000 base pairs) gene in humans were still required, and even three independently produced DNA sequences from the same gene in the same species might be considered overkill in a laboratory more concerned with disease, structure, and function.

In contrast, mitochondrial genes were known to be inherited maternally, to undergo rapid evolution compared to genes in the nucleus, to code for a limited number of essential genes controlling energy production in the cell, and to be free from genetic recombination, which would scramble the lineages of genes each generation in an evolving population. Building on early work of Wesley Brown, Douglas Wallace, and others, Cann, Stoneking, and Wilson published results of their analysis of a small but significant number of human mtDNAs [7].

We suggested that humans, first, shared a recent (about 200,000 years before present) common maternal ancestor, post-dating most occurrences of H. erectus but predating the recognized origin of H. s. sapiens and that, second, in all probability that ancestor was from sub-Saharan Africa. It appeared to us that instead of evolving from isolated archaic ancestors on many continents, modern people stemmed from some pool of African ancestors who colonized new areas. Groups of these people must have left, and some later returned to recolonize Africa. Our expectation was that if archaic lineages in Asia made significant maternal genetic contributions to the gene pools of modern Asians, the diversity (implying time to achieve that diversity) among Asians would equal that of Africans. Unless these absent lineages were missed due to incomplete sampling or the action of natural selection, we couldn't understand how to accommodate these data with direct continuity models favored by many paleontologists. The inferred migration rates would have to be higher than many population geneticists would be comfortable with, much greater than one migrant per generation, because the time scales were so short.

African ancestry for the root of the tree was inferred because the most divergent lineages were found among Africans and their descendents. The time was estimated from the mutation rate, and measurements of the pairwise mtDNA distances between two average humans taken at random. This paper was considered highly controversial by many anthropologists. Our wish was to engage a community of researchers in the real computational problems that exist in the analysis of large pairwise data matrices.

Although a large number of morphological similarities between archaic Asians and modern people from Asia had been identified by previous generations of physical anthropologists, we also questioned the biological significance of these similarities. Eventually, we have come to regard them as illusions, convergence due to similar environmental constraints. Natural selection has produced the same functional suite of characters in response to the same environmental conditions many times in humans.

Human paleontologists remained divided. Some found evidence in their own analysis of African characteristics and support for the replacement of older populations[8]. Others questioned every assumption of the mtDNA study, championed regional continuity, and then dismissed maternal genetic studies as irritating, irrelevant, or illogical at best [9].

A SHORT SUMMARY OF THE CONTROVERSY

The utility of MtDNA for evolutionary research is tied to an accurate understanding of its tempo of change, and the standard error of that estimate. Resolution of clock questions comes from having a large enough data set to begin statistical analysis, as well as a broad understanding of the molecular consequences of some mutations. Second, in order to present an accurate representation of the evolutionary process of mitochondrial change (a reconstruction of human maternal lineages and their evolution in a phylogenetic tree), it is important to have confidence in the rate of lineage extinction and the method used to infer a common ancestor at the root of the tree. Third, assumptions about the ability of human groups to exchange genes, the maternal history of Afro-Americans, the arrival at genetic equilibrium conditions, and the time to achieve that equilibrium under different levels of migration, assortative mating, or selection need to be discussed. Fortunately, new methods for rapid production and analysis of mtDNA sequences appeared, so that we will not have to wait long for the answers.

Excellent reviews of the organization and evolution of animal mitochondrial genomes exist, and this paper will not attempt to duplicate their breadth or scope. Most molecular diversity uncovered in these genes is probably neutral with respect to natural selection. Molecular anthropologists tried to address criticism of methods, data collection, and analysis of that data with an increased research effort, not rhetoric. Four major, new studies help raise our confidence that the African origin model for modern humans is the correct one.

The easiest criticism to answer was the problem posed by the substitution of Afro-Americans for "authentic" African donors. The original African Mother hypothesis was dismissed by some on the grounds that the sample of Afro-Americans constituting the majority of the "African" donors was inappropriate. These individual lineages may have considerable numbers of "non-African" genes for historic reasons, and it was reasoned that they could not be considered representatives of the gene pools of Africans still living in Africa.

Vigilant and coworkers reported[10] new information from several ethnic sub-Saharan African populations, notably the first detailed mtDNA studies of the !Kung, Yorubans, Hadza, Herero, Naron, and of Pygmies. They utilized individual hairs and the Polymerase Chain Reaction (PCR) to enzymatically amplify rare DNA sequences present in cells of the hair to a sufficient copy number for DNA sequencing. Their results confirm the deep genetic divisions within maternal African gene pools, found also in Afro-Americans, and support placing the root among Africans on the basis of existing diversity alone. No one disputes the claim that most of Africa is still uncharacterized genetically, and we are eager to investigate the connections between demography and lineage survival among modern African populations [11,12].

Another criticism concerned the proper calibration for the overall rate of mtDNA change in primates, and potential errors in estimating that rate. Two widely different rates often appear in the literature, and one which is much lower. The lower rate implied more ancient ancestry, and therefore encouraged anthropologists to think about a genetic division in <u>H. erectus</u>

populations, rather than between <u>H. sapiens</u> populations. The one percent to 0.7 percent often-used rate actually reflects a measure of distance along a single lineage since divergence from a common ancestor, rather than the distance between two lineages separated by that ancestor, which was the basis of the two to four percent rate used in the 1987 report.

Kocher and Wilson [13] address that issue, as well as one concerning the question of rooting the human tree. They produced hypervariable control region mtDNA sequences from three common chimpanzees, one pygmy chimp, and then aligned them with fourteen human sequences. This comparison led to the insight that humans are tediously similar from a chimpanzee geneticists' perspective. Specialists argue about the absolute rate of change, which depends on having the correct alignment of sequence. In the hypervariable region, that rate may be difficult to be certain of, due to gaps in the sequence, and some mutations could be counted more than once. Whatever the absolute rate of change, relative to the chimp-human divergence, the intra-human distances are about 1/40th the total length of the tree. If humans and chimps have last shared a common ancestor five million years ago, then the last common maternal ancestor of humans lived roughly 200,000 years ago.

Kocher and Wilson's study provided data that also allowed a great ape species to be used as an outgroup, for purposes of rooting the human phylogenetic tree. When chimpanzee mtDNA is used in this manner, it is clear that highly divergent African lineages are present at the base of the human mtDNA tree. This would be expected if, indeed, they represent the ancestral human population pool from which all other maternal lineages arose. Other African lineages are dispersed throughout the tree, demonstrating that as some groups stayed in Africa, other peoples left only to later return. This knowledge about the fluidity of African peoples and their geography may be important in the consideration of how frequent major expansions, such as that of the Bantu peoples, might have altered the face of the African continent in the past 100,000 years.

Recent summaries of hypervariable mtDNA sequences and additional information about dispersal of maternal lineages from Africa appears in Vigilant et al.'s newest report on sub-Saharan mtDNAs[14]. Publication of this data set led to a heated exhange in the same journal, where researchers pointed out that misuse of the parsimony-based approach resulted in an inadequate sample of trees being evaluated, some which were equally short with Asian-derived lineages at the base. A distance-based approach supports the African-origin hypothesis, but not at a statistically significant level[15]. Thus, the deepest branches of the human mtDNA tree are likely to remain contentious for some time until new methods in tree computations allow us to calculate branch lengths as well as affinities from the same matrix[16].

Third, questions of stochastic loss of lineages, bottlenecks, and the need to think about extreme population declines during migration have been the focus of Ward et al.'s [17] analysis of mtDNA variation in a single Amerindian tribe, the Nuu-Chah-Nulth. Much had been written previously about the extreme population bottleneck associated with the founding of the Americas, and the recent ages of archaeological evidence firmly supporting Paleoindian colonization. Schurr et al's [18] report had provided evidence from low-resolution mtDNA studies for limited population diversity in Amerindians, finding only four lineages that appeared to stem from a recent ancestor. More recent restriction mapping studies of mtDNA among Asians from the same researchers have sought to confirm the cohesiveness of Asian maternal lineages, and search for the roots of Amerindian origins [19,20].

These studies must now be contrasted with higher-resolution work of Ward and colleagues. Based on mtDNA hypervariable sequences, they demonstrate much greater diversity (28 maternal lineages) in a single tribe. These lineages shared a common maternal ancestor who dates to 60,000 years ago, and suggests that either the initial colonization of the Americas was more ancient than the 20,000-18,000 year bp estimate commonly cited, or that the colonization involved large numbers of people carrying many already distinct maternal lineages with them. Satoshi Horai's work now in press identifies these 4 clades in Amerindian control regions, and the question will hinge on effective population sizes, the total length of the tree and the proportional distance to monophyly in each cluster.

I personally suspect that there were no significant genetic population bottlenecks in the founding of the Americas, at least not the type we think of when working with endangered species. Thousands of founders, not just hundreds, probably streamed through Beringia. Different tribes were differentially impacted by disease, the gifts of later colonists. The limited diversity of genetic markers from blood samples probably reflects the massive disease epidemics that killed most Native Americans, and not the low numbers of genetic founders. This assertion could be tested by skeletal DNA analyses. The choice to depict limited genetic diversity found today in the descendants of the first Americans as a consequence of distance and migration may simply reflect our unexamined biases about what technologiy these people were capable of carrying.

Studies such as this demonstrate the need to have better mathematical models for understanding factors leading to the loss of lineages in populations. Rapidly expanding populations would not be expected to loose lineages, because every female might leave a large number of offspring, and some of these should be women. Most lineages would appear closely related, and have few distinct mutations. An older population approaching genetic equilibrium conditions (large size, stabilized for migration, stable mutation rate, and no extreme types of mate choice) should show a different spectrum of diversity than a newly founded or rapidly expanding one. An equilibrium population would have a few old, highly divergent lineages, a majority of middle-aged ones (with moderate genetic diversity) and a few new ones, with very little change [21]. If the same experimental standards are met, mtDNA studies could provide information about population history based solely on the shape of curves describing the distribution of genetic variants in the group studied.

When natural selection is the primary force operating on the expansion of one lineage to the exclusion of all others, it could appear that one population had replaced the other. It is tempting to try and identify selective agents which could operate in this way. Some researchers have attempted to identify the evolution of spoken language as a possible reason for mitochondrial lineage replacement. It is hard to imagine how to test such a hypothesis, because the links between phenotypic traits and mitochondrial mutations are rare, mostly confined so far to maternally inherited neural disorders [22].

We also have no evidence for the process of selection operating to completely sweep mitochondrial diversity out of a population, except in the unnatural context of human Hela cells in tissue culture. In the extreme where cancer cells are maintained, many cell lines are established from different donors and different cell lines can become overrun with contaminating Hela cells. In fact, one way to detect cell-line contamination is to screen for them for the presence of Hela mtDNA (Wesley Brown, pers. com.). Cancerous cells have escaped the body's ability to control their total proliferation, and so do not really provide us with an appropriate model for thinking about levels of natural selection that would normally be operating.

Finally, a possible migration corridor out of Africa for the rapidly expanding population which would replace earlier peoples has been identified by Di Rienzo's study of Mediterranean, Middle East, and North African populations[21]. They advance the hypothesis that the Nile Valley may have been the migration corridor used by our ancestors in their spread from Africa, based on the distributions of mtDNA sequence differences in modern Egyptians. Rapidly expanding populations continue to require non-equilibrium population genetic models, and the short time scale over which humans expanded cannot accommodate the level of migration between populations over the distances implied by continuity models. Redating of Middle-Eastern fossils, particularly Qafza in Israel, intensifies with need to examine mtDNA sequences from Neandertal skeletal remains. The mystery of what caused populations to expand 100,000 years ago, either due to a new culturally based, or a biologically based mixture of these adaptations, continues. Qafza cave fossils show us that archaic and modern people lived in the same region for over 50,000 years, with little evidence of mixing. A major task is to now quantify how much admixture is necessary to generate the morphological complexes identified in modern and skeletal populations from the same region.

THE CRUCIAL IMPORTANCE OF DEMOGRAPHIC STUDIES IN BIOLOGICAL ANTHROPOLOGY

Lifetime reproductive success is the ultimate basis of population continuity. Primatologists are able to measure this important biological variable in only a few wild-ranging species, but some important principles emerge. In one study of vervet monkeys, 57% of all infants die in the first year of life, and 69% of these deaths were due to predation [23]. Only 27% of females live to breeding age, and the extreme variation in infant survival means that even females who survive to breed over a number of years cannot be assured of leaving **any** offspring, either male or female. Baboon populations defy simple generalizations [24], but a bias towards greater numbers of female offspring born to high-ranking females has been noted. A study of African agropastoralists in southwest Kenya, the Kipsigis, has identified differences in female fertility and infant survival as the primary determinants of variance in lifetime reproductive success of these females [25].

Religious and social beliefs are important in promoting large families in modern African populations. We have no reason to suspect that such factors were any less important to African women 200,000 years ago. To focus on the issue of population replacement, and jokingly model it as the action of aggressive killer females, misses the point that Darwin made in 1859. Demographic studies which identify aspects of fertility [26] lay the groundwork for understanding how population replacement actually works in an anthropological context.

Variance in female reproductive success must have been extreme among archaic Asian maternal lineages, such that none survive today. Whether the variance was due to high infant mortality, differential fecundity, or combinations of other factors is impossible to say. If infant mortality is consistently biased against female survival, the speed at which female lineages are lost in a population, will increase. Even with systematic female infanticide, one would still predict that continuity models should allow the identification of the ancestral archaic maternal lineage. No others would be found, since few female lineages would survive long enough to undergo significant diversification. So, extreme infanticide would just reduce the relative number of old lineages, but not the presence of an old lineage. It appears that regional continuity adherents must continue to claim bad sampling strategies make us miss the surviving archaic lineages among Asians. The thousands of donors studied now argue against this.

Every astute observer of adolescent humans knows that reproductive strategies between males and females are not always congruent. One important strategy where males and females appear to differ involves distances traveled for improving access to resources. Longer dispersal distances related to females are well known for animal populations [27]. Koenig [28] identifies an interesting trend in contemporary Americans, where females migrate further geographic distances from birth to marriage than males. Alternatives to such patterns may be plentiful among mammals, yet female dispersal to virilocal residence in societies practicing patrilineal inheritance of accumulated resources is common in nonindustrialized human societies. How important this potential bias has been in detecting patterns of human genetic change is yet to be evaluated. If mother tongues depend on mothers, female genetic lineages which advance geographically ahead of male lineages may differentially transmit intrusive linguistic elements important to their sphere of activites in their new residences.

EFFECTIVE POPULATION SIZE AND PERSISTENCE

Geneticists are intimately involved in efforts to save endangered species, and growing numbers of them have become familiar with the general concepts important to the management side of conservation science. Along with food webs, communities, and edge-effects crucial to the design of preserves, population size has been examined from many angles. It contributes to the probability of long term survival of any local group, and for that reason, becomes the foci of intense study. Geneticists have for a long time made the important distinction between census size, and breeding size, or effective population size in inbreeding studies. Demographic studies in anthropology also commonly focus on the exact proportion of non-reproductive age individuals which exist in any age-structured population. Did the early modern humans have more non-reproductive individuals in their populations than archaic peoples which preceded them?

Population sizes fluctuate in nature, with direct biological consequences. However, only recently has information been available to demonstrate concretely that population size and persistence in nature of large bodied animals are directly correlated. The intuition of biologists was that this would be the case, but little direct confirmation was available. Paleontologists had seen that in the Holocene extinctions of North America, large bodied mammals were more likely to disappear than small bodied ones. Joel Berger's longitudinal study of bighorn sheep persistence in the western US clearly shows now that 100% of the populations with size 50 or less are extinct in 70 years. Only those populations with size greater than 100 have a reasonable probability of surviving for longer than 70 years [29]. One would estimate that the effective population size for some of these groups is probably on the order of 10-30, a common number that arises in discussions with ethnographers about indigenous peoples

The contribution of conservation biologists to human evolutionary studies is clear. African mothers were important, and one reason for the persistence of their lineages may be that their matrilineal kin groups had larger sizes than others. Even if infant mortality made lifetime reproductive success hard to predict, a larger population base would ensure against the stochastic loss of your lineage over time. Long range dispersal of female kin to multiple virilocal residences would be an excellent management strategy to minimize the chances of stochastic loss of any given lineage, and is in fact an effective species management practice to insure long-term viability and persistence of endangered taxa. Population geneticists suggest that the mtDNA data are consistent with a population size of 5000-10,000 individuals, which fits the endangered species model [30]. We may be blinded to the importance of chronically small sized populations and local extinctions, simply because the global problems caused by the unimpeded growth of recent humans have been the experience of most academics. It is not clear that a population of this size would constitute a genetic bottleneck, because most nuclear genetic variability would be still maintained.

ANCIENT DNAS AND MODERN PEOPLE

It is unfortunate that our ability to conduct socially sensitive research with the help of aboriginal peoples is so limited. Biological research on human skeletal remains acquired under aggressive or at best, non-neutral conditions has been a focus of concern and the point of heated the communication between laboratory biologists, archaeologists, and tribal elders.

Research questions about the role of infectious disease, social stratification, and displacement and replacement of the aboriginal populations in the Pacific by colonial powers will be limited to deductions from indirect scientific analyses in the absence of the ability to extract DNA from ancient donors and their pathogens. Only by comparing ancient sequences to those of their descendents will we truly understand the pace and impact of recent cultural shifts in the biology of modern humans. This is in contrast to the nebuluous situation which formerly existed for the study of dense pre-industrial societies, genetic markers for infectious disease, and population replacement.

I suggest we need to work towards the development of trust which comes when minority students are educated and trained to conduct research of interest to their peoples. J.B.S. Haldane serves as a striking reminder of this ethic. There is both the professional and moral challenge implied by the perceptions of certain individuals that no direct medical advances towards the prevention, diagnosis, or treatment of disease have come from the study of Native American skeletal remains [31]. This statement made by a former Assistant Surgeon General of the United States has now been extended to include the remains of other indigenous peoples who are protesting for the return of the bones of their ancestors from museums and university teaching archives. As scientists, we must demonstrate this perception is false, and precisely how it is false, if we think scientific studies of ancient remains have validity for the lives of modern peoples descended from these ancient donors.

The drive to develop DNA technology that is applicable to direct analysis of fossil materials and disease organisms therein is intense, and is also subject to enormous pressures for early claims of success before appropriate quality control is performed. The agreed upon rules for what constitutes proof of ancient DNA versus modern contaminants, what it will look like under various states of preservation and increasing age, and how best to preserve newly found fossil materials for the technology of the future are just being discussed by specialists [32]. Anthropologists should expect that some spectacular claims will be made, and that there will be dissent among these specialists until research results are replicated by other laboratory groups. Not only can contamination be a problem, but also the artifacts associated with partially amplifed DNA sequences, which later recombine in the test tube, complicate the universal acceptance of potentially valid experiments .

Ancient DNA studies are moving towards non-destructive sampling techniques which could potentially lesson the objections of aboriginal peoples towards this type of research. Instead of powdering a small section of bone, we can sometimes extract DNA from it by soaking a bone in buffer. If religious beliefs require preservation of intact bones, this alternative may be more socially acceptable to some groups. Again, Haldane stands as a model for religious tolerance among scientists, both believers and non-believers.

Our global society has become more familiar with DNA typing, and perhaps as a consequence, many non-scientists will become more curious about the biological information which resides in their DNA sequences. The first impact of this knowledge may cause us to plan our families more carefully, limit our diets, or avoid certain known health hazards. Molecular anthropologists hope that non-scientists will also be curious about our past, and learn that the history of our human family is recorded in the patterns of our genes, as surely than it resides in the cusps of our teeth or the ridges on our brows.

REFERENCES

1. M. Nei. "Molecular Evolutionary Genetics," Columbia University Press, New York (1987).
2. D. L. Hull. "Science as a Process," University of Chicago Press, Chicago (1988).
3. H. Erlich (ed.) " PCR Technology," Stockton Press, New York (1989).
4. S. Horai, R. Kondo, K. Murayama, S. Hayashi, H. Koike, and N. Nakai, Phylogenetic affiliation of ancient and contemporary humans inferred from mitochondrial DNA, *Phil. Trans. R. Soc. Lond.B* 333:409 (1991).
5. R. Cann, DNA and human origins, *Ann.Rev.Anthropol.* 17: 127 (1988).
6. J. S. Wainscoat, A.V.S. Hill, A.L. Boyce, J. Flint, M. Hernandez et al. Evolutionary relationship of human populations from an analysis of nuclear DNA polymorphisms, *Nature* 319:491 (1986).
7. R. L. Cann, M. Stoneking, and A.C. Wilson, Mitochondrial DNA and human evolution, *Nature* 325: 31(1987).
8. C. B. Stringer and P. Andrews, Genetic and fossil evidence for the origin of modern humans, *Science* 239: 1263 (1988).
9. M. Wolpoff, Multiregional evolution: the fossil alternative to Eden, *in* : "The Human Revolution," P. Mellars and C. Stringer, eds., Edinburgh University Press, Edinburgh (1989).
10. L. Vigilant , R. Pennington, H. Harpending, T. D. Kocher, and A.C. Wilson, Mitochondrial DNA sequences in single hairs from a southern African population, *Proc. Nat. Acad. Sci.* USA 86: 9350 (1989).

11. A. C.Wilson and R.L. Cann, The recent African genesis of humans, *Sci* . *Amer* . 266:68 (1992).

12. J. C. Caldwell, and P. Caldwell, High fertility in sub-Saharan Africa, *Sci. Amer.* 262:118 (1990).

13. T. D. Kocher. and A.C.Wilson, Sequence evolution of mitochondrial DNA in humans and chimpanzees: control region and a protein coding region, *in* :"Evolution of Life: Fossils, Molecules, and Culture" S. Osawa and T. Honjo, eds., Springer-Verlag, Tokyo (1991).

14. L. Vigilant, M. Stoneking, H. Harpending, K. Hawkes, and A.C. Wilson, African populations and the evolution of mitochondrial DNA, *Science* 253: 1503 (1991).

15. S.B. Hedges, S. Kumar, K. Tamura, and M. Stoneking, Human origins and the analysis of mitochondrial DNA sequences, *Science* 255:737 (1992).

16. J. Felsenstein, Estimating effective population size from samples of sequences:ineffiency of pairwise and segregating sites as compared to phylogenetic estimates, *Genet. Res. Camb.* 59:139 (1992).

17. R. H.Ward, B. L. Frazier, K. Dew-Jager, and S. Paabo, Extensive mitochondrial diversity within a single Amerindian tribe, *Proc. Nat. Acad. Sci.* USA 88: 8720 (1991).

18. T.G. Schurr, S. W. Ballinger, Y-Y. Gan, J. A. Hodge, D.A. Merriwether, et al. Amerindian mitochondrial DNAs have rare Asian mutations at high frequencies, suggesting they derived from four primary maternal lineages, *Am. J Hum. Gen.* 46: 613 (1990).

19. A. Torroni, T. G. Schurr, C-C. Yang, E. Szathmary, R.C. Williams et al. Native American mitochondrial DNA analysis indicates that the Amerind and the Nadene populations were founded by two independent migrations, *Genetics.* 130: 153 (1992)

20. S. W. Ballinger, T. G. Schurr, A. Torroni, Y.Y. Gan, J.A. Hodge, et al. Southeast Asian mitochondrial DNA analysis reveals genetic continuity of ancient Mongoloid migrations, *Genetics* 130: 139 (1992).

21. A. Di Rienzo and A.C. Wilson, Branching pattern in the evolutionary tree for human mitochondrial DNA, *Proc. Nat. Acad. Sci.* USA 88: 1597 (1991).

22. L. I. Grossman, Mitochondrial DNA in sickness and in health, *Am J. Hum. Gen.*: 46: 415 (1990).

23. D. L. Cheney, R. M. Seyfarth, S. J. Andelman, and P. C. Le, Reproductive success in vervet monkeys, *in:* " Reproductive Success," T.H. Clutton-Brock, ed., University of Chicago Press, Chicago (1988).

24 J. Altman, G. Hausfater, and S.A. Altman, Determinants of reproductive success in savannah baboons, Papio cynocephalus, *in*:: "Reproductive Success," T.H. Clutton-Brock, ed., University of Chicago Press, Chicago (1988).

25. M..B. Mulder, Reproductive success in three Kipsigis cohorts, *in*:: "Reproductive Success," T.H. Clutton-Brock, ed., University of Chicago Press, Chicago (1988).

26.J. Wood, Fertility in anthropological populations,*Ann. Rev. Anthropol.* 19: 211 (1990)

27. G.E. Woolfenden and J. W. Fitzpatrick,"The Florida Scrub Jay: Demography of a Cooperative- Breeding Bird, "Monographs in Population Biology #20, PrincetonUniversity Press, Princeton (1984).

28. W.D. Koenig, Sex-biased dispersal in the contemporary United States, *Eth. and Sociobiol.* 10: 263 (1989).

29. J. Berger, Population size and persistence in bighorn sheep (<u>Ovis canadensis</u>, *Conser. Bio.* 4: 91 (1990).
30. D.L. Hartl and A.G. Clark, "Principles of Population Genetics, 2nd edition," Sinauer Associates Inc., Sunderland (1989).
31 .E.H.Ayau, Native Hawaiian burial rights, *in:* "Native Hawaiian Rights Handbook". M. K. MacKenzie, ed., Native Hawaiian Legal Corporation and Office of Hawaiian Affairs, Honolulu (1991).
32. G. Eglinton and G.B. Curry , (eds). "Molecules through time: fossil molecules and biochemical systematics," *Phil. Trans. Roy. Soc. Lond B.* 333: 315 (1991).

ACKNOWLEDGMENTS

I thank W. Brown, A. Di Rienzo, R. Higuchi, S. Horai, T. Kocher, K. Lum, T. Quinn, S. Pääbo, E. Prager, O. Rickards, G. Shields, M. Stoneking, B. Sykes, K. Thomas, and L. Vigilant for mtDNA sequence information prior to publication. Members of my laboratory, B. Feldman, C. Reeb, K. Lum, and B. Laxson helped me focus on the biological issues at stake in human mtDNA studies. H. Carson, L. Freed, J. Hunt, and C. Simon contributed insights in discussion of larger issues of evolutionary biology in humans. Funding came from the National Science Foundation, Hawaii-Bishop Institute, and the J.D. and Catherine T. MacArthur Foundation.

EVOLUTION OF CHEMICAL SIGNALS

R.L. Brahmachary [1], Mousumi Poddar Sarkar [1] and J. Dutta [2]

[1]Indian Statistical Institute, Calcutta 700 035, India

[2]Wetland Management Laboratory, Calcutta, India

INTRODUCTION

Chemical signals comprise a wide assortment of molecules that play various key-roles in organisms as diverse as *E. coli* and elephant. The evolution of these signals and certain aspects of the genetics of histocompatibility in mice and men are relevant to the theme of this conference. These have been discussed and certain new findings on chemical signals in the tiger mentioned.

A relevant theme is also the role of chemical signals in reproductive success and hence, in turn, in forming a selection mechanism. For example, the status of health including normal hormonal function is likely to be reflected in the urine. Given a choice, a female is likely to test and prefer the urine of a highly androus male. Many mammals evince a keen interest in the urine of the opposite sex and the testing is accomplished through sniffing and vomeronasal organ (see later).

In some animals selection of a mate does indeed take place through semiochemicals. An interesting study on bank voles in Great Britain was carried out some years ago (Godfrey, 1968; see also Stoddart, 1980). Within the species *Clethironomys britannicus* there are races, for example, the Edinburgh population and the Western *Kintyre* population. In islands like Skomer, different species have probably evolved. The mainland voles leave tiny trails of urine marking and the Skomer race (species) uses larger drops. It was found however that individuals chose sex partners from the same race by odour alone and hybridization in the laboratory leads to less viable animals. An even better report exists on mice, where, given a choice, both inbreeding and outbreeding are avoided, – mates are chosen from non-related individuals of the same strain (Gilder and Slater, 1978). These examples suggest that urine – borne signals do help the animals in partner selection.

Again, the immunological self-nonself recognition has evolved into a mechanism for partner selection through the agency of semiochemicals. It is now known that with the help of such signals in the urine mice can sniff out the degree of relatedness in the major histocompatibility complex (MHC) (Yamazaki et al., 1981). Congeneric mice have now been bred where the only difference lies in the MHC and experiments with such animals show that they test the urinary semiochemicals and select mates who share lesser amounts of MHC, so that MHC in the offspring would be more heterozygous (Jones and Partridge, 1983; Yamazaki et al., 1981).

ROLES OF CHEMICAL SIGNALS (SEMIOCHEMICALS)

Chemical signals are ubiquitous in the living world. Chemical cues orchestrate the architecture, development and differentiation of the embryo into an animal. From classical

Human Population Genetics, Edited by P.P. Majumder
Plenum Press, New York, 1993

experimental embryology it is known that early blastomeres, especially in the so-called mosaic egg, are aware of the presence or absence of neighbouring cells and develop accordingly. More recently some of the chemical interaction among the "Northern" (animal) and "Southern" (vegetal) cells of the embryonic "globe" (blastula of Xenopus) have been traced to chemical signals which are polypeptides, members of the TGF family (Green and Smith, 1990). In the nematode, *Caenorhabditis elegans*, chemical signals from anchor cells (AC) are known to act on vulval cells (Greenwald and Rubin, 1992). It has also been known for quite some time that free-living slime mold or myxamoebic cells "call" one another by emitting cAMP (which attract these cells) and the initial conglomerations attract bigger and bigger crowds which, after a certain stage, move together as a single "slug" (Law and Regnier, 1971)

Further, chemical signals like amino acids bind to 30 types of receptors, the equivalent of sensory organs of bacteria and act on 9 protein structures, the equivalent of "brain", causing these organisms to move rather like guided missiles, towards the putative source of food (Koshland, 1980; Paoni et al, 1983). Likewise, the leucocytes, scavangers and police force in the body, are attracted by a modified form of the amino-acid, methionine, namely n-formylmethionine; oligopeptides, from at least dipeptide onwards, containing n-formylmethionine and elaborated by bacteria are the chemical cues (Schiffman et al., 1975; Sukumar et al., 1985). Interaction of an attractant (formyl peptide) with the receptor (in neutrophils of rabbits) has also been studied (Becker et al., 1985). Similarly, a 14-amino-acid peptide (Resact) "calls" sea-urchin sperms to bind to the egg (Dangott, 1988). There are further aspects of these phenomena which are worth noting and we will return to this theme in the section on Evolution.

The exudates of tadpoles may inhibit the growth of smaller tadpoles (Richards, 1962; Akin, 1966). It is a well known fact that queen bees (or termites) prevent the development of other queens and the chemical aspect of this phenomenon has been studied (Wilson, 1973). Only recently, a similar state of affairs has been detected in mammals – in the naked mole rat (Gamlin, 1987) and dwarf mongoose (Attenborough, 1990). A bizarre example is that of the water-borne chemicals emitted by larva of a mosquito species (*Aedes sirrensis*) which cause a free-living protozoa, the initial prey of the larva, to redifferentiate into a different, parasitical form which can then attack and kill the larva (Washburn and Anderson, 1991).

An equally startling example has been reported by Raina et al. (1992) on ethylene as a chemical signal from corn (maize) that induces the female *Helicoverpa zea* moth to broadcast the sexual attractant so that males come and copulate and eggs are laid on ripening plants (to ensure larval food).

PHEROMONES

The most well-known semiochemicals are pheromones. Up until comparatively recently, insect pheromones, especially the sexual attractants, had been intensively studied (Shorey, 1976). It is now common knowledge that Bombykol can attract male moths from a distance of 1 or 2 km; a good inkling of this fact was recorded by Fabre in the last century. This, therefore, is an immense extension of the "calling" semiochemical of slime molds which acts over a range of millimetres or centimeteres. Likewise, worker bees, after being attracted by the queen bee pheromone, call other workers over a distance with the help of pheromones exuded from Nassanov glands (Wilson, 1973). Insects on land and in air, as well as mammals, make use of volatile air-borne molecules as pheromones, which are of up to 300 Mol. Wt. Bigger molecules cannot be volatile and so an upper limit of size is fixed by physical constraints. By contrast, other classes of molecules may be utilized in the aquatic medium. Amino-acid which is not particularly volatile is, as already mentioned, an odorant or attractant for bacteria and from here up to the emergence of an aquatic vertebrate, namely fish, we can follow a nearly linear course of development. Fish, too, sense odorants dissolved in liquid media and some of these can be amino-acids. In fact, the amino-acids binding to receptors of the olfactory rosette of fish have been studied (Cagan and Zeiger, 1978). Meanwhile, air-borne olfactory molecules began to play a very important role in the lives of insects who left water. For pollinating flowers as well as for locating the sexually mature partner, the sense of smell and sexual attractants developed in the world of insects. Most of these odour molecules, perfume of flowers or sexual semiochemicals, have to be volatile, i.e. within the molecular weight range 300. (This is valid for terrestrial mammals as well.)

There is however a connecting link between mammals and fish in the form of vomeronasal organ (VNO) or Jacobson's organ. Discovered as early as in 1703 by Ruysch (see Schilling, 1970), and thoroughly studied by Jacobson in the early nineteenth century, the functional aspects of this organ remained mysterious. Only recently the possible function of this organ has been suggested (Schilling, 1970; Estes, 1972; Hart, 1983). Non-volatile pheromones seem to be transferred to the VNO by the tongue. The VNO occurs in various vertebrates and mammals, ranging from snakes to elephants. In some mammals, such as the cetacea, which have invaded the watery environment, the chiroptera and in primates, the VNO is absent or reduced. But in the embryonic stage even primates possess this organ and very recently the presence of VNO in adult humans has been noticed (Moran, 1991).

In the embryonic stage even a mammal (in the amniotic fluid) behaves like a fish responding to liquid-borne odorants. Only at birth, the system is suddenly switched off and the air-borne olfactory mechanism switched on. This is very clearly borne out in the rat (Pedersen et al., 1983). Pregnant rats, on the last day of gestation, were injected with $2 - [14_C]$ deoxy - D - glucose which was earlier used for studying "post natal, developing olfactory pathways". While the main olfactory pathway arises from receptors on the olfactory epithelium in the nasal cavity, the <u>accessory</u> olfactory system originates from receptors of the VNO. The uptake of the tracer in the accessory olfactory bulb and its absence in the main olfactory bulb indicates that *in utero*, the former is active. Thus, it is the VNO which enables the embryo to sense odor molecules in the liquid ambient through the accessory system and so is the equivalent of olfactory florette in fish. So far we can still follow a continuity of development from fish or even from bacteria. An evolution of receptor cells is also evident (Vinnikov, 1982). The original form of olfactory receptors is a cell with flagella and microvilli. Olfactory cells in the next stage are microvillous only, and in higher vertebrates these are flagellate only. In the teleosts or bony fishes, the diverticulum of olfactory sac, the equivalent of VNO, has microvillous cells and the VNO cells of higher vertebrates are also microvillous.

SOURCES OF SEMIOCHEMICALS IN MAMMALS

Urine and faeces are two primitive as well as common source of mammalian pheromones. this is not as surprising as it sounds. Excretory products as they are, they reflect the metabolic stage of the animals; in short, their status of health and, perhaps, the mark of individuality. An animal under stress, just as many a human being, urinates or defaecates. Later this familiar smell is reassuring and the animal is loth to enter into a neighboring clean cage, say. So cleaners may have to learn, at their peril, that many animals prize their faecal heaps/urine-soiled corners, etc. The cleaner is in fact trying to rob the animals of their valuable "property".

Apart from urine and faeces, a number of specialized glands have evolved to exude semiochemicals (Birch, 1974; Albone, 1984). These have been presented in table 1.

Table 1. Some of the specialised glands supposed to secrete semiochemicals in mammals

anal	More than 100 species
tarsal	
metatarsal	various antelopes and deer
caudal	
temporal	elephant
chin gland	rabbit
morillo	capybara
forehead gland	sugar glider
axillary (armpit)	human male
vaginal secretion	human female [1,2]
vaginal secretion	female monkey [3]

1. Michael et al. (1974); **2.** Maclintock et al. (1971); **3.** Curtis et al. (1971)

Two very good reviews were published some years ago wherein the ethological aspects of (pheromonal) marking have been carefully described (Ewer, 1968, 1973). This author emphasized the fact that in all carnivores the frequency of "scent marking" dramatically increases during mating period and that she was not aware of any exception to this rule. This stands to reason, for members of each sex must blatantly advertise their presence or, at least one sex must do so and the other must be keyed up to respond to that. This self-advertisement must be one of the principal functions of semiochemicals on which the reproductive success of the species, and hence its future, heavily leans. Theoretically, it also follows that just as coupling or mating is about to start, the necessity for advertisement no longer exists, at least for the time being.

An equally important urge is of course to always maintain a territory. It may be noted that a pet dwarf mongoose was lost and found after a few days in an emaciated, starving condition. When brought home and given food it first went around, "marking" the usual spots in its territory and only after that began eating (Ewer, 1973).

Urination itself seems to have evolved into two different modalities. Scent-marking or scent-urination in ungulates has been described with special reference to feral goats (Coblentz, 1976). Many goat species, urinate on some part of their own body which has long hair. This may carry one's own "olfactory flag" which may repel or intimidate another male and attract females. In Catalina Island, during the breeding period or just before that, no male feral goat was observed to urinate in the normal manner, instead, they "squirted" the urine on the long hair of chest, neck and beard.

The male eland is known to rub his forehead on the ground after urination and then he apparently transfers this scent to tree trunks (Leuthold, 1977). The male Nile Lechwe urinates on his chest and neck and then marks the female by putting his neck on her rump (Leuthold, 1977).

Scent urination, ethologically distinct from the process of mere excretion, has also been noticed in carnivores like wolf, tiger, lion, etc. (see later in tiger). The dominant male or female in the wolf pack was seen to squirt a jet of urine apparently for marking convenient spots in their territory (Ewer, 1973). Raised leg urination (RLU) has been adopted as a term to distinguish it from ordinary urination (Peters and Mech, 1977). A later study describes four types of urination in a captive wolf pack in a large enclosure (Asa et al., 1985). These are RLU, FLU (flexed leg urination), STU (standing urination), SQU (squatting urination). Only the dominant male and female indulged in RLU and FLU, respectively.

These, therefore, can be presumed to be sources of semiochemicals which indicate dominance within that territory. Since antiquity man must have closely watched the importance of urine in the animal world and he may still have an unconscious innate tendency to prize urine marking. In Essex, England, the male of the privileged class could urinate in broad daylight on the front wheel of his horse-drawn carriage and later, on the wheels of his motor car (Willis, 1985). In two communities in India and Africa, respectively, the bridegroom and the bride, respectively, used to be made to smell the urine of the partner (Choudhury, 1986).

A pattern of faeces location was also noted in the wolf pack described above. The scats of the dominant and the second dominant (i.e. alpha and beta) male were more frequently found in the feeding area, than those of others. Anal gland marking on scats were most frequent in the alpha male while the beta male took the second position. Again, the European lynx and wild cat bury their faeces but, at the boundary of the territory, these are not buried but deposited on prominent spots wshich are also urinated upon (Lindeman, see Ewer, 1968). It thus seems that these are sources of semiochemicals which indicate territoriality. That the spots where faeces are accumulated serve a function in scent marking among carnivores has been the subject of a more recent review (Macdonald, 1980). Brown hyenas may defecate in "latrines" and they also scent mark on grass stalks, shrubs etc. which may be as high as 2900/ hyena/year (Gorman and Mills, 1984). The latrines lie mostly near the boundary and the anal scent markings mostly in the inner part of the territory. A comparative study of brown hyenas and spotted hyenas in different geographical regions and under different conditions of population density leads to the conclusion that "scent marking strategies will not be species specific, but instead will be determined by the different ecological pressures acting on particular populations". When the territory is small, the border line or perimeter can be regularly marked, but in larger areas this is not feasible. So the hyenas there take recourse to "hinterland marking" – latrines scattered inside the area, along the pathways, etc. In hyneas, urine is not used for scent marking (Kruuk, 1972).

The badger marks its sett and the paths running along the range borders with the help of a lipid-rich exudate from its sub-caudal gland (Gorman et al., 1984). Dominant males of a clan scent mark most (Kruuk et al., 1984) (On snow this is easily visible and seems to be mixed with anal gland secretions.) All badgers in a clan mark each other but the dominant males do so most often and especially on females. Cubs are avidly marked by adults. It is thus a sign of community bondage, a sort of kibbutz, with characteristic mark or dress. Badgers are not unique in exhibiting these bahaviour patterns. Many other species do so (Ewer, 1968). Young rabbits, tree shrews, aardwolf, monogoose are marked by the adults with the help of chin glands, sternal or chest glands, and cheek glands. In the sugar glider, the males of a clan mark ther females wth exudates from the frontal gland. A female marked by a foreign male is attacked by males of her own clan.

Another agency for markings is that of saliva (Ewer, 1968). The sugar glider as well as some other marsupials and the tenrec deposit saliva on external objects. They may also transfer saliva borne odour on their own body. The hedgehog "self anoints" with saliva; so do certain members of the Lorisidae (Schilling, 1970). A curious case is that of "hand stand posture". The giant African rat (*Cricetomys gambianus*) and the female Afrcan bush dog defeacate normally as well as "standing" on the front legs, with the hind part resting against a tree. It is believed that at that time they also add the marking from a caudal or anal gland which is transferred to the faeces (Kleiman, 1966, Ewer, 1968). The mongoose may also adopt the "hand stand posture" (Gorman et al., 1974).

From the above descriptions, far from exhaustive as they are, there emerges a picture of a multi-faceted function of semiochemicals. These can indicate sex, dominance, clan membership, territoriality, etc. In addition, there is an aspect of semiochemicals that is closely associated with reproductive biology, well known as Whitten effect, Lee-Boot effect and Bruce effect. There is also the class of semiochemicals that forge the links of recognition between the mother and young.

It is the concept of territoriality which is perhaps the most complex and challenging in the context of mammalian semiochemicals. That a very large number of animals, including man, have a very strong attachment to their own home region is now part of common knowledge and the phrase "territorial imperative" has become a cliche (Ardrey, 1967).

According to the definitions in ethological literature, a territory is an area that is actively defended but a home range is not. These expressions may not be absolutely satisfactory and mechanisms for spacing without entailing fighting have necessarily evolved. It is too much of a risk to fight, – heavily armed predators like lions, tigers, leopards, even if they are winners, will in the long run die of the comparatively minor injuries, for they would be unable to predate. According to some, there is an inbuilt inhibitor in these so called fierce animals (Lorenz, 1953) but, this is not fully true. Lions and tigers (according to recent observations in Africa and India) may be killed by their conspecifics but roarings and semiochemicals as warnings and inhibitors certainly reduce the head-on collisions. A home range or territory need not be a continuous area and defending need not be an all out fight. An example of interest is the description of a female cat which had staked out a territory of 64 acres within which certain areas were regularly visited and others rarely or not at all. (Leyhausen, see Ewer, 1973). She chased females from her territory. Either here or in the case of larger cats, this defending of the territory need not lead to a fight, the intruder generally respects the "righteous indignation" of the "owner" and retreats. Its ventures are opportunistic sallies, in order to probe the alertness and strength of the owner, rather than a predetermined will to conquer. This is not without its counterpart in the human world, – whether in the proverbial castle of a home of the Englishman or a hotel room, prior occupation is respected just as in the animal world. Again, a known neighbour at the boundary evokes less agitation than an unknown in both the worlds. "It is neither necessary nor justifiable to suppose that the animals read and analyse the information in human terms: all that is required is that they should respond appropriately to different characteristics of the mark" (Ewer, 1973). Thus, an intruding male shows caution in the territory of another male (unless scent marks are too old) and eager interest in the territory of a female.

The concept of territory in animals is at least as fluid and variable as in the human society; in the latter we notice various gradations from an ancestral castle where a family has been living for a couple of centuries to a hotel or a dormitory where several people are spending a couple of nights sharing everything except just a bed, perhaps. While watching the hippos, almost literally spanning the river Rwindi in Zaire, the senior author would have been tempted to believe that these animals have no sense of territory at all, had he not also known and watched the nocturnal feeding areas of adult hippos on land and their beaten tracks along which they may scatter their faeces, by beating with the tail. Only careful observation shows that the dominant rat(s) liberates a substance that fluoresces under the ultraviolet ray and his (their) markings are scattered over the place in the largest number, others urine mark only in a few spots (Ropartz, 1977). The dominant ones thus "grab" the territory, which with naked eye observations of urine stains would seem to indicate a communal territory.

Male cats sharing junctions of roads in their territory may wait like railway trains on tracks, waiting for the "down" or green signal, when another cat is passing along through crossroads (Leyhausen, 1964). Even here a certain degree of exclusiveness is evident. Again, several different types of territories were known quite some time ago (Ewer, 1973). For example, the Uganda Kob defends a circular patch of territory for mating, not for feeding; Grant's gazelle defend their territorial space and Impala, their moving "territory" of a harem. The mountain goat may do likewise but not when the area lies under deep snow.

SELF-NONSELF RECOGNITION

The ability to distinguish self and non-self is implicit in any territorial context and so is the mother-infant recognition through semiochemicals. There is a general tendency of avoiding extreme inbreeding (and extreme outbreeding) in animal populations. Mechanisms have evolved to ensure this at the semiochemical level. It even seems that in a sense, the animal, given the choice, smells out histocompatibility "reflected" in urine and "decides" against mating with shared histocompatibility genes (Yamazaki et al., 1981).

In invertebrates like sponges and gorgonian corals, an aspect similar to vertebrate histocompatibility has been studied (Avise and Neigel, 1983; Theodor, 1970). These organisms reproduce both sexually and asexually, and autografts lead to fusion or acceptance but grafts between individuals growing in distant spots (and therefore, presumably, unrelated) are rejected. Even more important is the study on *Botryllus*, an ascidian (i.e., at the boundary line of invertebrates and vertebrates). Here the colony fusion is controlled by a single highly polymorphic genetic region, and, in addition the fusibility (or, compatibility) locus "affects self fertilization by sperm-egg incompatibiltiy" (Scofield et al., 1982).Since here the "histocompatibilty genes control both somatic tissue recognition and response as does the vertebrate MHC and gamete behaviour at fertilization (the) findings support ... that vertebrate histocompatibility genes evloved from gametic self-nonself recognition" (Scofield et al., 1982).

The "supergene" of histocompatibility has been well studied in mice and men. These supergenes, MHC or HLA are located on chromosomes 17 and 6 in mice and men, respectively, and adjacent to that is another supergene - the T/t system (Hammerberg and Klein, 1975; Lyon et al., 1979). The MHC region is subdivided into parts like H_2K, H_{20}, Ir, etc. The MHC determine graft-rejection, cell-cell interaction, immune response, control of complement activity and even the levels of plasma testosterone (Goldstein and Cagan, 1981). As Ir genes have a high response to particular antigens, heterozygous Ir or MHC would be more effective against mortality. Thus, mate choice involving unrelated partners would be better from the point of view of survival value. Furthermore, the fact that about 75% of the human embryos fail to come to term may at least partly be due to histocompatibility assortments of the parents (Jones and Partridge, 1983). They state that "the foetus is nature's transplant" which, because of the foreign genes of the father, first elicits a challenge in the mother's system and this "boosts foetal antigens and protects the foetus from a later attack by the mother's lymphocytes" (Jones and Partridge, 1983). This later attack in fact led to abortions in three women who shared much of the MHC genes of their husbands and so the first attack was weak. Then, in the course of their next pregnancy, they were treated with foreign lymphocytes which apparently elicited a strong early challenge and all three mothers gave birth to healthy babies. These authors also point out that the evolutionary basis of this aspect may lie in the *Botryllus*, where fertilization between individuals sharing the same histocompatibility genes usually fails.

The t/t supergene system is also very complex (Glucksohn-Waelsch and Erickson, 1970). Historically, Dovrovalskaia in 1927, discovered the semidominant mutation T (brachyury) which, when heterozygous, leads to shorttailed mice, but is lethal if homozygous. A combination of a t-allele and a non-t allele has a selective advantage over non-t bearing sperm. Interestingly, a male heterozygous for two different complementing t - alleles, may produce non-functional sperm. The homozygous lethality, "a costly invention " may have evolved to ensure generation of a superior sperm, - t mutants (Hammerberg and Klein, 1975).

This t complex is now known to cover the proximal one-third of chromosome 17 and includes T locus and MHC (Herrman et al., 1986). It was also found that major testicular cell proteins are specified by a mouse T/t complex. The MHC t/T supergene system thus encodes several agents that play a role in reproductive biology and survival. Interestingly, the body odour gene locus is closely related to it (see later).

INBREEDING AND OUTBREEDING

What has been descrbed above may be discussed in connection with inbreeding/outbreeding. It is indeed well known that too close, consanguineous marriages lead to inbreeding depressions, accumulation of genetic defects, etc. in human populations. Mechanisms to avoid self-pollination and close inbreeding in plants and animals have been noted by classical biologists. An exception is the Syrian hamster. Millions of healthy hamsters live in laboratories and pet shops all over the world, although these were bred from a single family of wild caught hamsters.

In man, excessive inbreeding may be avoided with the help of elaborate cultural traditions. Female chimpanzees begin to avoid related males after puberty (Pusey, see Jones and Partridge, 1983). It has also been noted that children raised together in the kibbutz, like a large family of brothers and sisters, on coming of age do not choose their wives/husbands from the same kibbutz-reared group.

In animals with well developed olfactory powers, the non-kin selection may well depend on semiochemicals in urine or body odour. The cases of bank voles and congeneric mice differing only in MHC have already been described.

PHEROMONES OF THE TIGER

The principal source of tiger pheromone is the marking fluid sprayed upwards and backwards through the urinary channel and at first wrongly believed to be mixed with anal gland secretion (Schaller, 1967). We have endeavoured to dispel this confusion (Brahmachary and Dutta, 1981, 1987) and have studied the various volatile constituents of this fluid, likely candidates for pheromones, (Brahmachary and Dutta, 1981, 1987; Brahmachary et al., 1990, 1991, 1992). The volatile molecules are a number of amines, free fatty acids, and a highly unstable, presumably acetylated, Nitrogen-containing compound. This last is particularly interesting because it imparts the very characteristic, musky odour of the tiger and leopard. This seems to be the same substance which imparts a pleasant aroma to certain varieties of Indian fragrant rice in the unboiled state and is different from, though similar to, 2-acetyl-1-pyrroline, the only identified component of the smell of boiled, fragrant rice (Buttery et al., 1982). We have succeeded in synthesizing this new compound though with a very poor yield, but the exact formula has not yet been established (Brahmachary et al.,1990a; Poddar-Sarkar et al., 1992).

So far as we are aware, tigers and leopards are the only two mammalian species biosynthesizing this aroma substance. Domestic cat, African lion, goat, cattle and Indian camel have actually been tested by us (unpublished) and to our knowledge this substance has not been recorded in any other mammal though the urine of the wolverine and the Canadian lynx is said to be musky (personal communication).

Fragrant rice-water did indeed elicit flehmen gesture from an adult leopard while the control (the same or similar piece of cloth or paper moistened in water and presented a short while before and again a short while after the experimental (i.e., rice-water moistended) piece did not. Many different strong odours induce the flehmen gesture in the tiger and leopard but it is worth noting that strong smelling butyric acid and valeric acid (natural constituents of MF) did not elicit flehmen from cubs or the adult leopard although they strongly reacted by audibly sniffing. Further bioassay with tigers and leopards is being planned but the preliminary results suggest that this aroma may be a perceptible volatile component of the MF and therefore a putative pheromone. The fact that the same aroma molecule is biosynthesized in living beings as different as the tiger and a member of the graminae family is of interest *per se* but at present the evolutionary implication of this, if any, is unknown. "Boar-taint" (androsterone), a male pheromone in the boar and the human male is known to be present in certain vegetables like celery (Clause and Hoppen, 1979). The function of boar-taint and of the aroma in celery or rice is anybody's guess.

Quite early in the course of our investigations it became clear that the volatiles of which the aroma is a very representative example, quickly dispersed on separtion from the non-volatile lipids as a result of steam distillation (Brahmachary and Dutta, 1979). The quantity of lipids in tiger MF may be as high as 2 g/Litre. Thus, a tiger may throw away several grams of lipid per day but this apparent wastage makes sense; the lipid as fixatives cause the volatiles to linger for days or weeks and otherwise the tiger could never meaningfully mark its domain.

THE CHEMICAL BASIS OF INDIVIDUALITY WITH SPECIAL REFERENCE TO THE TIGER

The uniqueness of an individual is evident at visual, auditory, physiological and biochemical levels. Not only is every face different from another, every one has a recognizable voice, every individual has also a characteristic fingerprint which fact has been used in forensic science for more than a century. The use of police dogs in hounding out criminals on the basis of their distinctive body odour is also well known.

As early as in 1756 Withof stated that every individual human being exudes a distinctive odour (Schildeknecht and Ubl, 1986) and this idea has now culminated into the finding that the distinctive body odour and histocompatibility are determined by neighbouring genes (Goldstein and Cagan, 1981).

Again, biochemical individuality in the excretory products, specially in urine, is a theme discussed by Garrod in 1882 (Williams, 1956). It is impressive, in retrospect, to reread his words, implying that alkaptanurea etc. are "extreme examples of variations of chemical behaviour which are probably everywhere present in minor degrees" and that "just as no two individuals of a species are absolutely identical in bodily structure" the biochemical processes in their systems also vary. The idea has been well supported by more recent findings on the patterns of specific excretory products in rats and human beings; certain amino acids, vitamin etc. seem to be excreted in particular individuals in characteristic, nearly constant amounts which are largely independent of diet (Williams, 1956). Even when the same diet is fed to rats or in milk-fed human babies, individuals exhibit different, characteristic amounts of those substances excreted in the urine.

An impressive example is a series of star-shaped patterns, a sort of biochemical tagmark for an individual that can be drawn up on the basis of relative ratios of 41 different substances in urine, saliva, etc. (Strauss, 1960). Presumably, the police dog works on some such molecular basis.

The semiochemicals of territorial animals — and most animals are so, though in varying degrees — must, theoretically, encode the messages for self-non-self and presumably moreover the distinction between a regular neighbour and an intruder. Brownlee's findings on black tailed deer can be very aptly presented as "All females carried a flag coloured some shade of red, all males a blue flag so that sex is immediately obvious and with a little more care one can say, but I have never seen exactly that shade before, he/she must be a newcomer" (Ewer, 1973). The semiochemicals of mongoose, a series of carboxylic acids, are maintained in characteristic ratios in each individual and mongoose can be trained to distinguish the odour of other specific mongoose (Gorman et al., 1974, Gorman, 1976). An interesting earlier example is that of Rasa, very smartly presented in the following manner (Ewer, 1973). Cheek gland secretion of a dwarf mongoose elicits fear in another but anal sac secretion causes only curiosity; "one letter is a personal signature, the former an anonymous threat. Normally, the signature is added to the threat. That animals actually discriminate among conspecifics has been indicated in several species, like Tamarin monkey (Epple et al., 1981), wolves and dogs (Dugmore and Evans, 1989) cats (Passanisi and MacDonald, 1986) but only Gorman (1976) established a clear molecular basis on which the animal might work, namely that the ratios of the carboxylic acids are unique for every one of 24 mongoose.

Eleven free fatty acids have now been quantitatively estimated for three tigers. Considering these values, the "Euclidean distance" between the tigers has been measured. The distance turns out to be 2.0 between the mother and son, 7.3 between the mother and the third tiger (female) and 6.7 between this last and son. Tiger studbook data show that the third tiger is more closely related to the son (from paternal side) than to the mother. Though less convincing than Gorman's data (1976), these results support the theory that semiochemicals encode individuality. All these data offer pitfalls for interpretation, especially because replication results may vary due to intrinsic changes and partly due to experimental errors.

EVOLUTION OF CHEMICAL SIGNALS

The mechanism of reception of chemosignals evolving from unicellular organisms to mammalian olfactory system reveals a highly conservative scheme. A common receptor gene family encodes olfactory receptor and sperm cell receptors and sperm chemotaxis (Parmentier et al., 1992). The infrastructure of using cyclic nucleotides can be noticed in olfaction as well as sperm-egg binding (Anholt, 1986). Chemical evolution is evident from single amino acids as signals for bacteria, to di-, tri- and tetra-peptides for egg-sperm signalling as has been described. Again, among pheromones, pyrazines occurring in the insect world are detectable even in the primitive mammal *Tupaia* (Stralendorff, 1987). Pyrazines, in fact, are widely distributed in warning-coloured insects and plants (e.g., red and green pepper, etc.) and have also been detected in coyote urine and rabbit faecal pellets (Rothschild, 1989). Free fatty acids seem to be a conservative class of molecules evolving from *Tupaia* (Stralendorff, 1986, 1987) and traceable in the mongoose, tiger, monkey and human species. Interestingly, androstenone or boar-taint occurs both in the boar (i.e. the male of the species) and the human male, functioning as sexual attractant/stimulator, definitively in the former species and probably in the latter while the presence of this substance in celery-like plants (Clause and Hoppes, 1979) has so far not been functionally correlated. Likewise, that aroma common to both tiger and rice may act as a tiger pheromone but its relevance to the life cycle of the rice plant is unknown.

ADDENDUM

Together with Dr. P. Bhattacharya we have now surmised a possible structure of the new aroma molecule on the basis of periodate treatment which immediately destroys the aroma. Both 2-acetyl-1-pyrroline (2AP) and a 3-keto-1-pyrroline structure would be periodate sensitive. Since the new molecule is not 2AP, the second possibility is strengthened.

REFERENCES

Akin, G.C., 1966, Self-inhibition of growth in *R. pipiens* tadpoles, *Physiol. Zool.* 39:341.

Albone, E., 1984, "Mammalian Semiochemicals," John Wiley, New York.

Anholt, R.R.H., 1987, Primary events in olfactory reception, *Trends Biochem. Sci.*, 12:58.

Ardrey, R., 1967, "The Territorial Imperative," Collins, London.

Asa, C.S., Mech, L.D., Seal, U.S., 1985, The use of urine, faeces and anal gland secretion in scent marking of wolf, *Anim. Behav.* 33:1034.

Attenborough, D., 1990, "The Trials of Life," Collins/BBC, London.

Avise, J.C., Neigel, J.E., 1982, Population biology aspects of histocompatibility polymorphisms in marine invertebrates, *in* "Genetics: New Frontiers," V.L. Chopra, B.C. Joshi, R.P. Sharma, H.C. Bansal, eds., Oxford and IBH Publishing Co., New Delhi.

Becker, E.L., Freer, R.J., Toniolo, C., Balaram, P., 1985, The specificity of chemotactic formyl peptide receptor of rabbit neutrophils, *in* "Membrane Receptors and Cellular Regulations," Alan R. Liss, New York.

Birch, M.C., 1974, "Pheromones," North Holland, Amsterdam.

Brahmachary, R.L., 1986, Ecology and chemistry of mammalian pheromones, *Endeavour*, New Ser. 10:65.

Brahmachary R.L., Dutta, J. , 1979, Phenylethylamine as a biochemical marker of tiger, *Zf. Naturforsch.* 34C:632.

Brahmachary R.L., Dutta, J., 1981, On the pheromones of tigers: experiments and theory, *Amer. Nat.* 118:561.

Brahmchary, R.L., Dutta, J., 1987, Chemical communication in the tiger and leopard, *in*: "Tigers of the World," R.L. Tison, U. Seal, eds., Noyes, New Jersey.

Brahmachary, R.L., Poddar-Sarkar, M., Dutta, J., 1990, The aroma of rice and tiger, *Nature* 344:36.

Brahmachary, R.L., Dutta, J., Poddar-Sarkar, M., 1991, The marking fluid of tiger, *Mammalia*, 55:150.

Brahmachary, R.L., Poddar-Sarkar, M., Dutta, J., 1992, Chemical signals in vertebrates VI, Plenum Press, New York. *(In press)*

Brown, D.S., Johnston, R.E., 1983, Individual discrimination on the basis of urine in dogs and wolves. *in* "Chemical Signals in Vertebrates III," D. Muller-Schwarze, R.M. Silverstein, eds., Plenum Press, New York.

Buttery, R.G., Ling, L.C., Juliano, B.O., 1982, 2-acetyl-1-pyrroline, — an important aroma component of cooking rice, *Chem. Industry* 4 December: 958.

Cagan, R.H., Zeiger, W.N., 1978, Biochemical studies of olfaction: binding specificity of radiolabelled stimuli to an isolated olfactory preparation from rainbow trout, *Proc. Natl. Acad. Sci., U.S.A.* 75:4679.

Choudhury, D., 1986, Letter, *New Scient.* No. 1526:87.

Clause, R., Hoppen, H.O., 1979, Androstenone in vegetables like celery, *Experientia*, 35:1674.

Coblentz, B.E., 1976, Functions of scent urination in ungulates with special reference to feral goats, *Amer. Nat.* 110:549.

Curtis, R.F., Ballantine, J.A., Keverne, E.B., Bonsall, R.W., Michael, R.P., 1971, Identification of primate sexual pheromones and properties of synthetic attractants, *Nature* 232:396.

Dangott, L.J., 1988, News on sea-urchin chemoattractant, *Nature*, 334:649.

Dugmore, S.J., Evans, C.S., 1989, Discrimination of conspecific chemosignals by female ring-tailed lemur, *in* "Chemical Signals in Vertebrates V", D. Macdonald, ed., Oxford University Press, Oxford.

Epple, G., Golob,N.F., Cebul, Mary-Scot, Smith, A.B., 1981, Communication by scent in some callitrichidae, *Chem. Senses* 6:377.

Estes, R.D., 1972, The role of VNO in mammalian reproduction, *Mammalia* 36:315.

Ewer, R.F., 1968, "Ethology of Mammals," Elek Science, London.

Ewer, R.F., 1973, "Carnivores," Weidenfeld & Nicholson, London.

Gamlin, L., 1987, Rodents join the commune, *New Scient.* 30 July:40.

Gilder, P.M., Slater, P.J.B., 1978, Interest of mice in conspecific male odours is influenced by degree of kinship, *Nature* 274:364.

Glucksohn-Waelsch, A., Erickson, R.P., 1970, The T-locus of the mouse *in* "Current Topics in Developmental Biology," A.A. Moscona, A. Monroy, eds., Academic Press, New York.

Goldstein, N.I., Cagan, R.H., 1981, The major histocompatibility complex and olfactory receptors, *in* "Biochemistry of Taste and Olfaction," R.H. Cagan, M.R. Kare, eds., Academic Press, New York.

Gorman, M.L., 1976, A mechanism for individual recognition by odour in *Herpestes, Anim. Behav.* 24:141.

Gorman, M.L., Nedwell, D.B., Smith, R.M., 1974, An analysis of contents of anal scent pockets of *Herpestes, J. Zool. Lond.* 172:389.

Gorman, M.L., Mills, M.G.L., 1984, Scent marking strategies in hyaenas, *J. Zool Lond.* 202:535.

Gorman, M.L., Kruuk, H., Leitch, A., 1984, Social functions of the sub-caudal scent gland secretion of the European badger, *J. Zool. Lond.* 204:549.

Green, J.B.A., Smith, J.C., 1990, Graded changes in dose of an activin homologue elicit stepwise transitions in embryonic cell fate, *Nature*, 347:391.

Greenwald, I., Rubin, G.N., 1992, Making a difference: the role of cell-cell interactions in establishing separate identities for equivalent cells, *Cell* 68:271.

Hammerberg, C., Klein, J., 1975, Linkage disequilibrium between H_2 and T complexes in chromosome 17 of the mouse, *Nature*, 258:296.

Hart, B.L., 1986, Flehmen behavior and vomeronasal organ function *in* "Chemical Signals in Vertebrates IV," D. Duvall, D. Muller-Schwarze, R.M. Silverstein, eds., Plenum Press, New York.

Jones, J.S., Partridge, S.A., 1983, Tissue rejection: the price of sexual acceptance? *Nature* 304:484.

Kleiman, D., 1966, Scent marking in the canidae, *Symp. Zool. Soc. Lond.*, No. 18:167.

Koshland, D.E., 1980, Biochemistry of sensing and adaptation, *Trends Biochem. Sci.* 5:297.

Kruuk, H., 1972, "The Spotted Hyena," University of Chicago Press, Chicago.

Kruuk, H., Gorman, M., Leitch, A., 1984, Scent marking with the sabcaudal gland by the European badger, *Anim. Behav.* 32:899.

Law, H., Regnier, F.E., 1971, Pheromones, *Ann. Rev. Biochem.*, 40:533.

Leuthold, W., 1977, "African Ungulates," Springer, Berlin.

Leyhausen, P., 1964, The communal organisation of solitary mammals, *Symp. Zool. Soc. Lond.* No. 14:249.

Lorenz, K., 1953, "King Solomon's Ring," Methuen, London.

Lyon, M.F., Evans, E.P., Jarvis, S.E., Sayers, I., 1979, t-haplotypes of the mouse may involve a change in intercalary DNA, *Nature* 279:38.

McClintock, M., 1971, Menstrual synchrony and suppression, *Nature*, 229:244.

Macdonald, D.W., 1980, Patterns of scent marking with urine and faeces amongst carnivore communities, *Symp. Zool. Soc. Lond.* 45:107.

Michael, R.P., Bonsall, R.W., Warner, P., 1974, Human vaginal secretions: volatile fatty acid contents, *Science* 186:1217.

Mills, M.G.L., Gorman, M.L., Mills, M.E.J., 1980, The scent marking behaviour of the brown hyena, *S. Afr. J. Zool.* 15:240.

Moran, D.T., 1991, Vomeronasal (Jacobson's) organ in man, Abstr., *in* "Chemical Signals in Vertebrates VI," Plenum Press, New York. *(In press)*

Paoni, N.F., Maderis, A.M., Koshland, D.E.,Jr., 1981, Chemical sensing by bacteria, *in* "Biochemistry of Taste and Olfaction," R.H. Cagan, M.R. Kare, eds., Academic Press, New York.

Parmentier, M., Libert, F., Schumans, S., Schiffman, S., Lefort, A., Eggericks, D., Ledont, C., Mollereau, C., Gerard, C., Perrot, J., Grootegoed, A., Vassart, G., 1992, Expression of members of the putative olfactory receptor gene family in mammalian germ cells, *Nature* 355:453.

Passanisi, W.C., Macdonald, D. W., 1989, Group discrimination on the basis of urine in a farm cat colony, *in* "Chemical Signals in Vertebrates V," D.W. Macdonald, ed., Oxford University Press, Oxford.

Pedersen, P.E., Stewart, W.B., Greer, C.A., Shepherd, G.M., 1983, Accessory olfactory bulb *in utero*, *Science* 221:478.

Peters, R., Mech, L.D., 1978, Scent marking in wolves, *in* "Wolf and Man," R.L. Hall, H.S. Sharp, eds., Academic Press, New York.

Poddar-Sarkar, M., Sarkar, N., Dutta, J., Brahmachary, R.L., 1992, A Method of synthesizing the aroma of certain varieties of fragrant Indian rice, *IRRN* 17:5.

Raina, A.K., Kingan, T.G., Mattoo, A.K., 1992, Chemical signals from host plant and sexual behaviour in a moth, *Science* 255:592.

Richards, C.M., 1958, The inhibition of growth in crowded *R. pipiens* tadpoles, *Physiol. Zool.* 31:138.

Rothschild, M., 1989, Moths and memory, *Endeavour* New Ser. 13:15.

Ropartz, Ph., 1977, Chemical signals in agonistic and social behaviour of rodents *in* "Chemical Signals in Vertebrates I," D. Mueller-Schwarze, M.M. Mozell, Plenum Press, New York.

Schaller, G., 1967, "The Deer and the Tiger," University of Chicago Press, Chicago.

Schiffman, E., Corcoran, B.A., Wahl, S.M., 1975, N-formylmethionyl peptides as chemoattractants for leucocytes, *Proc. Natl. Acad. Sci. USA* 72:1059.

Schildeknecht, M., Ubl, J., 1986, Ecochemistry of life, *Interdiscipl. Sci. Rev.* 11:153.

Schilling, A., 1970, "Memoires du Museum National d'Histoire Naturelle," Tome LXI:203.

Shorey, H.H., 1976, "Animal Communication," Academic Press, New York.

Stoddart, M.D., 1976, "Mammalian Odours and Pheromones," Edward Arnold, London.

Stralendorff, F., 1986, Urinary signalling pheromone in tree shrews, I, *J. Chem. Ecol.* 12:99.

Stralendorff, F., 1987, Partial chemical characterization of urinary signalling pheromone in tree shrews, *J. Chem. Ecol.* 13:655.

Strauss, B. S., 1960, "Chemical Genetics," Saunders, Philadelphia.

Sukumar, M., Raj, P.A., Balaram, P., Becker, E.L., 1985, A highly active chemotactic peptide analog, *Biochem. Biophys. Res. Communs.* 128:339.

Theodor, J.L., 1970, Distinction between "self" and "not-self" in lower invertebrates. *Nature* 227:690.

Vinnikov, Y.A., 1982, "Evolution of Receptor Cells," Springer, Berlin.

Washburn, J.O., Anderson, J.R., 1991, Mosquito semiochemical alters morphogenesis of protozoa, *Nat. Hist.* July:30.

Williams, R.J., 1956, "Biochemical Individuality," John Wiley, New York.

Willis, R., 1985, Pot pourri: male urination in the United States, *Anthrop. Today* 1 (No. 1):25.

Wilson, E.O., 1973, "Insect Societies," Harvard University Press, Mass.

Yamazaki, K., Yamaguchi, M, Beauchmp, G.K., Bird, J., Boyse, F.A., Thomas, L., 1981, Chemosensation: an aspect of the uniqueness of the individual, *in:* "Biochemistry of Taste and Olfaction," R.H. Cagan, M.R. Kare, eds., Academic Press, New York.

SEGREGATION ANALYSIS USING THE PROPERTIES
OF BINOMIAL DATA

C. C. Li

Department of Human Genetics
Graduate School of Public Health
University of Pittsburgh
Pittsburgh, PA 15261

INTRODUCTION

The first chapter of my first book (1948) was about segregation analyses based entirely on the pioneer work of Haldane (1932, 1938). These two papers still constitute my basic knowledge of segregation analysis today, I am not ashamed to say. These two papers also got me thinking about the properties of the Mendelian segregation data which are essentially binomial in nature. A diagram illustrating the Mendelian segregation pattern for an autosomal locus with dominance appeared in the first chapter of the 1948 book and is reproduced here (reluctantly) as Fig. 1 for the convenience of later reference. (This figure may also be found in several textbooks on human genetics).

Having Fig. 1 in mind, we may suppose (hypothetically) that we have a collection of fifty families of normal parents each with four children, but invariably the first child is "affected" by some condition and the remaining three are normal. The frequency of the affected children is apparently 1/4, but we can hardly call it a Mendelian phenomenon. The point is thus clear. Mendelian segregation data have several other properties that must be satisfied in addition to the segregation value of 1/4 for a recessive trait in the offspring of heterozygous parents (Aa x Aa).

In this paper I limit myself to review briefly my attempts to utilize some of the other properties of Mendelian segregation in order to have a fuller appreciation of the Mendelian phenomenon. I regret that these attempts were made too late (1964 and after) to get the benefit of Haldane's criticism.

Human Population Genetics, Edited by P.P. Majumder
Plenum Press, New York, 1993

Fig. 1 Segregation of a simple recessive trait (solid symbol) in families of three children when both parents are heterozygous.

The Conventional Method

Before we proceed, it may be well to say a few words about the well known conventional segregation analysis. We note from Fig. 1 that there are 48 recessives (solid symbols) among the total 64 × 3 = 192 children; the segregation "ratio" (fraction) is 48/192 = 1/4 correctly.

However, the first 27 sibships without any recessive members are not distinguishable from the normal families in the population and hence they are unobservable. Only the 37 sibships with at least one recessive member are observable. Among these 37 observed sibships, there are still 48 recessives as before, but there are only 37 × 3 = 111 children, yielding a raw segregation ratio 48/111 = .432 which is too large, obviously.

The conventional method is to keep the number of recessives (48) as observed but enlarge the observed total children (111) to its theoretical total number (192). This is easily done by noting that the unobservable and thus missing sibships constitute $(3/4)^3$ = 27/64 of the theoretical number of total sibships and thus also of the theoretical number of children. In other words, the observed total number of children is only $1 - (3/4)^3$ of the theoretical total:

$$theoretical\ total\ [1-27/64] = observed\ total$$

$$theoretical\ total = \frac{obs.\ total}{1-27/64} = \frac{111}{37/64} = 192$$

The corrected segregation ratio is 48/192 = 0.25.

Now we may summarize the conventional method of finding the correct segregation ratio in more general terms. Let the segregation probability for dominants be q and that for recessives be p, where q + p = 1. Let us concentrate on the observed n_s sibships of "size" s (= 3 in Fig. 1), so that $t_s = n_s \times s = 37 \times 3 = 111$ in Fig. 1 is the total number of children in the n_s sibships. Finally, let r_s be the total number of recessives (48 in Fig. 1) among these t_s children. The estimate of p by the method of maximum likelihood is

$$\hat{p} = \frac{r_s}{\left(\dfrac{t_s}{1 - q^s}\right)} \tag{1}$$

The meaning of this expression is as explained numerically in the preceding paragraphs. A more popular form of (1) is

$$\frac{r_s}{p} = \frac{t_s}{1 - q^s} \tag{1a}$$

Expressions (1) and (1a) are true for any fixed size (s) of sibships. Pooling the observed data on sibships of various sizes, we obtain the pooled estimate of p:

$$\frac{r_2 + r_3 + r_4 + \cdots}{p} = \frac{t_2}{1 - q^2} + \frac{t_3}{1 - q^3} + \frac{t_4}{1 - q^4} + \cdots$$

i.e.

$$\frac{R}{p} = \sum_s \left(\frac{t_s}{1 - q^s}\right) \tag{2}$$

where $R = \Sigma r_s$. This estimation equation is sometimes known as the Haldane's equation (1932, 1938).

A straight forward application of the maximum likelihood method will also give us the "weight" or "information" (reciprocal of the variance of the estimate). The weight for the n_s sibships of size s is

$$W_s = \frac{n_s \, s}{pq} \left[\frac{1 - q^s - spq^{s-1}}{(1-q^s)^2} \right] \tag{3}$$

If we divide this quantity by n_s (i.e. omit n_s in the expression), the remaining quantity is then the weight per sibship of size s, which may be tabulated once and for all. The total weight for the pooled estimate from sibship of all sizes is simply $W = \Sigma W_s$. The variance of the pooled estimate is $V = 1/W$.

We mention the weight (3) here, as we shall use it as the standard value, against which the weights of other estimates may be compared. If there were no truncation of the sibships without any recessives, the weight of the estimate per sibship would be simply s/pq. It may be shown that $(1 - q^s)^2$ is always greater than $1 - q^s - spq^{s-1}$, so that the fraction within brackets in (3) is always less than unity, indicating the loss of information on p due to truncation of the binomial distribution.

Subdivision of Data by First Appearance Time

We have mentioned that a true Mendelian segregation not only produces 1/4 recessive offspring but these recessives must occur at random as exhibited in Fig. 1. One way to investigate the problem of random occurrence is to reclassify the "data" of Fig. 1 by the recessive's first appearance time. If the first child is a recessive, we say the appearance time is t = 1, regardless of the phenotypes of the remaining members of the sibship. If the first child is dominant and the second child is recessive, we say t = 2, regardless of the phenotypes of the remaining members of the sibship; etc. This type of classification may be better illustrated by sibships of size 4 (Li, 1966, 1970). The essential properties of this type of classification, however, may be seen from sibships of size 3 with little arithmetic labor.

Fig. 2 gives the classification of the 37 sibships of Fig. 1 with at least one recessive according to the recessive's first appearance time. Thus, the data are divided into two portions, one up to the first recessive and one after the first recessive.

Consider the 16 sibships with $t = 1$ in the left column of Fig. 2. The number of children after the first recessive is $16 \times 2 = 32$. The number of recessives among these 32 children is $3 + 3 + 2 = 8$, so the segregation ratio is $8/32 = 1/4$ correctly. Next, consider the 12 sibships with $t = 2$. The number of children after the first recessive is $12 \times 1 = 12$, of which 9 are dominants and 3 are recessives; yielding the segregation ratio $3/12 = 1/4$ again. There are no children after the first recessive when $t = 3 = s$.

These preliminary observations enable us to see some general properties of the children after the first appearance. For sibships of size s and first appearance time t, the distribution of the remaining children is given by the complete expansion of $(q + p)^{s-t}$. Example: $s = 3$ and $t = 1$, the distribution of the remaining children is $(3/4 + 1/4)^2$ with terms 9:6:1, which is exactly the case in the left column of Fig. 2. For $s = 3$ and $t = 2$, the distribution of the remaining children is $(3/4 + 1/4)^1$ with terms 3:1. Since those are complete expansions of $(q + p)^{s-t}$, a simple count of total children and recessives will give us an unbiased estimate of the segregation ratio. Since this is true for any given t, the children after the first appearance may be combined to yield a single estimate of p.

Let f_t be the observed number of sibships with first appearance time t. In Fig. 2, $(f_1, f_2, f_3) = (16, 12, 9)$ and $16 + 12 + 9 = 37 = n_s$, because it is the same 37 sibships in both Fig. 1 and Fig 2. The total number of children after t is $T_0 = 16 \times 2 + 12 \times 1 + 9 \times 0 = 44$. For sibships of size s,

no. of sibships	t=1	no. of sibships	t=2	no. of sibships	t=3	Total
9	●○○	9	○●○	9	○○●	27
3	●●○	3	○●●			9
3	●○●					
1	●●●					1
16	●xx	12	○●x	9	○○●	37
f_1		f_2		f_3		n

Fig. 2 Classification of the observed 37 sibships of Fig. 1 by the first appearance time (t) of a recessive. × = any phenotype, dominant or recessive.

$$T_o = f_1(s-1) + f_2(s-2) + \ldots + f_{s-1} \qquad (4)$$

The pooled estimate from all children after t is simply

$$p_o = \frac{total\ recessives\ after\ t}{total\ children\ after\ t} = \frac{R_o}{T_o} \qquad (5)$$

Since the estimate is based on simple counts from complete binomial distribution, its variance is pq/T_o and its weight is

$$W_o = \frac{T_o}{pq} \qquad (6)$$

Another estimate of p may be obtained from the distribution of t as shown in the bottom line of Fig 2. In other words, we now use the information preceding the first appearance of a recessive: The children preceding t are, of course, all dominants ("failures" in the study of "waiting time" before the first "success" happens). The distribution of t for sibships of size s is as follows:

1st appearance	t=1,	t=2,	t=3,	...,	t=s,	Total
obs numbers	f_1,	f_2,	f_3,	...,	f_s	$N=N_s$
"probability"	p,	qp,	q^2p,	...,	$q^{s-1}p$,	$1 - q^s$

This is a truncated geometric distribution which terminates at $t = s$. The "probabilities" shown above sum to $1 - q^s$. In using the method of maximum likelihood, we normalize the probabilities by dividing each term by $1 - q^s$, so that they would add up to unity.

The number of dominants before the first recessive is $m = 12 + 2 \times 9 = 30$ in Fig. 2. In general,

$$m = f_2 + 2f_3 + \ldots (s - 1)f_s \qquad (7)$$

A straightforward application of the maximum likelihood method leads to the estimation equation:

$$\frac{m}{n} = \frac{q}{p} - \frac{sq^s}{1-q^s} \qquad (8)$$

where m/n is the average (per sibship) waiting time before a recessive appears. If the geometric distribution were not truncated (s indefinitely large), the estimation equation becomes $m/n = q/p$, which is the standard result for geometric distributions. Then the factor $sq^s/(1-q^s)$ in (8) represents the amount of decrease in the average waiting time, because of the truncation of the distribution's tail with long waiting times. In our numerical example (Fig. 2), $s=3$, $n=37$, $m=30$, and the equation (8) becomes

$$\frac{30}{37} = \frac{q}{p} - \frac{3q^3}{1-q^3}$$

The solution will be found to be $p = 1/4$. In general, we shall call it p_t, the solution of (8) based on the distribution of t.

The maximum likelihood method also gives us the weight (the reciprocal of variance) of the estimate p_t as given by (8):

$$W_t = \frac{n}{p^2} + \frac{m}{q^2} - nsq^{s-2} \left[\frac{s-1+q^s}{(1-q^s)^2} \right] \qquad (9')$$

Now we have two estimates of p. The first one is p_o based on data after the first recessive; and the second estimate is p_t based on data before the first recessive. The weighted average of the two estimates is

$$\bar{p} = \frac{w_o\, p_o + w_t\, p_t}{w_o + w_t} \qquad (10)$$

with weight $w_o + w_r$.

We naturally wonder how efficient this average estimate is. In order to compare $w_o + w_t$ with $w = w_s$ given in (3), the weight for the fully efficient estimate, we have to put w_o and w_t in terms of $n = n_s$ and (p,q) only. To do this, we may use (8) to eliminate m in (9') which thus becomes

$$W_t = \frac{n}{p^2q} \left[1 - \frac{s^2 p^2 q^{s-1}}{(1-q^s)^2} \right] \qquad (9)$$

We also have to rewrite T_o in (4), the total number of children after the first recessive. There are a total of ns children. Each sibship has a first recessive; so there are n first recessives. The number of dominants before the first recessive is m. So the number of children after the first recessive is

$$T_o = ns - n - m = n(s-1) - m \qquad (11)$$

where $m = n\,[q/p - sq^s/(1-q^s)]$ by (8). With these substitutions in W_o and W_t, it may be shown that

$$w_o + w_t = w_s \qquad (12)$$

i.e.
$$\qquad (6) + (9) = (3)$$

The last expression shows there is no loss of information in dividing the data into two parts by the first appearance time (of recessives). The estimate $p_o = R_o/T_o$ is based on simple counts of total children and recessives after the first appearance regardless of randomness of occurrence. The estimate p_t is based on the distribution of the first appearance time (t = 1, 2, ...) which depends on the randomness of occurrence of the recessives. These are two different properties of Mendelian segregation. The method outlined in this section enables us to study these properties separately without losing information. An example of application may be found in Li (1965), using the data on Ellis-van Creveld syndrome.

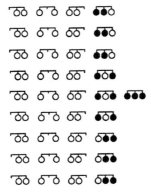

Fig. 3 The discarding of the singleton recessives from the 27 sibships of Fig. 1. After discarding, each of these sibships has only 2 normal children (size 2, no recessives).

Discarding the Singleton Recessives

Let us look back at Fig. 1 once more. A consideration of the missing (unobserved) sibships without any recessive members leads to the estimate (1). This method amounts to enlarge the observed number of children (t_s) to its theoretical total $t_s/(1 - q^s)$. The method to be described in this section is just the opposite; we shall try to decrease the numerator (r_s) of the estimate.

Limit our attention to the observed 37 sibships with at least one recessive member (Fig. 1). One may recall that $t_s = 37 \times 3 = 111$, $r_s = 48$, and $48/111 = .432$, being much too large than the true segregation ratio. We note there are 27 sibships, each with only one recessive member. If we discard such recessive singletons from these 27 sibships, the situation would become that shown in Fig. 3. Each of these 27 sibships becomes one of size 2, with two normal members.

Now we (Li and Mantel, 1968) assert that the data in Fig. 3 gives us directly the segregation ratio; no further calculation is needed. A simple count shows that the total number of children in Fig. 3 is 84 and the number of recessives is 21; and $21/84 = 1/4$. For those who have already obtained $t_s = 111$ and $r_s = 48$ from Fig. 1, the new estimate may be written as

$$\acute{p} = \frac{r_s - j_s}{t_s - j_s} \tag{13}$$

$$= \frac{48 - 27}{111 - 27} = \frac{21}{84} = \frac{1}{4}$$

where $j_s = 27$ is the number of singleton recessives.

An algebraic demonstration is in order to show that (13) is generally true for any s. The simplest way to do this is to write out the binomial distribution and indicate the corresponding observed numbers (t_s, r_s, j_s), as shown in Table 1.

Table 1. Binomial distribution $(q + p)^s$ of sibships of size s and the corresponding observations

Recessive in a sibship r	Frequency of sibships f	No. of recessives fr	Correspondence to observation
0	q^s	0	(unobserved)
1	spq^s	spq^{s-1} . . . . j_s singletons	
2	-	-	-
.			
.			
.			
s	p^s	sp^s	

Complete total	1.00	sp	
*Incomplete total	$1 - q^s$	sp. . . . . r_s recessives	
Incomplete children	$s(1 - q^s)$ . . . . t_s children		

* without the first row where r = 0.

From Table 1, we see that the estimation equation (13) corresponds to

$$\frac{sp - spq^{s-1}}{s(1-q^s) - spq^{s-1}} = \frac{p(1-q^{s-1})}{1-q^{s-1}} = p$$

so that (13) is an estimate of p. Gart (1968), independently, reached the same conclusion by a different approach.

Since (13) is true for any s, sibships of all sizes may be pooled to yield an overall estimate of p:

$$\hat{p} = \frac{\Sigma r_s - \Sigma j_s}{\Sigma t_s - \Sigma j_s} = \frac{R - J}{T - J} \tag{14}$$

where the summations are with respect to s.

A further property of Fig. 3 may be mentioned in passing. If the sibships with singleton recessives are not usable (for any reason at all), then we may ignore the 27 sibships on the left of Fig. 3 and regard the remaining 10 sibships as the usable data, from which we wish to obtain an estimate of p. In this situation, we may discard the doubleton recessives in the 9 sibships which then consist of 9 sibships with one normal child each. Then, a simple count shows there remains 12 children, 3 of them are recessives, and 3/12 = 1/4 is an estimate of p. This example is to show that the method of discarding certain recessives is applicable to a binomial distribution truncated at any point.

The method of estimating the segregation probability by discarding the singleton recessives is so simple (there being no equation to solve) that Nicholas (1982) thinks, and I concur, that it is the only method that should be used in practice. The variance of $\hat{p}$ = (r-j)/(t-j) is , however, more complicated than the variance of the estimate obtained by maximum likelihood method. Skipping the details, we merely give the final result on a per-sibship basis $(n=1)$:

$$Var(\hat{p}) = \frac{pq}{s} \cdot \frac{(1-q^s)[1-q^s + (s-2)pq^{s-1}]}{(1-q^{s-1})^2}$$

(15)

Its reciprocal is the weight (w) per sibship, which has been tabulated by Li and Mantel (1968, Table 5 for segregation problems and Table 6 for general problems where the binomial distribution is without the first term). Example: for $s = 3$ and $p = 1/4$, the tabulated weight is 7.37, so the total weight of the 37 sibships in Fig. 1 is

$$W = 7.37 \times 37 = 272.69$$
$$Var(\hat{p}) = 1/W = .003667$$

(15a)

Davie (1979) gave an approximate variance formula for large samples in terms of observed numbers:

$$Var(\hat{p}) = \frac{(r-j)(t-r)}{(t-j)^3} + \frac{2j_2(t-r)^2}{(t-j)^4}$$

(16)

In our present numerical example (Fig. 1) $t = 111$, $r = 48$, $j = j_1 = 27$ sibships with singleton recessives, and $j_2 = 9$ sibships with doubleton recessives. Substituting:

$$Var(\hat{p}) = \frac{21 \times 63}{(84)^3} + \frac{18(63)^2}{(84)^4} = .003667$$

(16a)

in agreement with (15a).

How does this method compare with the maximum likelihood estimate? Putting $n_s = 1$ in (3), we obtain the weight per sibship for the maximum likelihood estimate (which has been tabulated by Li, 1961, p. 66). For example, when $s = 3$ and $p = 1/4$, the weight is 7.48, larger than 7.37, the weight for the estimate obtained by discarding singles. The relative efficiency is $7.37/7.48 = 98.5\%$. For $s = 4$ at $p = 1/4$, the relative efficiency is $11.66/11.948 = 97.6\%$. For $s = 2$, it is fully efficient, because the maximum likelihood estimate (1),

$$\frac{r}{p} = \frac{t}{1-q^2}$$

reduces to

$$\hat{p} = \frac{2r-t}{r} = \frac{r-j}{t-j}$$

on account of the fact $t = r + j$ for $s = 2$.

The three examples above seem to suggest that the relative efficiency (singles method/maximum likelihood) decreases with s. It is, however, not the case. The minimum efficiency is 96.8% when $s = 6$ and p not too different from 1/4. Then the relative efficiency rises again, reaching 99% when $s = 15$ and $p = .25$, and 99.6% when $p = .30$ (Li and Mantel, 1968, Table 7, p. 69). In view of its high efficiency and the tabulated weight per sibship, the singles method is a highly practical one.

Table 2. Distribution of sibships with 4 children from Aa × Aa parents and subdivision into subsets of dominants and recessive children.

Type and frequency of sibships of size 4		Subsets of children		Subset total
dom, rec (d, r)	frequency f	dom fd	rec fr	expansion of $4(q+p)^3$
(4, 0)	q^4	$4q^4$	0	
(3, 1)	$4pq^3$	$12pq^3$	$4pq^3$	$4q^3$
(2, 2)	$6p^2q^2$	$12p^2q^2$	$12p^2q^2$	$12pq^2$
(1, 3)	$4p^3q$	$4p^3q$	$12p^3q$	$12p^2q$
(0, 4)	p^4	0	$4p^4$	$4p^3$
Total	1.00	$4q$	$4p$	$4 = s$

Self-contained Subsets of Data

We continue to consider sibships of a fixed size s. As usual, let r = number of recessive members in a sibship, and d = s - r is the number of dominants in a sibship. If $f_r = f$ is the frequency of sibship with r recessives, then fr is the number of recessive members in such sibships. In almost all occasions we do not bother to write out the number of dominants explicitly because if we know the number of recessives, we would also know the number of dominants. However, if we write out the number of dominants explicitly, we would discover new properties that we could use to estimate the segregation ratio. To illustrate these properties, we give the details in Table 2 for sibship size s = 4.

In Table 2, we classify the children into subsets by slanted lines. Example: the dominants of the (4,0) sibships and the recessives of the (3,1) sibships belong to the same subset. The dominants of the (3,1) sibships and the recessives of the (2,2) sibships belong to the same subset; etc. The first thing one notices is that the two entries of the same subset always have the same numerical coefficients. This is always true for any value of $s = d + r$, as shown below:

sibship		Children	
dom rec		dominants	recessives
s-r, r		$(s-r)\cdot\begin{pmatrix}s\\r\end{pmatrix}p^r q^{s-r}$	----------
s-r-1, r+1		----------	$(r+1)\cdot\begin{pmatrix}s\\r+1\end{pmatrix}p^{r+1} q^{s-r-1}$

Table 3. Numerical version of Table 2

Type and frequency of sibships, s=4			Children		
dom rec (d, r)	frequency f		dom fd	rec fr	Subset total
(4, 0)	[81]		[324]	0	
(3, 1)	108		324	[108]	[432]
(2, 2)	54		108	108	432
(1, 3)	12		12	36	144
(0, 4)	1		0	4	16
Complete Total 256			768	256	1024
Incomplete Total* 175			444	148	592

* = total without the first number of each column, as marked by [...].

Writing out the binomial coefficients and canceling, we see that

$$(s-r) \cdot \binom{s}{r} = (r+1) \cdot \binom{s}{r+1} = s \cdot \binom{s-1}{r} \qquad (17)$$

which is also the coefficient of the subset total (Table 2, last column). The subset totals of children are distributed as $s(q+p)^{s-1}$.

Since the coefficients of the two entries of a subset (17) are equal, it is easy to see that the two quantities are always in the ratio

$$p^r q^{s-r} : p^{r+1} q^{s-r-1} = q : p \qquad (18)$$

In other words, in every subset, the numbers of dominates and recessives are in the ratio $q : p$ and thus every subset permits an estimate of p directly. Such a subset may be called a self-contained subset (Li, 1986).

Table 3 gives a numerical version of Table 2, so that we can see more readily the various possible estimates of p that can be made from the subsets of children. The first example is the case where the sibships without any recessive children are not observed, indicated by [81] in Table 3. So, the first subset is incomplete, although the 108 singleton recessives from the (3,1) sibships are observed. Now, we discard the incomplete subset and estimate p from the remaining three complete subsets. The estimate then is from the incomplete total of Table 3 $\hat{p} = 148/592 = 1/4$ This is equivalent to the method of discarding singleton recessives described in the previous section. For, if we count the total number of children observed (including the 108 singleton recessives), we would have found $t = 1024 - 324 = 700$, $r = 256$, and $j = 108$; and the estimate is

$$\hat{p} = \frac{r-j}{t-j} = \frac{256-108}{700-108} = \frac{148}{592} = \frac{1}{4}$$

The method of discarding singleton recessives may be viewed as a special case of discarding incomplete or defective subsets.

Some investigators call the (3,1) or (s-1, 1) sibships "sporadic" which may consist of other cases than pure segregation. If we omit the (3,1) sibships from consideration, an estimate of p may be made from the remaining two complete subsets:

$$\hat{p} = \frac{recessives}{total} = \frac{36+4}{144+16} = \frac{40}{160} = \frac{1}{4}$$

Some other investigators would like to omit the (0,4) or (0,s) sibships which may be due to other causes than pure segregation. Finney (1949) treated this case (double truncation) by the method of maximum likelihood. By our method, we would simply delete both the first and the last subsets and estimate p from the middle two complete subsets:

$$\hat{p} = \frac{recessives}{total} = \frac{108 + 36}{432 + 144} = \frac{144}{576} = \frac{1}{4}$$

Besides, we may have three separate estimates of p, one from each complete subset of Table 3, and see if they are more or less in agreement.

One important point to remember is that if one number is missing or defective, we discard the subset which contains the defective number, not just the defective number itself.

Sibships of varying sizes may be combined, because the two numbers of a subset are in the ratio q : p for any value of s.

As far as estimation is concerned, the method of self-contained subsets is the simplest, with no equation to solve. The variance of such an estimate, however, has not been investigated, except for the case of discarding the first subset (equivalent to discarding the singleton recessives) and the case of double truncation. Li (1986) raised a number of general questions about the variance of such an estimate and he seems willing to settle on approximate formulas.

SUMMARY AND DISCUSSION

Several methods of segregation analysis, developed in the post-Haldane period, have been reviewed. These methods are applicable for cases of complete ascertainment. The method involving the first appearance time of a recessive is to put explicit emphasis on the randomness of the occurrence of recessives. The method of discarding singleton recessives is to avoid the correction factor $(1 - q^s)$ and arrive directly at an estimate without solving complicated equations. The method of constructing self-contained subsets of children is useful when the missing or defective observation occurs anywhere in the distribution of sibships, not just the missing first term.

Although these methods, as reviewed here, apply to cases of complete ascertainment only, similar properties also exist when the ascertainment is incomplete. For instance, Li, Chakravarti and Holloran (1987), based on Davie (1979), showed that self-contained subsets of data may also be constructed with incomplete ascertainment. Both the probability of ascertainment and the segregation ratio may be estimated in the same simple manner.

We must realize that the probability of ascertainment is not a genetic parameter. Rather, it is an attempt to describe the way the investigators operate; the latter varies with time, place, experience, circumstances, personnel, and funding. In other words, it is a complex of social factors that contribute to the incompleteness of ascertainment. Most workers use a single value (denoted by π) for the ascertainment probability which is obviously inadequate. But the models involving varying values of ascertainment probabilities are also arbitrary and inadequate, in view of the unlimited possibilities of missing or ascertaining a case. I thought the best way to overcome the effects of social factors is to strive for complete ascertainment by establishing adequate systems of registration and reporting by health agencies.

REFERENCES

Davie, A.M. 1979. The singles method for segregation analysis under incomplete ascertainment. *Ann. Hum. Genet.* 42:507-572.

Finney, D.J. 1949. The truncated binomial distribution. *Ann. Eugen.* (London) 14:319-328.

Gart, J.J. 1968. A simple nearly efficient alternative to the simple sib method in the complete ascertainment case. *Ann. Hum. Genet.* 31:283-291.

Haldane, J.B.S., 1932. A method for investigating recessive characters in man. *J. Genet.* 25:251-255.

Haldane, J.B.S., 1938. The estimation of the frequencies of recessive conditions in man. *Ann. Eugen.* (London) 8:255-262.

Li, C.C., 1948. An introduction to population genetics. *Peking Univ. Press.*

Li, C.C., 1961. Human Genetics, Principles and Methods. McGraw-Hill Book Co., New York.

Li, C.C., 1964. Estimate of recessive proportion by first appearance time. *Ann. Hum. Genet.* 28:177-180.

Li, C.C., 1965. Segregation of the Ellis-van Creveld syndrome as analyzed by the first appearance method. *Am. J. Hum. Genet.* 17:343-351.

Li, C.C., 1966. A new method of studying Mendelian segregation in man. Proc. Symp. on mutation in population, pp. 155-166. Prague: Publ. House, Czech. Acad. Sci.

Li, C.C., 1970. The incomplete binomial distribution. In Mathematical Topics in Population Genetics (K. Kojima, Ed.) Springer-Verlag, Berlin.

Li, C.C., 1986. A method of subdividing genetic data into self-contained subsets. *Ann. Hum. Genet.* 50:259-270.

Li, C.C., Chakravarti, A., and Halloran, S.L., 1987. Estimation of segregation and ascertainment probabilities by discarding the single probands. *Genet. Epidem.* 4:185-191.

Nicholas, F.W., 1982. Simple segregation analysis: A review of its history and terminology. *J. Heredity* 73:444-450.

SOME RECENT DEVELOPMENTS IN THE THEORETICAL ASPECTS OF SEGREGATION ANALYSIS

Robert C. Elston

Department of Biometry and Genetics
Louisiana State University Medical Center
1901 Perdido Street
New Orleans, LA 70112

INTRODUCTION

Segregation analysis can be defined as the statistical methodology used in the analysis of family and pedigree data to determine the mode of inheritance of a particular phenotype, especially with a view to uncovering Mendelian segregation at a single locus. For the past two decades multiparameter models have been available to perform likelihood-based analyses of quantitative or qualitative traits measured on members of sibships or extended pedigrees. By the end of the 1970's two such models were well established: the transmission probability — or generalized major gene — model and the mixed model (reviewed by Elston, 1981). In this paper I shall describe these two models and how they have evolved over the last decade. For simplicity, the various models will be described in the context of randomly sampled nuclear family data, the complications of extended pedigrees and non-random sampling being briefly dealt with later. Under each model specific statistical hypotheses can be tested about segregation parameters, and I shall compare and contrast these tests for the purpose of detecting major gene segregation. Although the recombination fraction is in fact a segregation parameter, and linkage to a marker has distinct advantages in the detection of major gene segregation, I shall restrict myself to situations in which information on polymorphic genetic markers or any other type of measured genotype information is not available; this is the situation for which those methods of analysis that are usually called segregation analysis were developed. I shall not go into computational details, except to explain the rationale underlying certain approximations used in the likelihood calculations. Finally, I shall discuss certain theoretical aspects of data transformation that do not appear to be well understood.

Human Population Genetics, Edited by P.P. Majumder
Plenum Press, New York, 1993

MODELS FOR RANDOMLY SAMPLED NUCLEAR FAMILY DATA

In the case of all the models that will be described, we can write likelihood expressions for the data on all members of a nuclear family, and hence, as a special case, for the data on the parents alone. Thus, dividing the former by the latter, we have the likelihood for the offspring phenotypes conditional on their parents' phenotypes. It has sometimes been suggested that inferences should be based on this conditional likelihood, rather than on the unconditional likelihood. This has merit if there has been selection of the families for study through the phenotypes of the parents, or if for any other reason the sample distribution of parental phenotypes is not the same as that modelled in the likelihood. For example, it has been shown that the effect of wrongly assuming random mating among the parents, when in fact there is assortative mating, is trivially small if inferences are based on the conditional likelihood (MacLean et al., 1975); and it has been suggested that the conditional likelihood approach is less sensitive to temporal trends that might induce certain systematic differences between parents and offspring (Lalouel et al., 1983). Barring such situations, however, inferences should be based on the full unconditional likelihood, which is more informative.

The likelihoods of all the genetic models that will now be described contain the following three components: the population distribution of major genotypes, the offspring distribution of major genotypes conditional on their parents' major genotypes, and the phenotypic distribution conditional on major genotype. This last distribution may differ between parents and offspring, and a major difference among the models lies in how the offspring phenotypic distribution is allowed to depend on the parental phenotypes other than through major genotypes. Although it is not an essential assumption for any of the models, for simplicity I shall assume that there are just three major genotypes, determined by two alleles at one autosomal locus. All the models are often implemented with this assumption.

The Transmission Probability Model

Under the assumption just stated, the transmission probability model (Elston and Stewart, 1971) supposes that the population is made up of not more than three types of persons, whom we can label AA, AB and BB, and that conditional on type the individuals' phenotypes are independently distributed. If there is Mendelian segregation, the types correspond to major genotypes. Let y denote phenotype and u denote type (u = AA, AB, BB), and let the subscripts m, f and i denote, respectively, mother, father and the ith offspring (i = 1, 2,..., n). Let $P(u)$ be the population probability of type u, $P(u_i \mid u_f, u_m)$ the *transition* probability that parents with types u_f and u_m have a child with type u_i, and $P(y \mid u)$ the conditional probability (mass or density) function of y given u. Assuming random mating, the likelihood of the transmission probability model for data on a nuclear family is then

$$L_{tp} = \sum_{u_f} P(u_f) \, P(y_f | u_f) \sum_{u_m} P(u_m) \, P(y_m | u_m)$$

$$\cdot \prod_i \left[\sum_{u_i} P(u_i | u_f \, u_m) \, P(y_i | u_i) \right]. \tag{1}$$

The first line of this expression gives the likelihood for the two parents' phenotypes and the second line gives the likelihood for the offspring phenotypes, assumed independent conditional on their parents' types. The probability functions depend on unknown parameters. In particular, the $3^3 = 27$ transition probabilities $P(u_i | u_f, u_m)$ are parametrized in terms of three *transmission* probabilities, $\tau_u = P(\text{individual with type } u \text{ transmits } A \text{ to a child})$, as follows:

$$
\begin{aligned}
P(AA | u_f, u_m) &= \tau_{u_f} \tau_{u_m} \\
P(AB | u_f, u_m) &= \tau_{u_f}(1 - \tau_{u_m}) + \tau_{u_m}(1 - \tau_{u_f}) \\
P(BB | u_f, u_m) &= (1 - \tau_{u_f})(1 - \tau_{u_m}).
\end{aligned}
\tag{2}
$$

The population probabilities $P(u)$ are usually made functions of a single parameter, q, interpretable as a gene frequency if the types are genotypes, and Hardy-Weinberg equilibrium proportions are assumed:

$$P(AA) = q^2, \quad P(AB) = 2q(1-q), \quad P(BB) = (1-q)^2. \tag{3}$$

Although this is a testable hypothesis, i.e. it is possible to test the null hypothesis $P(AB) = 2\sqrt{P(AA) \, P(BB)}$ under a model of arbitrary type probabilities that sum to one, the power of such a test is in practice often very low. In the case of a continuous trait the conditional probability functions $P(y | u)$ are often assumed to be normal densities whose means (only) depend on u, while in the case of a dichotomous disease trait they are simple Bernoulli probabilities. In each case there is no fundamental difficulty in letting the functions depend on other covariates, such as age and sex, or in adapting them for situations in which the phenotype of interest is a disease with variable age of onset.

Under the transmission probability model, the test for major gene segregation corresponds to testing the joint null hypothesis

$$
\begin{aligned}
\tau_{AA} &= 1 \\
\tau_{AB} &= \tfrac{1}{2} \\
\tau_{BB} &= 0 \, .
\end{aligned}
\tag{4}
$$

However, because acceptance of a null hypothesis is not good evidence that the hypothesis is true, one should also be able to reject the null hypothesis that there is no transmission, i.e. that the distribution of offspring types is the same in all sibships. This can be formally tested as the null hypothesis:

$$\tau_{AA} = \tau_{AB} = \tau_{BB} \, , \tag{5}$$

which allows the distribution of types to differ between parents and offspring. If these three transmission probabilities are equal to τ, under random mating the distribution of offspring types is, independent of the parental types:

$$P(AA) = \tau^2, \quad P(AB) = 2\tau(1 - \tau), \quad P(BB) = (1 - \tau)^2.$$

Comparing this to (3), we see that if we assume Hardy-Weinberg equilibrium proportions and wish to test simultaneously that there is no transmission and that the distribution of types is the same among both parents and offspring, the null hypothesis is:

$$\tau_{AA} = \tau_{AB} = \tau_{BB} = q. \tag{6}$$

It should be noted that for a continuous trait, even under the assumption of normality conditional on type, none of these tests assumes that the population distribution of the phenotype is normal, but rather a mixture of normals. Because of the flexibility of such a distribution, it is unlikely that non-normality of the trait *per se* would lead these tests to suggest the spurious existence of a major gene for a continuous phenotype. However, the model can only allow for the presence of parent-offspring or sibling correlations by having inequality of the transmission probabilities, and so there is the possibility that polygenic inheritance or an environmentally caused sibling correlation could give the appearance of major gene segregation.

The Mixed Model

The mixed model can most easily be understood by first considering the polygenic analogue of (1). Denoting the continuous polygenotype g, the likelihood for data on a nuclear family under polygenic inheritance can be written, replacing u by g and summation by integration,

$$L_p = \int_{-\infty}^{\infty} P(g_f) \, P(y_f \mid g_f) \int_{-\infty}^{\infty} P(g_m) \, P(y_m \mid g_m)$$

$$\cdot \prod_i \left[\int_{-\infty}^{\infty} P(g_i \mid g_f, g_m) \, P(y_i \mid g_i) \, dg_i \right] dg_m \, dg_f \, . \tag{7}$$

The population probability function $P(g)$ is taken to be a normal density with mean 0 and (additive genetic) variance σ_g^2, and the transition probability function $P(g_i \mid g_f, g_m)$ is a normal density with mean $(g_f + g_m)/2$ and variance $\sigma_g^2/2$. The mixed model assumes the phenotype y depends on both a major genotype u and an independent additive polygenotype g, and the corresponding likelihood is thus just a combination of (1) and (7):

$$
L_m = \sum_{u_f} P(u_f) \int_{-\infty}^{\infty} P(g_f) \, P(y_f \mid u_f, g_f) \sum_{u_m} P(u_m) \int_{-\infty}^{\infty} P(g_m) \, P(y_m \mid u_m, g_m)
$$

$$
\cdot \prod_i [\sum_{u_i} P(u_i \mid u_f, u_m) \int_{-\infty}^{\infty} P(g_i \mid g_f, g_m) \, P(y_i \mid u_i, g_i) \, dg_i] \, dg_m \, dg_f \,,
$$

(8)

in which $P(u)$ is defined as in (3) and the $P(u_i \mid u_f, u_m)$ are fixed Mendelian transition probabilities corresponding to the transmission probabilities (4). In the case of a continuous trait, $P(y \mid u, g)$ is assumed to be normally distributed with mean $\mu_u + g$, where μ_u is the overall mean for genotype u, and with (environmental) variance σ_e^2. In the case of a dichotomous disease trait, on the other hand, it is assumed that the conditional probability of disease, given u and g, is $\Phi(\{g - t_u\}/\sigma_e)$, where $\Phi(\bullet)$ is the standard cumulative normal distribution function and t_u is a major genotype-specific parameter that is interpretable as a threshold on a liability scale. In the latter case the genetic and environmental variances σ_g^2 and σ_e^2 are not separately estimable. In both cases these variances can be parametrized in terms of $H_p = \sigma_g^2 / (\sigma_g^2 + \sigma_e^2)$, the heritability of the non-major gene components or polygenic heritability, together with a further variance parameter in the case of continuous traits.

Morton and MacLean (1974) utilized the fact that the polygenic likelihood (7) can be rewritten so that only one integration is necessary, as follows. First, note that under the polygenic model the offspring phenotypes depend on the parental phenotypes only through the midparental polygenotype $v = (g_f + g_m)/2$, and that conditional on v the offspring phenotypes are independent. Thus the joint density of y_f, y_m, v and the y_i can be written as

$$
P(y_f) \, P(y_m) \, P(v \mid y_f, y_m) \prod_i P(y_i \mid v),
$$

and hence we can write

$$
L_p = P(y_f) \, P(y_m) \int_{-\infty}^{\infty} P(v \mid y_f, y_m) \prod_i P(y_i \mid v) dv.
$$

Thus the mixed model likelihood (8) can be equivalently written as

$$
L_m = \sum_{u_f} P(u_f) \, P(y_f \mid u_f) \sum_{u_m} P(u_m) \, P(y_m \mid u_m)
$$

$$
\cdot \int_{-\infty}^{\infty} P(v \mid y_f, y_m, u_f, u_m) \prod_i [\sum_{u_i} P(u_i \mid u_f, u_m) \, P(y_i \mid v, u_i)] \, dv,
$$

(9)

in which, for a continuous trait, the appropriate normal densities are substituted. Using the notation

$$\phi(w, \sigma^2) = \frac{1}{\sqrt{2\pi}\,\sigma} \exp(-w^2/2\sigma^2), \tag{10}$$

these densities are [see Boyle and Elston (1979) for their derivations]:

$$P(y \mid u) = \phi(y - \mu_u, \sigma_g^2 + \sigma_e^2)$$

$$P(v \mid y_f, y_m, u_f, u_m)) = \phi(v - H_p\{y_f - \mu_{u_f} + y_m - \mu_{u_m}\}/2, H_p\sigma_e^2/2)$$

$$P(y \mid v, u) = \phi(y - \mu_u - v, \sigma_g^2/2 + \sigma_e^2).$$

Analogous substitutions are made for a dichotomous trait, i.e., if $y = 1$ represents presence of disease and $y = 0$ represents absence of disease,

$$P(y \mid u) = \left[\Phi(-t_u / \sqrt{\sigma_g^2 + \sigma_e^2})\right]^y \left[\Phi(t_u / \sqrt{\sigma_g^2 + \sigma_e^2})\right]^{1-y}$$

$$P(y \mid v, u) = \left[\Phi(\{v - t_u\} / \sqrt{\sigma_g^2/2 + \sigma_e^2})\right]^y \left[\Phi(\{t_u + v\} / \sqrt{\sigma_g^2/2 + \sigma_e^2})\right]^{1-y}.$$

Because in this situation σ_g^2 and σ_e^2 are not separately estimable, we can arbitrarily assume $\sigma_g^2 + \sigma_e^2 = 1$, in which case $\sigma_g^2/2 + \sigma_e^2 = 1 - H_p/2$.

As originally proposed by Morton and MacLean (1974), the model was extended to allow for discrete covariates in the form of liability classes, and for the possibility of simultaneously using information on a correlated continuous trait. The population prevalence of a dichotomous trait (or prevalences — one for each liability class in the presence of covariates) was assumed to be known. Furthermore, the model included an additional sibling environmental component. This was done by letting $v = (g_f + g_m)/2 + b$, where b is an environmental component common to all offspring in the same sibship, normally distributed across sibships with mean 0 and variance σ_b^2. The likelihood of this model for a continuous trait, for example, is then obtained by substituting into (9):

$$P(y \mid u) = \phi(y - \mu_u, \sigma_g^2 + \sigma_e^2)$$

$$P(v \mid y_f, y_m, u_f, u_m) = \phi(v - H_p\{y_f - \mu_{u_f} + y_m - \mu_{u_m}\}/2, H_p\sigma_e^2/2 + \sigma_b^2) \tag{11}$$

$$P(y \mid v, u) = \phi(y - \mu_u - v, \sigma_g^2/2 + \sigma_e^2 - \sigma_b^2).$$

However, instead of including a common sibling environmental component, later versions of the mixed model made the ratio of σ_g^2 to the total variance (including the major gene component), a parameter denoted H by Morton and MacLean, generation-dependent.

Under the mixed model, the test for major gene segregation corresponds to testing the null hypothesis

$$q = 0, \tag{12}$$

or equivalently

either $\mu_{AA} = \mu_{AB} = \mu_{BB}$,   in the case of a continuous trait

or $t_{AA} = t_{AB} = t_{BB}$,   in the case of a dichotomous trait,

so that the major gene component is significant if this null hypothesis is rejected. This model has the advantage that the presence of polygenic inheritance or of an environmentally caused sibling correlation is very unlikely to suggest the spurious existence of a major gene, and in fact it is possible to test for the presence of these components explicitly. However, the model assumes that all transmission from one generation to the next is Mendelian, either major gene or polygenic, and any departure from polygenic inheritance can give the appearance of major gene segregation. Furthermore, any non-normality in the population distribution of a continuous phenotype, such as platykurtosis (Eaves, 1983) or skewness, will tend to be interpreted as a major gene.

The Unified Model

Because the transmission probability model and the mixed model are complementary in their strengths and weaknesses, an overall model that subsumes both is particularly appealing. Such a model was introduced as the unified model by Lalouel et al. (1983). The likelihood of this model is essentially the same as that of the mixed model, (8) or (9), with the exception that the transition probabilities $P(u_i \mid u_f, u_m)$ are expressed in terms of arbitrary transmission probabilities in the unit interval. Under this model we infer the existence of a major gene, with or without multifactorial background effects, provided we can accept hypothesis (4) and reject hypotheses (5) and/or (6) and (12). If all these conditions are met except acceptance of hypothesis (4), we conclude that there is transmission of a major effect, but that this transmission is not simply one-locus Mendelian.

It is possible for an analysis under the unified model to suggest the spurious existence of a major gene, in the case of a continuous trait, if there is any polygenically determined non-normality. For this reason transformations of the data should be explored in this situation, as discussed later. This is not necessary for a dichotomous trait; with sufficient data a segregation analysis under the unified model can provide strong evidence of major gene segregation when the phenotype being analyzed is dichotomous, provided care is taken to perform a complete analysis including a test of hypothesis (4) (McGuffin and Huckle, 1990).

Regressive Models

Bonney (1984, 1986) proposed four classes of regressive models for the genetic analysis of family data. These models allow for familial dependencies other than through major gene transmission by the simple expedient of making the conditional probability functions $P(y \mid u)$ dependent on the phenotypes of antecedent family members. The likelihood of a class A model is very similar to (1), and so is relatively easy to compute. It does not assume random mating and assumes that the offspring phenotypes are independent only conditional on the phenotypes and types of their parents. Thus (1) changes to

$$L_A = \sum_{u_f} P(u_f) \, P(y_f | u_f) \sum_{u_m} P(u_m | u_f) \, P(y_m | y_f, u_f, u_m)$$

$$\cdot \prod_i \, [\sum_{u_i} P(u_i | u_f, u_m) \, P(y_i | y_f, y_m, u_f, u_m, u_i)] \, . \tag{13}$$

In the case of a continuous trait it is assumed that, conditional on the type of each person in the family, the phenotypes are multinormally distributed. It follows that the sibling correlation is completely determined by the spouse and parent-offspring correlations. Thus $P(y_m | y_f, u_f, u_m)$ is the univariate normal density of y_m, conditional on u_m, after regressing out the effect of $y_f - \mu_{u_f}$; and $P(y_i | y_f, y_m, u_f, u_m, u_i)$ is the analogous density of y_i after regressing out the effects of $y_f - \mu_{u_f}$ and $y_m - \mu_{u_m}$. In the case of a dichotomous trait, $y = 0$ or 1 (unaffected or affected, respectively), it is assumed that each phenotypic probability function is of the logistic form $e^{y\theta}/(1 + e^\theta)$, where the logit θ is a linear function of covariates and a baseline parameter that reflects the person's type. The dependencies among relatives are allowed for by introducing extra terms into θ. The logit for $P(y_m | y_f, u_f, u_m)$, θ_m, contains a term that depends on y_f, the affection state of the father, and the logit for $P(y_i | y_f, y_m, u_f, u_m, u_i)$, θ_i, contains two extra terms, one dependent on y_f and the other on y_m. Thus, denoting the type-specific baseline parameter β_u, in the absence of covariates these logits are

$$\theta_m = \beta_{u_m} + \delta_f(y_f)$$

$$\theta_i = \beta_{u_i} + \delta_f(y_f) + \delta_m(y_m),$$

where $\delta_j(1)$ and $\delta_j(0)$, $j = f$ or m, are parameters that increase or decrease the logit according as to whether the jth parent is affected or not. We can think of these extra terms as regressing out the effects of the father and mother, as in the continuous case, but a difference should be noted. Whereas in the continuous case multinormal theory can be used to regress out the residuals $y_f - \mu_{u_f}$ and $y_m - \mu_{u_m}$, there are no analogous residuals in the case of a dichotomous trait and the logistic function. For this reason, the logits for the spouse and the offspring are made dependent on only the phenotypes, and not the types, of the antecedent family members.

The class B, C and D regressive models allow for dependencies among the sibs over and above any dependency that can be attributed to their parents' phenotypes. In a class B model each sib's phenotype can depend on both the parents' phenotypes and the phenotypes of the *first* (one or several) sib(s) in the sibship, while in a class C model each sib's phenotype can depend on both the parents' phenotypes and the phenotypes of the *preceding* (one or several) sib(s) in the sibship. The class D models assume only that, conditional on their types, the sibs' phenotypes are equally correlated. Because in these models the sibs' phenotypes are no longer independent, conditional on the phenotypes and types of their parents, the likelihood can no longer be written in the same form as (1) and (13). Rather, Bonney (1984) expressed the likelihood for the children as a nested sum, involving much more computation. The likelihood of a class D model, for example, was written as

$$
L_D = \sum_{u_f} P(u_f) \, P(y_f | u_f) \sum_{u_m} P(u_m | u_f) \, P(y_m | y_f, u_f, u_m)
$$

$$
\bullet \sum_{u_1} P(u_1 | u_f, u_m) \, P(y_1 | y_f, y_m, u_f, u_m, u_1)
$$

$$
\bullet \sum_{u_2} P(u_2 | u_f, u_m) \, P(y_2 | y_f, y_m, y_1, u_f, u_m, u_1, u_2)
$$

$$
\vdots
$$

$$
\bullet \sum_{u_n} P(u_n | u_f, u_m) \, P(y_n | y_f, y_m, y_1, y_2, ..., y_{n-1}, u_f, u_m, u_1, u_2, ..., u_n). \qquad (14)
$$

In the case of a continuous trait, the probability function for the ith offspring in this expression, $P(y_i | y_f, y_m, y_1, y_2, ..., y_{i-1}, u_f, u_m, u_1, u_2, ..., u_i)$, is the normal density of y_i, conditional on u_i, after regressing out the effects of $y_f - \mu_{u_f}$, $y_m - \mu_{u_m}$, $y_1 - \mu_{u_1}$, $y_2 - \mu_{u_2}, ..., y_{i-1} - \mu_{u_{i-1}}$, i.e. the residuals of the parents and of all the preceding sibs. It should be noted that this probability function depends on the types of all the sibs up to the ith, which causes the nested sum to appear in (14). [Just as in expressions (1) and (13) the second line is nested within the summations of the first line, so in (14) the expression on each line is nested within the summations of previous lines]. It is possible to avoid this nested sum, regaining a likelihood of the form (13), by using for the ith child an approximate probability function that does not depend on the types of the preceding sibs, i.e. of the form

$$
P(y_i | y_f, y_m, y_1, y_2, ..., y_{i-1}, u_f, u_m, u_i). \qquad (15)
$$

This is taken to be the density of y_i, conditional on u_i, after regressing out the effects of the residuals $y_f - \mu_{u_f}$, $y_m - \mu_{u_m}$ and $y_j - \bar{\mu}$ for $j < i$; $\bar{\mu}$ can be one of several weighted averages of the genotypic means μ_u, the weights being possibly dependent on y_j, u_f, and u_m, but not on any u_j (Demenais et al., 1990). Approximate likelihoods of this form are much faster to compute, especially for large sibships.

In the case of a dichotomous trait, a class D model is formulated by including in the logit for the ith offspring, θ_i, terms that depend on the affection states of all preceding sibs together with the terms that allow for the affection states of the parents. As before, these terms do not depend on the sibs' (or even parents') types, so that the probability function for the ith child is of the form that is a special case of (15), and so the class D model likelihood can be expressed in the form (13). The logit θ_i is increased by the same amount, say $\delta_b(1)$, for each preceding sib who is affected; similarly a (usually negative) term $\delta_b(0)$ is added to the logit for each preceding sib who is unaffected. Thus, in the absence of covariates, the logits for the first three sibs in a family in which both the parents are unaffected, the first two sibs are affected and the third sib is unaffected are:

$$\theta_1 = \beta_{u_1} + \delta_f(0) + \delta_m(0)$$
$$\theta_2 = \beta_{u_2} + \delta_f(0) + \delta_m(0) + \delta_b(1)$$
$$\theta_3 = \beta_{u_3} + \delta_f(0) + \delta_m(0) + 2\delta_b(1).$$

Thus, fixing their types at u_1, u_2, and u_3, the joint probability for the three sibs is the multiple logistic:

$$P(y_1 \mid y_f, y_m, u_1)\ P(y_2 \mid y_f, y_m, y_1, u_2)\ P(y_3 \mid y_f, y_m, y_1, y_2, u_3)$$

$$= \frac{e^{\theta_1}}{1 + e^{\theta_1}} \cdot \frac{e^{\theta_2}}{1 + e^{\theta_2}} \cdot \frac{1}{1 + e^{\theta_3}}. \tag{16}$$

Now in the case of a normally distributed continuous trait, multinormal regression theory makes it easy to decompose $P(y_1, y_2, y_3)$ as the product of three univariate normal densities $P(y_1)\ P(y_2 \mid y_1)\ P(y_3 \mid y_1, y_2)$ with the resulting product independent of the order of the decomposition (i.e. the joint density does not depend on how the subscripts are permuted). In the case of the joint multiple logistic (16), however, the order of the sibs can change the corresponding probability. Suppose, for example, we condition on the previous sibs' phenotypes in an analogous fashion, but now the first sib is unaffected and the last two sibs are affected. The logits are then

$$\theta'_1 = \beta_{u_1} + \delta_f(0) + \delta_m(0)$$
$$\theta'_2 = \beta_{u_2} + \delta_f(0) + \delta_m(0) + \delta_b(0)$$
$$\theta'_3 = \beta_{u_3} + \delta_f(0) + \delta_m(0) + \delta_b(1)$$

and the joint probability for the three sibs is

$$\frac{1}{1 + e^{\theta'_1}} \cdot \frac{e^{\theta'_2}}{1 + e^{\theta'_2}} \cdot \frac{e^{\theta'_3}}{1 + e^{\theta'_3}}.$$

It is easy to verify that this is not identical to (16).

The tests for major gene segregation under a regressive model are exactly the same as under the unified model. However, they have the advantage of being performed under a more general model that can allow for a spouse correlation and/or a difference between father-offspring and mother-offspring correlations, which should tend to reduce both type I and type II errors.

Compound Regressive Models

To overcome the non-interchangeability of the sibs in class D logistic regressive models, and at the same time increase the speed of computation for class D normal regressive models, Bonney (1992) introduced compound regressive models. For continuous normal traits, compound and class D regressive models are mathematically identical, though formulated differently.

Consider a set of independent normally distributed random variables x_i, $i = 1, 2, ..., n$, with common mean 0 and common variance σ_x^2. Let

$$z_i = x_i + c,\qquad(17)$$

where c is a normally distributed random variable, independent of the x_i, with mean 0 and variance σ_c^2. Then the variance of z_i is $\sigma_z^2 = \sigma_x^2 + \sigma_c^2$, and the correlation between z_i and $z_{i'}$ ($i \neq i'$) is $\sigma_c^2 / (\sigma_x^2 + \sigma_c^2)$. Now using the notation (10), the joint density of c and the z_i is

$$\phi(c, \sigma_c^2) \prod_i \phi(z_i - c, \sigma_x^2),$$

so that we can write the joint (compound) density of the z_i as

$$\int_{-\infty}^{\infty} \phi(c, \sigma_c^2) \prod_i \phi(z_i - c, \sigma_x^2)\, dc,\qquad(18)$$

in which the compounding density is normal with variance σ_c^2. Now consider the likelihood of a class A model, (13), for a normally distributed continuous trait:

$$\begin{aligned}
L_A = \sum_{u_f} P(u_f)\, \phi(y_f - \mu_{u_f}, \sigma^2)\\
\cdot \sum_{u_m} P(u_m \mid u_f)\, \phi(y_m - \mu_{u_m} - b_{mf}\{y_f - \mu_{u_f}\}, \{1 - \rho_{fm}^2\}\sigma^2)\\
\cdot \prod_i [\sum_{u_i} P(u_i \mid u_f, u_m)\, \phi(z_i, \{1-\delta\}\sigma^2)],
\end{aligned}\qquad(19)$$

where σ^2 = the variance of the residuals from the major type mean

ρ_{fm} = the spouse correlation of the residuals

b_{mf} = the regression of the mother's residual on the father's residual (which equals

ρ_{fm} when the variances are equal)

$z_i = y_i - \mu_{u_i} - b_{if}(y_f - \mu_{u_f}) - b_{im}(y_m - \mu_{u_m})$, the ith offspring's residual after regressing out the effects of the parents' residuals

δ = the proportion of the residual variance that can be accounted for by regression on the parents' residuals. Bonney (1984) showed that $\delta = (\rho_{fo}^2 + \rho_{mo}^2)/(1 + \rho_{fm})$, where ρ_{fo} and ρ_{mo} are respectively the father-offspring and mother-offspring residual correlations.

It is then a simple matter to allow for additional sibling correlation, over and above that due to having the same parents, by compounding over a normally distributed random variable c with variance σ_c^2. Thus the likelihood of a class D model, identical to (14) when the probability density functions are normal, can be written as the following modification of (19):

$$L_D = \sum_{u_f} P(u_f)\ \phi(y_f - \mu_{u_f}, \sigma^2)$$

$$\bullet \sum_{u_m} P(u_m|u_f)\ \phi(y_m - \mu_{u_m} - b_{mf}\{y_f - \mu_{u_f}\}, \{1 - \rho_{fm}^2\}\sigma^2) \qquad (20)$$

$$\bullet \int_{-\infty}^{\infty} \phi(c, \sigma_c^2) \prod_i [\sum_{u_i} P(u_i|u_f, u_m)\ \phi(z_i - c, \{1 - \delta\}\sigma^2 - \sigma_c^2)]\ dc.$$

In (19) the z_i are independent with variance $(1 - \delta)\sigma^2$. On the other hand in (20), the last line of which is analogous to (18), the $z_i - c$ are independent as in (16) with variance $(1 - \delta)\sigma^2 - \sigma_c^2$. Thus the variance of z_i remains the same, $(1 - \delta)\sigma^2$, but now the z_i have common correlation $\sigma_c^2/(1 - \delta)\sigma^2$. In the class A likelihood (19) the residual variance σ^2 is divided into just two parts:

$\delta\sigma^2$ = variance accounted for by regression on parents' residuals

$(1 - \delta)\sigma^2$ = variance not accounted for by regression on parents' residuals.

In the class D likelihood (20), however, the residual variance is divided into three parts:

$\delta\sigma^2$ = variance accounted for by regression on parents' residuals

σ_c^2 = variance common to offspring not accounted for by regression on parents' residuals

$(1 - \delta)\sigma^2 - \sigma_c^2$ = residual variance unique to individual offspring.

Pooling the first two of these three parts, which represent the total variance common to siblings, we see that the residual sibling correlation, i.e. the correlation between $y_i - \mu_{u_i}$ and $y_{i'} - \mu_{u_{i'}}$, $i \neq i'$, is $\rho_{ss} = (\delta\sigma^2 + \sigma_c^2)/\sigma^2 = \delta + \sigma_c^2/\sigma^2$. Solving this for σ_c^2, we see that we can substitute $\sigma_c^2 = (\rho_{ss} - \delta)\sigma^2$, where $\delta = (\rho_{fo}^2 + \rho_{mo}^2)/(1 + \rho_{fm})$, into (20) for the likelihood to be expressed in terms of the correlations of interest. Alternatively, when $\rho_{fm} = 0$ we can write the third line of (20) as

$$\int_{-\infty}^{\infty} \phi(c, 1) \prod_i [\sum_{u_i} P(u_i | u_f, u_m) \phi(z_i - \sigma_c c, \{1-\delta\}\sigma^2 - \sigma_c^2) \, dc$$

$$= \int_{-\infty}^{\infty} \phi(c, 1) \prod_i [\sum_{u_i} P(u_i | u_f, u_m) \phi(z_i - \sigma_c c, \{1 - \rho_{ss}\}\sigma^2) dc,$$

the form Bonney (1992) uses, which is mathematically identical to it.

It should also be noted that when there is random mating and $b_{if} = b_{im}$ (= b., say), the likelihood (20) becomes

$$L_D = \sum_{u_f} P(u_f) \, \phi(y_f - \mu_{u_f}, \sigma^2) \sum_{u_m} P(u_m) \, \phi(y_m - \mu_{u_m}, \sigma^2)$$

$$\bullet \int_{-\infty}^{\infty} \phi(c, \sigma_c^2)$$

$$\bullet \prod_i [\sum_{u_i} P(u_i | u_f, u_m) \, \phi(y_i - \mu_{u_i} - b.\{y_f - \mu_{u_f} + y_m - \mu_{u_m}\} - c, \{1 - \rho_{ss}\}\sigma^2)] dc,$$

while if we substitute the density functions (11) into (9) we obtain

$$L_m = \sum_{u_f} P(u_f) \, \phi(y_f - \mu_{u_f}, \sigma_g^2 + \sigma_e^2) \sum_{u_m} P(u_m) \, \phi(y_m - \mu_{u_m}, \sigma_g^2 + \sigma_e^2)$$

$$\bullet \int_{-\infty}^{\infty} \phi(v - H_p\{y_f - \mu_{u_f} + y_m - \mu_{u_m}\}/2, H_p\sigma_e^2/2 + \sigma_b^2) \qquad (21)$$

$$\bullet \prod_i [\sum_{u_i} P(u_i | u_f, u_m) \, \phi(y_i - \mu_{u_i} - v, \sigma_g^2/2 + \sigma_e^2 - \sigma_b^2)] dv.$$

These two likelihoods are identical if we equate

$$\sigma^2 = \sigma_g^2 + \sigma_e^2$$

$$\sigma_c^2 = H_p\sigma_e^2/2 + \sigma_b^2$$

$$b. = H_p/2$$

$$\text{and} \quad \rho_{ss} = (\sigma_g^2/2 + \sigma_b^2)/(\sigma_g^2 + \sigma_e^2).$$

Thus in the case of continuous traits measured on members of nuclear families, the mixed and unified models are special cases of the class D regressive, or compound regressive, models. On the assumption of random mating, homoscedasticity and equal parent-offspring correlations, the models differ only in that the mixed and unified models parametrize the parent-offspring and sibling correlations in terms of polygenic heritability and possibly an environmental sibling variance component.

In the case of dichotomous disease traits Bonney (1992) retains a logistic form for the phenotypic probability functions, rather than the cumulative normal form of the mixed model, but includes in each offspring's logit a baseline risk variable c, analogous to c in (20), which is assumed to follow some compounding distribution across sibships. Using the transformation

$$R = e^c / (1 + e^c),$$

Bonney makes the suggestion, among others, that a beta distribution be assumed as the compounding distribution of R.

The finite polygenic mixed model

Fernando et al. (1993) have recently proposed an alternative formulation of the mixed model that, like the compound regressive models, has both theoretical and computational advantages over the mixed and unified models. In this case the difference lies solely in how the polygenic component is modelled. We have noted that in the usual polygenic model the transition probability function $P(g_i \mid g_f, g_m)$ is assumed to be a normal density whose variance, $\sigma_g^2/2$, is independent of g_f and g_m; in other words, the within-sibship polygenic variance is a constant that does not depend on the parental polygenotypes. It can be shown that this assumption is correct if the polygenic component is in fact the sum of the effects of an infinite number of segregating loci each of which is infinitesimally small. When there is a finite number of "polygenic" loci, however, the within-sibship variance would be expected to be smaller when the parents have extreme phenotypes. Suppose, for example, that each parent is homozygous at every polygenic locus (which occurs with probability zero if there is an infinite number of loci); in this case the within-sibship polygenic variance is necessarily zero. On the other hand, the offspring polygenic variance will be largest when all the parental loci are heterozygous, in which case the parental phenotypes are likely to be less extreme.

In the finite polygenic mixed model it is assumed that the polygenic component is determined by a set of k diallelic loci with additive effects. One of the alleles at each locus adds a positive effect to the polygenic component, the same for all loci, while the other adds a negative effect, again the same for all loci. The magnitudes of these positive and negative effects are determined by the common allele frequencies p and 1 - p at each locus, the value of k and the population variance of the polygenic component, σ_g^2. The polygenic component can take on only one of 2k + 1 values, as each person must have either 0, 1,..., or 2k alleles adding to it (and conversely either 2k, 2k - 1,..., or 0 alleles subtracting from it). There are 3^k polygenotypes possible, and hence 3^{3k} transition probabilities (some of which will be zero) for the polygenic component. However, if we assume that the polygenic components of two sibs are independent conditional on the polygenic components of their parents, then we need only consider $(2k + 1)^3$ transition probabilities. Suppose, for example, we let the polygenic component be determined by k = 5 diallelic loci, so that it can take on only 2k + 1 = 11 values. Then in the finite polygenic mixed model we add three genotypes for the major gene and have a total of $3 \times 11 = 33$ "genotypes" to sum over in calculating the likelihood (1), instead of $3 \times 3^5 = 729$ genotypes. Stricker et al (1993) have shown in a specific situation that using this approximation and setting $k = 5$ and $p = \frac{1}{2}$ gives likelihoods and maximum likelihood estimates that are very close to those obtained under the mixed model.

The finite polygenic mixed model can be extended to have non-Mendelian transmission probabilities at the major locus, as in the unified model, and the same tests then apply for major gene segregation. In addition, it is possible to vary the parameters k and p to find which values fit the data best.

EXTENDED PEDIGREES AND NON-RANDOM SAMPLING

Models for General Pedigrees

In theory all the same segregation analysis models can be used for the analysis of randomly sampled pedigrees as can be used for randomly sampled complete nuclear families. In practice this is not possible except by using approximations, because for some of the models the amount of time required to compute a likelihood increases exponentially with the size of the pedigree or with the number of pedigree members whose phenotypes are missing. We define a pedigree to be a collection of related persons such that every person in the pedigree has either both parents or neither parent also in the pedigree. This can cause the phenotype y to be missing for certain members of the pedigree (whether a nuclear family or not), and hence materially increase the computer time needed for certain models. For all models the computational time can become prohibitive if the pedigree contains many and/or large loops. (A pedigree is said to contain a loop if it is possible, on a pedigree chart as usually drawn, to follow a continuous line starting from one person through one or more other persons back to the original starting person).

Provided it does not contain a loop, there is no difficulty in computing the likelihood of the transmission probability model or the finite polygenic mixed model in the case of an extended pedigree: the computational time required is no more than that required to calculate the likelihood for all the nuclear families it contains. We make a distinction between pedigree members who have no parents in the pedigree, whose contribution to the likelihood is analogous to the first line of (1), and pedigree members who have both parents in the pedigree, whose contribution is analogous to the second line of (1) (Elston and Stewart, 1971; Lange and Elston, 1975; Cannings et al., 1978; Fernando et al., 1993). The probability functions $P(y \mid u)$ are set equal to unity for each person with a missing phenotype.

The likelihoods of the mixed and unified models cannot be calculated for large pedigrees without approximation. Lalouel and Morton (1981) have recommended that one divide the pedigree into its constituent nuclear families and assume that, conditional on the parental phenotypes in each case, these nuclear families are independent. The closer the phenotype is to determining the genotype, the better this approximation will be. Clearly it is not possible to condition on any phenotype that is missing, and for this situation Morton et al. (1984) integrate the likelihood over the missing phenotypes. Whereas this approach is appropriate for independent nuclear families, it worsens the approximation involved when it is used for the component nuclear families of an extended pedigree. The effect of the approximation should be small in large samples on the values of maximum likelihood estimates obtained, but could be large on their estimated standard errors, and hence also on tests of hypotheses. Hasstedt (1982) has proposed an approximation to calculate the likelihood of the mixed model on a large pedigree, without dividing it into its component nuclear families, that is similar to the use of a weighted average of genotypic means in (15) in order to simplify the likelihood of a class D regressive model. This approximation has been shown to be good in specific situations.

In the case of regressive and compound regressive models, no approximation is needed so long as the pedigree contains no members with missing phenotypes. Bonney (1992) has indicated how one can allow for persons with no parents in the pedigree to have missing phenotypes, and a person with no offspring in the pedigree can simply be eliminated if his or her phenotype is missing. However, a computational problem occurs if any person who has both parents and offspring in the pedigree has a missing phenotype. The usual approximation in this situation is to assume that the residuals of the parents of the person with missing phenotype are independent of the residuals of that person's offspring. It should be noted in this connection that all the class A, B, C and D regressive models assume that, conditional on the phenotype and type of the intervening parent, the phenotypes of grandparents are independent of the phenotypes of their grandchildren. This assumption implies, for example when the residual mother-offspring and father-offspring correlations are equal, ρ_{po}, that the residual correlation between k^{th} degree unilineal relatives is ρ_{po}^k, whereas under the mixed model it would be $\rho_{po}/2^{k-1}$. Thus for pedigree structures that contain second and higher degree relatives, the mixed model is no longer a special case of the class D regressive models. However, the difference between these two quantities is always less than 1/16 provided $\rho_{po} \leq 1/2$.

Models for Non-random Sampling

There has long been a well-established general theory to allow for the ascertainment of sibships via probands, i.e. persons of extreme phenotype who, *each independently of all such persons,* cause their family to be included in the sample studied. (It is not uncommon to encounter misuse of the word proband. For example, if a sibship is ascertained because it contains two affected individuals, these individuals are often wrongly called "probands." In this case the ascertainment event is multiplex and it would be better to use a term such as "proband set"). Briefly, we let $\pi(y)$ be the probability that a person with phenotype y becomes a proband and express the appropriate likelihood (i.e. conditional on the sibship being ascertained, which is equivalent to it containing at least one proband) as a function of the $\pi(y_i)$. In the classical model for a dichotomous disease trait, $\pi(y)$ takes on only two values: $\pi(0) = 0$ or $\pi(1) = \pi$. Thus the probability that a sibship with r affected sibs contains at least one proband is $1 - (1 - \pi)^r$. When $\pi = 1$, this probability is independent of r. Li (this conference) has reviewed recent advances for this situation. When $0 < \pi < 1$, the probability that a sibship is ascertained is a concave function of r (i.e. the function increases more slowly as r increases), and as $\pi \rightarrow 0$, it becomes proportional to r. These two situations are known respectively as multiple and single ascertainment, because they correspond respectively to situations where there may be more than one or only one proband in each sibship sampled. Although it is not then a probability, it is mathematically possible for π to take on a negative value, in which case $1 - (1 - \pi)^r$, although negative, corresponds to a situation in which the probability that a sibship is ascertained is a convex function of r, i.e. it increases faster as r increases (Haldane, 1938; Elston and Bonney, 1986; George and Elston, 1991). In this situation, which may be realistic for mild diseases, it is not possible to identify probands. Elston and Yelverton (1975) suggested various ways of modelling $\pi(y)$ when y is continuous, or in the case of a disease with variable age of onset, in the context of analyzing nuclear family data.

Whereas there is no theoretical difficulty in extending all the segregation analysis models that have been described in this paper for randomly sampled sibships to randomly sampled extended pedigrees of arbitrary complexity, there are serious theoretical difficulties in extending to pedigree data the general model that has just been described for ascertainment via probands. The special case of single ascertainment ($\pi \rightarrow 0$) was essentially solved by Cannings and Thompson (1977) and by Elston and Sobel (1979). The former considered a sequential sampling scheme in which, after the initial ascertainment event, only the phenotypes of persons already sampled determine which additional pedigree members are studied, so that it is implicitly impossible for the same pedigree to be multiply ascertained. The latter assumed the existence of a proband sampling frame, such that pedigree members drawn from outside this sampling frame have zero probability of becoming probands, whatever their values of $\pi(y)$. (Alternatively, we can consider $\pi(y)$ to be undefined for persons outside the proband sampling frame). Then, by restricting membership in this sampling frame to those pedigree members involved in the original ascertainment event, the conditional likelihood becomes identical to that proposed by Cannings and Thompson. Both approaches reduce to the usual solution for single ascertainment in the special case of sampling sibships via probands. However, in each case it is theoretically simple to allow for a multiplex ascertainment event, such as "at least two affected persons in a sibship," provided the ascertainment is single, i.e. there cannot be more than one such ascertainment event in each pedigree.

Prior to 1980, two proposals had been made that allowed $\pi(y) > 0$ in extended pedigree data. The first of these is to ignore the fact that ascertainment is through probands and simply analyze the data as though they constitute a random sample of pedigrees. Provided the proportion of probands is small among all those in the sample with a similar extreme phenotype, estimates of the transmission probabilities and penetrance parameters will not have large biases, while parameters such as heritability and gene frequency will be severely overestimated (Elston and Rao, 1978; Elston, 1979; Majumder, 1985). The second is to assume that if *any* one (or more) of the pedigree members identified as being in the proband sampling frame becomes a proband, then the *same* pedigree structure is sampled, regardless of the position of the proband(s) in the pedigree (Elston and Sobel, 1979). Since then another solution to the problem of multiple ascertainment in pedigrees has been put forward by Lalouel and Morton (1981). As part of their proposal to calculate the pedigree likelihood of the mixed model by dividing it up into nonindependent nuclear families, they suggested that each of these be conditioned on, in addition to the parental phenotypes: (1) the fact that the sibship contains at least one proband, if in fact it does contain one or more probands, and (2) the phenotypes of up to three pointers — probands outside the nuclear family. Each component of the nuclear family (father, mother or the set of offspring) is taken to have at most one pointer, the closest proband outside the nuclear family, each proband being assigned only to the closest component of the nuclear family. Certain further simplifying assumptions are made, and then, on the assumption of conditional independence, the pedigree likelihood is taken to be the product of these conditional nuclear family likelihoods. All of these proposals make unrealistic assumptions whose effects have not been investigated in any detail. How to calculate appropriate pedigree likelihoods under multiple ascertainment is an area that deserves further investigation.

DATA TRANSFORMATION

It has been noted above that in the case of a continuous phenotype the scale of measurement, or equivalently the way the phenotypic measurements are transformed prior to analysis, can have an important effect on the outcome of a segregation analysis. A good example of this is afforded by Asamoah et al (1987), who analyzed serum dopamine-β-hydroxylase activity measured on 178 members of a six-generation pedigree under the transmission probability model. On the untransformed scale one would conclude that there is evidence of major gene segregation, while on the logarithmic scale one would conclude that there is no such evidence for a major gene. The reason for this result lies in the fact that unless the data points fall clearly into a number of discrete groups, fitting mixtures of normal distributions to a sample of data will often lead to quite different admixture proportions depending on the scale of measurement used. It is therefore important to investigate the effect of various transformations and to use an objective method to decide which is the most appropriate. The latter problem has led to arguments in the statistical literature [see Box and Cox (1982) and the literature cited therein]. I therefore conclude this paper with a simple explanation of the correct procedure in this situation.

Several families of transformations are available for segregation analysis (MacLean et al., 1976; George and Elston, 1988), based on variations of the original power transformation proposed by Box and Cox (1964), $x = (y^p - 1)/p$, where $x = \ln y$ when $p = 0$. Using this simple family as an example, a reasonable criterion would be to choose the maximum likelihood estimate of p, estimated jointly with all the other parameters that need to be estimated. In order to test one of the null hypotheses discussed above under a particular model, using the likelihood ratio criterion, p must be estimated twice — once under the hypothesis and once under the model. In most situations there will be a serious bias toward accepting the null hypothesis if, when maximizing the likelihood under the model, the value of p is held fixed at its maximum likelihood estimate under the hypothesis. However, because the value of p determines the scale of measurement, we cannot simply compare the two likelihoods obtained when different values of p are used to transform the data. The solution to this dilemma is to realize that in each case the likelihood must be appropriate for the data as measured on the same scale, say p = 1 which corresponds to untransformed data (Wilson et al., 1984). Let $\{y\}$ denote the pedigree data $y_1, y_2, ..., y_n$ on the original scale, and $\{x\}$ the data $x_1, x_2, ..., x_n$ after transformation. Then with the Box and Cox transformation, we can equate the likelihoods

$$L\{x\} = L\{y\} \prod_{i=1}^{n} \left| \frac{dy_i}{dx_i} \right| = L\{y\} \prod_{i=1}^{n} x_i^{p-1},$$

where the extra factor is the Jacobian of the transformation, and so

$$L\{y\} = L\{x\} \prod_{i-1}^{n} x_i^{1-p}. \tag{22}$$

Provided we always maximize $L\{y\}$ as given by (22), the likelihoods will all be on the same scale and can therefore be compared for different values of p.

An alternative way of interpreting (22) is to realize that if $P(x \mid u)$ and the other similar conditional densities are inserted into $L\{x\}$ as normal densities, then $L\{y\}$ as given in (22) is identical to the expression obtained for $L\{y\}$ on the assumption that the densities $P(y \mid u)$, etc., are power-normal. Thus the conditional densities are assumed to depend on three unknown parameters, which can be characterized as location, scale and skewness parameters. Similarly the generalized modulus power transformation (George and Elston, 1988) allows for conditional densities that depend on four unknown parameters that can be characterized as location, scale, skewness and kurtosis parameters.

REFERENCES

Asamoah, A., Wilson, A.F., Elston, R.C., Dalferes, E. Jr., Berenson, G.S., 1987, Segregation and linkage analyses of dopamine-β-hydroxylase activity in a six generation pedigree, *Am J Med Genet.* 27:613.

Bonney, G.E., 1984, On the statistical determination of major gene mechanisms in continuous human traits: regressive models, *Am J Med Genet.* 18:731.

Bonney, G.E., 1986, Regressive logistic models for familial disease and other binary traits, *Biometrics.* 42:611.

Bonney, G.E., 1992, Compound regressive models for family data, *Hum Hered.* 42:28.

Box, G.E.P., Cox, D.R., 1964, An analysis of transformation, *J Roy Statist Soc Ser B.* 26:211.

Box, G.E.P. Cox, D.R., 1982, An analysis of transformation revisited, rebutted, *J Am Statist Assoc.* 77:209.

Boyle, C.R., Elston, R.C., 1979, Multifactorial genetic models for quantitative traits in man, *Biometrics.* 35:55.

Cannings, C., Thompson, E.A., 1977, Ascertainment in the sequential sampling of pedigrees, *Clin Genet.* 12:208.

Cannings, C., Thompson, E.A., Skolnick, M.H., 1978, Probability functions on complex peidgrees, *Adv Appl Probab.* 10:26.

Demenais, F.M., Murigande, C., Bonney, G.E., 1990, Search for faster methods for fitting the regressive models to quantitative traits, *Genet Epid.* 7:319.

Eaves, L.J., 1983, Errors of inference in the detection of major gene effects on psychological test scores, *Am J Hum Genet.* 35:1179.

Elston, R.C., 1979, Likelihood models in human quantitative genetics, *in:* "Genetic Analysis of Common Diseases: Applications to Predictive Factors in Coronary Heart Disease," C.F. Sing, M. Skolnick, eds., AR Liss, New York.

Elston, R.C., 1981, Segregation analysis, *Adv Hum Genet.* 11:63.

Elston, R.C., Bonney, G.E., 1986, Sampling via probands in the analysis of family studies, *in:* "Proceedings of the XIIIth International Biometric Conference," July 27 - August 1, 1986, University of Washington, Seattle.

Elston, R.C., Rao, D.C., 1978, Statistical modeling and analysis in human genetics. Ann *Rev Biophys Bioeng.* 7:253.

Elston, R.C., Sobel, E., 1979, Sampling considerations in the gathering and analysis of pedigree data. *Am J Hum Genet.* 31:62.

Elston, R.C., Stewart, J., 1971, A general model for the genetic analysis of pedigree data, *Hum Hered.* 21:523.

Elston, R.C., Yelverton, K.C., 1975, General models for segregation analysis, *Am J Hum Genet.* 27:31.

Fernando, R.L., Stricker, C., Elston, R.C., 1993, The finite polygenic mixed model: An alternative formulation for the mixed model of inheritance. Submitted.

George, V.T., Elston, R.C., 1988, Generalized modulus power transformations, *Commun Statist.* 17:2933.

George, V.T., Elston, R.C, 1991, Ascertainment: An overview of the classical segregation analysis model for independent sibships, *Biometrical J.* 33:741.

Haldane, J.B.S., 1938, The estimation of the frequencies of recessive conditions in man, *Ann Eugen (London).* 8:255.

Hasstedt, S.J., 1982, A mixed model likelihood approximation on large pedigrees, *Comput Biomed Res.* 15:295.

Lalouel, J.M., Morton, N.E., 1981, Complex segregation analysis with pointers, *Hum Hered.* 31:312.

Lalouel, J.M., Rao, D.C., Morton, N.E., Elston, R.C., 1983, A unified model for complex segregation analysis, *Am J Hum Genet.* 35:816.

Lange, K., Elston, R.C., 1975, Extensions to pedigree analysis. I. Likelihood calculations for simple and complex pedigrees, *Hum Hered.* 25:95.

MacLean, C.J., Morton, N.C., Lew, R., 1975, Analysis of family resemblance. IV. Operational characteristics of segregation analysis, *Am J Hum Genet.* 27:365.

McLean, C.J., Morton, N.E., Elston, R.C., Yee, S., 1976, Skewness in commingled distributions, *Biometrics.* 32:695.

Majumder, P.P., 1985, Comparison of ascertainment-bias correction schemes for pedigrees ascertained through multiple probands, *Stat Med.* 4:163-173.

McGuffin, P., Huckle, P., 1990, Simulation of Mendelism revisited, *Am J Hum Genet.* 46:994.

Morton, N.E., MacLean, C.J., 1974, Analysis of family resemblance. III. Complex segregation analysis of quantitative traits, *Am J Hum Genet.* 26:489.

Morton, N.E., Rao, D.C., Lalouel, J.M., 1984, "Methods in Genetic Epidemiology," S. Karger, Basel.

Stricker, C., Fernando, R.L., Elston, R.C., 1993, Segregation analysis under an alternative formulation for the mixed model, *Genet Epidemiol.* In press.

Wilson, A.F., Elston, R.C., Siervogel, R.M., Weinshilboum, R., Ward, L.J., 1984, Linkage relationships between a major gene for catechol-o-methyltransferase activity and 25 polymorphic marker systems, *Am J Med Genet.* 19:525.

SEGREGATION ANALYSIS IN THE UNDERSTANDING OF COMPLEX DISORDERS: SOME RECENT DEVELOPMENTS

Partha P. Majumder

Anthropometry and Human Genetics Unit
Indian Statistical Institute
Calcutta 700 035, India

INTRODUCTION

In comparison with a genetic disorder which has no environmental contribution, which expresses itself at birth and which is determined by a completely penetrant allele at a single autosomal diallelic locus, genetics of many human disorders are more complex. The complexity can arise in a variety of ways. Although no claim is made that the following list is exhaustive, some of the more common causes of complexity of a disorder are :

1. Variable age of onset — individuals with the appropriate genotype do not manifest the disorder either at birth or at the same age later in life;

2. Incomplete penetrance — some individuals with the appropriate genotype manifest the disorder, while some others do not;

3. Phenotypic heterogeneity — all individuals of the same genotype do not manifest the same phenotype;

4. Allelic/genetic heterogeneity — different alleles, either at the same locus or at different loci give rise to the same phenotype;

5. Multiple loci — disorder is determined by the action of genes at more than one locus;

6. Environmental influence – environment acts jointly with genes or interacts with genotype in the manifestation of the disorder.

The causes listed above are not mutually exclusive.

In this paper, we shall initially discuss certain aspects of a multilocus epistatic model, with or without variable age of onset, and then proceed to describe some recent developments in computer-intensive methods that are useful in segregation analysis. We shall conclude with a discussion of a method useful for investigating the mode of inheritance of a complex disorder.

THE MULTILOCUS EPISTATIC RECESSIVE MODEL

This model was first explicitly considered by Li (1953) and was later developed more extensively by him (Li, 1987). Li's reconsideration in 1987 was prompted by David Lykken's Presidential Address to the Society for Psychophysiological Research (Lykken, 1982), in which Lykken drew attention to some behavioral traits which tend not to run in families,

Human Population Genetics, Edited by P.P. Majumder
Plenum Press, New York, 1993

but appear as isolated cases. Patterns of inheritance of such traits are drastically different from the inheritance pattern of a monogenic trait. While such traits may appear to be non-genetic because of their lack of familial aggregation, yet they may really be genetic. Li (1987) proposed a concrete genetic model for such traits. The model states that such a trait or disorder is determined by a particular combination of genes of a number of loci, thus forming a gene "configuration" or "constellation". Absence of any one gene in this configuration destroys the manifestation of the trait or disorder. More concretely, suppose we consider a large number (say, k) of epistatically interacting autosomal diallelic loci $(A_i, a_i); i = 1, 2, \ldots, k$. The model states that the gene configuration responsible for manifestation of the trait or disorder is $(a_1 a_1 a_2 a_2 \ldots a_k a_k)$; individuals of the remaining $3^k - 1$ genotypes do not manifest the trait.

Suppose q_i denotes the frequency of the allele $a_i (i = 1, 2, \ldots, k)$ in a population. If the population practices random mating, then the prevalence (δ) of the disorder in the population will be

$$\delta = \prod_{i=1}^{k} q_i^2. \tag{1}$$

If $q_i = q$ (for all $i = 1, 2, \ldots, k$), then

$$\delta = q^{2k}. \tag{2}$$

It is obvious that prevalence sharply decreases with increase in the number of loci for a fixed value of the gene frequency; also, that prevalance decreases with decrease in gene frequency for a fixed number of loci.

Since there are only two phenotypes (normal and affected), there will be three different types of matings in the population: normal × normal ($N \times N$); normal × affected ($N \times A$); and, affected × affected ($A \times A$). However, because individuals of the normal phenotype comprise $3^k - 1$ genotypes, $N \times N$ matings comprise a vast number of different genotypic matings. Li (1987) provides further details.

While the above results are a description of characteristics of the disorder among individuals in a population or among families drawn at random from the population, the mode of inheritance of such a disorder cannot be effectively studied unless families are ascertained through an affected individual.

When a family is sampled through an affected parent, no correction of any kind is necessary to take into account the effect of non-random sampling. The observed proportion of affected offspring in the sampled $N \times A$ families is compared with the expected proportion $S_1 = Q/(1 + Q)$, where $Q = \prod_{i=1}^{k} q_i$.

When families are ascertained through an affected child, the observed proportion of affected offspring is grossly upwardly biased because not only are these families capable of producing an affected child, but only those families who have actually produced an affected child, are being studied. A method of correcting for ascertainment-bias in such families has been provided in Majumder et al. (1989).

When the disorder is not expressed at birth, and the age of onset is variable, further complexity arises because individuals may be of the susceptible genotype but may not have expressed the disorder. Unless this variability in age of onset is taken into account, the segregation probability will be underestimated. An approximate likelihood method for dealing with variable age of onset has been developed by Majumder et al. (1988).

The methods of segregation analysis developed for this model were applied to family data on prelingual deafness for the case when the disorder presents itself at (or very near) birth (Majumder et al., 1989), and to family data on vitiligo for the case when the disorder has a variable age of onset (Majumder et al., 1988). Some other theoretical results relating to family data are also presented in Majumder and Nath (1992).

DISCRIMINATING BETWEEN SINGLE- AND MULTI-LOCUS MODELS

In this section, we shall primarily discuss how effectively single- and multi-locus recessive models can be discriminated using the technique of segregation analysis. But, before getting into the details of this, we shall discuss two pertinent points.

First, a major criticism of our analysis of family data on prelingual deafness that was raised (Kimberling et al., 1989; Marazita et al., 1989) concerned the sporadic proportion, and our handling of simplex families. In previous studies on prelingual deafness [for references see Majumder et al. (1989)], probands in most, if not all, simplex families were classified as sporadic (not genetic). We argued against this procedure, because this invariably leads to an overestimate of the true sporadic proportion. In fact, under a multilocus — even a two-locus — recessive model, the vast majority of $N \times N$ families ascertained through an affected offspring, are expected to be simplex, even though all of the probands in these families are genetic cases. This qualitative observation will be quantified in this section. Morton (1991) has further asserted that "if the unit of evaluation is the pedigree rather than the proband, the sporadic proportion is necessarily increased." We shall examine this assertion also.

Second, even before discussing the effectiveness of segregation analysis in distinguishing between single-locus and multi-locus models, it is desirable to investigate whether the problem, is at all statistically identifiable; that is, whether two distinct sets of parameter values cannot yield the same probability distribution (van der Genugten, 1977). Under a single-locus or the multi-locus epistatic recessive model, we have shown that the likelihood of observations on a nuclear family ascertained through an affected offspring is, except for a constant, a mixture of binomial distributions, the mixture proportions being mating probabilities (Majumder et al., 1989). In view of this result, it is easy to verify that the problem is statistically identifiable. Since it is identifiable for nuclear family data, it obviously remains to be identifiable for more extensive pedigree data.

To quantitatively investigate the ability to discriminate between one-locus and multi-locus recessive models, we have conduced a study involving computer simulation followed by segregation analysis. In this study, simulated pedigress of the structure given in Figure 1, were generated under a two-locus recessive model. The assumptions and parameters under which the simulated pedigrees were generated were : (i) founders were assumed to have been drawn from a population in Hardy-Weinberg equilibrium; (ii) the frequency of the disorder/trait in the general population was assumed to be fixed (1/1000 or 1/10000); (iii) the disorder/trait was assumed to be controlled by two unlinked diallelic (alleles A, a) autosomal recessive loci; (iv) the frequencies of the recessive alleles were assumed to be equal at each locus; (v) complete penetrance was assumed – that is, only individuals of genotype $aaaa$ manifested the trait/disorder with probability 1; (vi) Mendelian transmission was assumed; and, (vii) each simulated pedigree was assumed to have been ascertained through an individual who manifested the trait, and the proband was one of the individuals 26-30 in the pedigree (see Figure 1). Under these assumptions, when the population prevalence of the disorder is 1/1000, the frequency of the recessive allele turns out to be 0.1778 at each locus for the two-locus model (assuming equal gene frequencies at both loci), and 0.0316 if a single-locus recessive model is assumed.

Pedigrees were generated under these simulation conditions. Each simulation run comprised the generation of a batch of 10 pedigrees. Ten such runs were performed. The joint likelihood of each simulation run (batch of 10 pedigrees) was calculated under three models — non-genetic, single-locus autosomal recessive and two-locus autosomal recessive — using the computer program PAP (Hasstedt and Cartwright, 1981). Correction for ascertainment bias was performed by dividing the likelihood of a pedigree by the likelihood of the proband.

In Table 1 are presented the results of 10 simulation runs. For each run, we have also calculated the number of simplex pedigrees (that is, pedigrees with at least one affected relative of proband). It is seen that in the pooled set of 100 pedigrees, about 60% of the pedigrees are simplex. Therefore, even when only two epistatically interacting loci are involved, in the majority of the pedigrees, the proband is the only affected member even for a fairly common (prevalence = 1/1000) recessive disorder. We emphasize that none of these probands is a sporadic case. This point will be reconsidered in further detail later.

From the corrected $\log_{10}$-likelihood values presented it Table 1, it is seen that for the total set of 100 pedigrees, the two-locus model is correctly identified when compared to a non-genetic model (log-odds = 46.06) or a single-locus recessive model (log-odds = 18.90). To examine clearly how many pedigrees are necessary for identifying the correct model, we have plotted in Figure 2 the cumulative $\log_{10}$-likelihood values over the 10 simulation runs. It is seen from this figure that while a batch of 10 pedigrees is sufficient for favoring a genetic model to a non-genetic model, it is insufficient for discriminating between a single-locus and a two-locus recessive model. A set of twenty pedigrees is sufficient for distinguishing between one-locus and two-locus recessive models; twenty pedigrees provide a log-odds of about 5 in favor of the two-locus model.

To further investigate the characteristics of the aggregation of a multilocus recessive disorder in families, we generated 5 batches of 100 pedigrees each for two levels of population prevalence of the disorder: $\delta_1 = 1/1000$ and $\delta_2 = 1/10000$. The frequency distributions of the number of affected individuals in the nuclear family of the proband as also in the complete pedigree are given in Table 2. It is seen that the Mean ($\pm$ s.d.) numbers of nuclear families with the proband as the only affected member are, for $\delta_1 = 1/1000$ and $\delta_2 = 1/10000$, 78.2 (± 4.1) and 82.6 (± 4.3), respectively. In the complete pedigree, the corresponding figures are, 65.8 (± 2.3) and 79.6(± 4.6). Thus, we see that the vast majority of families — nuclear or pedigree — will have only one affected member, who is actually the proband, even when the disorder is controlled by just a pair of epistatically interacting recessive alleles. Obviously, to interpret the affected member as a sporadic case based on the nature of familial aggregation of the disorder is bound to be erroneous. We also see that the number of simplex families — nuclear or pedigree — increases with decrease in population prevalence of the disorder.

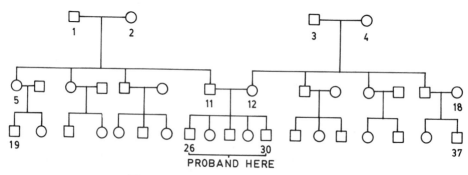

Figure 1. Structure of simulated pedigree

Table 1. Run-wise characteristics of simulated pedigree batches and results of segregation analysis

Run No.	No. of Pedigrees			Corrected log_{10} L under Model		
	Simplex	Multiplex	Total	Non-Genetic	One-Locus	Two-Locus
1	6	4	10	-18.15	-12.43	-10.50
2	7	3	10	-9.16	-11.55	-8.09
3	8	2	10	-9.16	-11.21	-8.34
4	6	4	10	-21.15	-9.47	-9.18
5	6	4	10	-12.15	-9.93	-7.97
6	6	4	10	-18.15	-12.21	-11.04
7	5	5	10	-15.15	-12.27	-10.66
8	5	5	10	-15.15	-9.85	-8.64
9	5	5	10	-24.15	-17.17	-14.98
10	7	3	10	-9.20	-9.32	-7.11
Total	61	39	100	-151.57	-115.41	-96.51

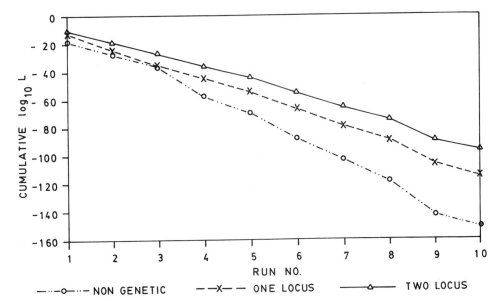

Figure 2. Cumulative log_{10}-likelihood values over 10 runs, each run comprising data on 10 simulated pedigrees, separately for three models

Table 2. Frequency distributions of number of affected individuals in complete pedigree and in proband's nuclear family obtained from batches of 100 simulated pedigrees for two levels of prevalence $\delta_1(=1/1000)$ and $\delta_2(=1/10000)$

Type of family	No. of affecteds	Batch (Run) No.									
		1		2		3		4		5	
		δ_1	δ_2	δ_1	δ_2	δ_1	δ_2	δ_1	δ_2	δ_1	δ_2
Complete	1	66	84	62	79	67	73	65	84	68	78
pedigree	2	25	14	30	20	24	24	23	13	24	15
	3	6	1	6	0	5	1	8	2	6	5
	4	3	0	2	1	2	2	3	0	1	2
	5	0	1	0	0	1	0	1	1	1	0
	6	0	0	0	0	1	0	0	0	0	0
Nuclear	1	77	86	80	82	72	76	79	87	83	82
family of	2	19	13	14	17	23	21	17	11	14	13
proband	3	4	1	4	1	4	1	2	2	2	4
	4	0	0	2	0	1	2	1	0	1	1
	5	0	0	0	0	0	0	1	0	0	0

Even though for a multilocus recessive disorder the vast majority of families ascertained through an affected individual will be simplex, segregation analysis enables the identification of the correct genetic model. We have already established this (Table 1 and Figure 2). However, to examine this issue more thoroughly when the population prevalence varies, we present appropriate results in Table 3. From the joint $\log_{10}$-likelihood values of 5 batches of 100 pedigrees each presented in this table, it is seen that, on an average, the non-genetic model can be rejected in favor of a genetic model with a $\log_{10}$-odds of about 22. The incorrect one-locus genetic model is also rejected in favor of the correct two-locus genetic model with a $\log_{10}$-odds of about 25. Further, even though under each model considered, the absolute values of the likelihood function are lower for lower values of disease prevalence, the relative $\log_{10}$-odds seem to change little with a change in disease prevalence.

In this study, no parameters were estimated under any model, because the primary purposes were to evaluate the characteristics of nuclear families and pedigrees under the two-locus recessive model, as also to quantify the usefulness of segregation analysis in discriminating between plausible models (especially when the majority of sampled families are simplex). Further, more extensive, studies along these lines are in progress.

Table 3. Joint $\log_{10}$-likelihood values for data on five batches of 100 pedigrees each under 3 models for two levels of prevalence

Prevalence	Model	Batch (Run) No.				
		1	2	3	4	5
1/1000	Non-genetic	-145.54	-148.54	-157.54	-130.54	-146.54
	One-locus	-125.61	-120.05	-127.93	-117.80	-127.92
	Two-locus	-102.16	-95.88	-106.74	-94.01	-99.64
1/10000	Non-genetic	-80.16	-92.16	-128.16	-124.16	-84.16
	One-locus	-71.46	-81.24	-80.70	-85.83	-79.87
	Two-locus	-45.07	-53.58	-58.22	-61.37	-50.20

COMPUTER-INTENSIVE METHODS

Elston (this volume) has described the various models that have been proposed for biological traits/disorders, especially quantitative ones. As is seen from this description, the proposed models have become more and more realistic by incorporation of the various familial and non-familial factors that are important in the determination of trait values. Consequently, the complexity of the models have also increased, especially in terms of the number of parameters. This, in turn, has resulted in an increase in computational complexity. In fact, in their classic paper, when Elston and Stewart (1971) proposed their algorithm, for computing likelihoods of pedigree data under specific genetic models, recognizing the computational complexity under the major gene-polygene model, they suggested restricting attention to the major gene component. Morton and MacLean (1974) later proposed an approximate solution to this computational complexity by polychotomizing the continuous scale into discrete classes. Cannings et al. (1976, 1978) then introduced the "peeling" algorithm in which they arranged terms in the Elston-Stewart algorithmic likelihood in a different order to speed up recursive computation of the likelihood function. However, Ott (1979) showed that the exact value of the likelihood of data on a pedigree under the mixed model (for a description, see Elston, this volume) cannot be calculated recursively, and that probabilities of all 3^N (where N denotes the number of individuals in the pedigree) genotypic combinations need to be calculated. He presented a method of calculating this likelihood using the EM-algorithm (Dempster et al., 1977), but had to restrict himself to 10 as an upper bound of N, because of computational complexity. Hasstedt (1982) suggested an approximate computation of likelihoods for mixed models, which has also been implemented in the Pedigree Analysis Package (Hasstedt and Cartwright, 1979).

The above discussion shows that computational complexity and computer time requirement have been two limiting factors in the analysis of data on pedigrees under the mixed model. Some recent developments in statistics, have provided computer-intensive methods to reduce computational complexity, though not necessarily computation time (which, however, is no longer a heavy constraint). Further, having obtained maximum likelihood estimates of parameters under genetic models, tests of hypothesis relied on certain statistics and their asymptotic distributions derived under many assumptions, some of which are now known to be false under specific genetic models, including the mixed model. Again, some recent computer – intensive methods have provided solutions to this problem. These methods are now described in the following subsections in the context of segregation analysis.

The Gibbs Sampler

This is a computer-intensive technique that has gained popularity in recent times. Although the roots of this technique can be traced back to the early 1950's (Metropolis et al., 1953), its recent popularity is primarily due to two papers by Geman and Geman (1984) and Gelfand and Smith (1990). Geman and Geman (1984) introduced this technique clearly in the context of image reconstruction and proved some important convergence theorems that provided theoretical justifications to the use of this technique. Gelfand and Smith (1990) later demonstrated the applicability of this technique in a wide variety of statistical contexts and showed how this technique is related to some other similar techniques, e.g., substitution sampling.

The Gibbs sampler algorithm is a sampling-based Monte Carlo method. Its attraction lies on the facts that it is conceptually simple and relatively easy to implement, even though it is not necessarily computationally efficient in comparison with some of the advanced numerical analytic techniques (Gelfand and Smith, 1990). In the paragraphs that follow, we shall illustrate the Gibbs sampler in the context of segregation analysis. But, first we shall provide a brief exposition to the technique in general statistical terms. [For details, see Cassella and George (1992).]

Suppose $X, Y_1, Y_2, \ldots, Y_k$ are random variables with joint probability density $f(x, y_1, \ldots, y_k)$. Then, the marginal density of X, $f(X)$, is given by :

$$f(x) = \int \int \cdots \int f(x, y_1, \ldots, y_k) dy_1 \, dy_2 \ldots dy_k. \tag{3}$$

The Gibbs sampler is a method for obtaining $f(x)$ when the integrations in (3) are difficult to perform either analytically or numerically. Actually, the Gibbs sampler is an algorithm to generate a sample $X_1, X_2, \ldots, X_m$ from $f(x)$. Thus, the marginal density of X or any population characteristic of this density can easily be obtained. The Gibbs sampler algorithm is based on properties of Markov chains (Hastings, 1970). If X and Y are two random variables, then

$$f_x(x) = \int f_{XY}(x, y) dy, \tag{4}$$

where $f_{XY}(x, y)$ denotes the joint density (possibly unknown) of X and Y, and $f_X(x)$ denotes the marginal density of X. The Gibbs sampler generates a sample from $f_X(x)$ by sampling from the conditional densities $f_{X|Y}(x|y)$ and $f_{Y|X}(y|x)$, which can often be explicitly derived. A sequence of random variables $\{Y_i', X_i'\}$, known as the "Gibbs sequence", is generated iteratively as follows. Starting with an initial value $Y_o' = y_o'$, X_i' is generated from the density $f(x|Y_i' = y_i')$, and Y_{i+1}' is generated from the density $f(y|X_i' = x_i')$. If there are three random variables X, Y and Z, so that

$$f_X(x) = \int \int f_{XYZ}(x, y, z) dz \, dy, \tag{5}$$

then, the "Gibbs sequence" comprises a sequence of random variables $\{Z_i', Y_i', X_i'\}$, generated as

$$X_i' \text{ from density } f(x|Y_i' = y_i', Z_i' = z_i'),$$

$$Y_{i+1}' \text{ from density } f(y|X_i' = x_i', Z_i' = z_i'), \text{ and}$$

$$Z_{i+1}' \text{ from density } f(z|X_i' = x_i', Y_{i+1}' = y_{i+1}'),$$

starting with two initial values $Y_o' = y_o'$ and $Z_o' = z_o'$. (Generalization for a larger number of random variables is straightforward.) Geman and Geman (1984) showed that as $i \longrightarrow \infty$, the distribution of X_i' converges to $f_X(x)$ at a geometric rate. The Gibbs sampler, it may be noted, uses the full set of univariate conditional distributions, which Besag (1974) proved is a sufficient set to uniquely determine the joint distribution (and hence all marginal distributions). When the number of variables is large, specification of the set of full conditionals is done by assuming that an individual full conditional distribution only depends on some "neighborhood" subset of variables. Details regarding the circumstances under which such "neighborhood specifications" uniquely determine the joint distribution are discussed in Geman and Geman (1984).

In the implementation of the Gibbs sampler, there are three major issues: choice of starting values, choice of a stopping value of i in the iterative process (that successive values of X_i in the Gibbs sequence are not independent needs to be borne in mind), and choice of the sample size m (based on which density or parameter estimation is to be done). No general rules can be given for any of these issues. As is generally true with all iterative algorithms, judicious choice of starting values will speed up covergence. Some strategies for choosing k and m are discussed in Gelfand et al. (1990) and Tanner (1991), although these strategies are not without caveats (Gelman and Rubin, 1991). [See also Ritter and Tanner (1992) for some methods that are useful in assessing convergence of the Gibbs sampler.]

In the context of segregation analysis, the Gibbs sampler has been thoroughly discussed by Guo and Thompson (1992). We shall base the subsequent discussion, including the notation, on this paper. Consider a quantitative trait. The data comprise observations on members of a pedigree consisting of n individuals. The model considered is the simple version of the mixed model (Morton and MacLean, 1974) which postulates the effect of a major diallelic locus with alleles A and a, with trait values μ_1, μ_2 and μ_3 given an individual's major-locus genotype as AA, Aa and aa, respectively; polygenic effect which follows a $N(0, \sigma_a^2)$ distribution and an individual environmental effect which follows a $N(0, \sigma_e^2)$ distribution. The observations on the members of the pedigree are denoted as $(y_1, y_2, \ldots, y_n) = \underset{\sim}{y}$. The problem is to evaluate the likelihood of the data $\underset{\sim}{y}$ under this mixed model. The parameters are $(p, \mu_1, \mu_2, \mu_3, \sigma_a^2, \sigma_e^2) = \underset{\sim}{\theta}$, where p denotes the population frequency of the allele A. The Elston-Stewart algorithm is the standard approach adopted for evaluation of the likelihood. Because the major-locus genotypes and polygenotypes are unobservable, estimation of $\underset{\sim}{\theta}$ is a "missing-data" problem that can generally be handled using the EM-algorithm (Dempster et al., 1977). Ott (1979) used the EM approach and showed that in estimating parameters, the joint conditional distribution of major locus genotypes and polygenotypes is required. He derived that:

$$f_{\underset{\sim}{\theta}}(\underset{\sim}{G}, \underset{\sim}{a} \mid \underset{\sim}{y}) = \frac{f_{\underset{\sim}{\theta}}(\underset{\sim}{y} \mid \underset{\sim}{G}, \underset{\sim}{a}) \cdot P_{\underset{\sim}{\theta}}(\underset{\sim}{G}) \cdot f_{\underset{\sim}{\theta}}(\underset{\sim}{a})}{\sum_{\underset{\sim}{G}} \int_{\underset{\sim}{a}} f_{\underset{\sim}{\theta}}(\underset{\sim}{y} \mid \underset{\sim}{G}, \underset{\sim}{a}) \cdot P_{\underset{\sim}{\theta}}(\underset{\sim}{G}) \cdot f_{\underset{\sim}{\theta}}(\underset{\sim}{a}) d \underset{\sim}{a}}, \tag{6}$$

where $f_{\underset{\sim}{\theta}}(\underset{\sim}{G}, \underset{\sim}{a} \mid \underset{\sim}{y})$ is the conditional distribution of major-locus genotypes $\underset{\sim}{G}$ and polygeno-types $\underset{\sim}{a}$ given the trait values $\underset{\sim}{y}$ of the n members in the pedigree, $P_{\underset{\sim}{\theta}}(\underset{\sim}{G})$ is the probability of the major-locus genotypes $\underset{\sim}{G}$, $f_{\underset{\sim}{\theta}}(\underset{\sim}{a})$ is the distribution of polygenotypes $\underset{\sim}{a}$ and $f_{\underset{\sim}{\theta}}(\underset{\sim}{y} \mid \underset{\sim}{G}, \underset{\sim}{a})$ is the conditional distribution of trait values $\underset{\sim}{y}$ given $\underset{\sim}{G}$ and $\underset{\sim}{a}$. Ott (1979) pointed out that there is no practical way to evaluate the denominator of equation (6) for large $(n > 10)$ pedigrees.

Guo and Thompson (1992) have accomplished the estimation of the conditional distribution $f_{\underset{\sim}{\theta}}(\underset{\sim}{G}, \underset{\sim}{a} \mid \underset{\sim}{y})$ using the Gibbs sampler. For implementing the Gibbs sampler, one needs to evaluate, for every individual $i (i = 1, 2, \ldots, n)$ in the pedigree, $P_{\underset{\sim}{\theta}}(G_i \mid \underset{\sim-i}{G}, \underset{\sim}{a}, \underset{\sim}{y})$ and $f_{\underset{\sim}{\theta}}(a_i \mid \underset{\sim-i}{a}, \underset{\sim}{y}, \underset{\sim}{G})$, where $\underset{\sim}{G} = G_i \cup \underset{\sim-i}{G}$, G_i, $\underset{\sim}{a} = a_i \cup \underset{\sim-i}{a}$, G_i is the major-locus genotype and a_i is the polygenotype of the $i - th$ member in the pedigree. By defining the "neighborhood" corresponding to the $i - th$ individual as the set of parents (if present in the pedigree), spouses (if any) and offspring (if any), Guo and Thompson (1992) derived explicitly the expressions for the local conditional distributions of $P_{\underset{\sim}{\theta}}(G_i \mid \underset{\sim-i}{G}, \underset{\sim}{a}, \underset{\sim}{y})$ and $f_{\underset{\sim}{\theta}}(a_i \mid \underset{\sim-i}{a}, \underset{\sim}{y}, \underset{\sim}{G})$, thereby making the generation of G_i and a_i straightforward.

The Gibbs sampler on pedigrees in thus specified as follows (Thompson and Guo, 1991):

(a) Find any genotypic configuration over the pedigree that is consistent with phenotypic observations.

(b) Take a random permutation of all individuals in the pedigree.

(c) For each individual, in the order indicated by (b), update the genotypic configuration by replacing the genotype of the individual by a realization from the local conditional distribution given his/her own phenotype, and current genotypic estimates for his/her parents, spouses and offspring.

(d) Repeat (b) and (c) for as long as required, storing and/or using the configuration achieved (for example) after the $400th$ scan, and every subsequent $20th$ scan of the pedigree.

[For algebraic details and for results based on simulated pedigrees comprising up to 230 members in 6 generations, Guo and Thompson (1992) may be consulted.] Thus, evaluation of likelihoods of data on complex pedigrees under fairly general models is no longer a limiting factor in the analysis of complex disorders/traits.

Bootstrap

This computer-intensive procedure proposed by Efron (1979) has now become a standard statistical technique. It is most useful in situations where the distribution of a statistic of interest cannot be derived algebraically. It, therefore, has found wide applicability in the context of testing statistical hypotheses. In segregation analysis, comparison of models is done by the use of the likelihood-ratio (LR) statistic and its asymptotic chi-square distribution (Wald, 1941; Morton, 1982). It has, however, recently been demonstrated that under models which are mixtures of normal distributions, the distribution of $-2\ell n$ LR may not follow a chi-square distribution asymptotically (Ghosh and Sen, 1985; Hartigan, 1985). Unfortunately, the likelihood of a pedigree under the mixed model is a mixture of normal distributions. Therefore, testing the validity of the mixed model against a polygenic alternative by use of the $-2\ell n$ LR statistic and the corresponding chi-square distribution may be invalid. While other techniques may be developed, one can continue to use the statistic $-2\ell n$ LR for this purpose, but since the chi-square approximation cannot be used, the empirical distribution of this statistic has to be obtained. The (parametric) bootstrap procedure is useful for obtaining the empirical distribution of the statistic. This approach was suggested in the general statistical setting by Williams (1970), and adopted in the context of segregation analysis by Schork and Schork (1989). Let $X = (X_1, X_2, \ldots, X_N)$ denote data on phenotypes on members of N pedigrees. We consider testing the null hypothesis $H_0 : h(x) = f(x, \theta)$ against $H_1 : h(x) = g(x, w)$, where $h(x)$ is the probability density from which the data are drawn. $f(x, \theta)$ and $g(x, w)$ may belong to separate families, and θ and w may be vector-valued. If $\hat{\theta}$ and $\hat{w}$ denote the maximum likelihood estimates of θ and w, respectively, then Williams (1970) proposed directly simulating the distribution of $T_N = \sum_{i=1}^{N} \log\{g(x_i, \hat{w})/f(x_i, \hat{\theta})\}$, assuming that $\theta = \hat{\theta}$. That is, for a sufficiently large integer M, M sets $\{(x*_{1k}, \ldots, x^*_{Nk}), k = 1, 2, \ldots, M\}$ of artificial data are drawn from the population with density $f(x^*, \hat{\theta})$. From the $k - th$ set, the m.l.e.'s $(\hat{\theta}^*_k, \hat{w}^*_k)$ and $T^*_{Nk} = \sum_{i=1}^{N} \log\{g(x^*_{ik}, \hat{w}^*_k)/f(x^*_{ik}, \hat{\theta}^*_k)\}$ are computed. The hypothesis H_0 is rejected at nominal level α if T_N computed from the actual data exceeds the $[M\alpha]$ largest of the $\{T^*_{Nk}\}$. Schork and Schork (1989) conducted a simulation study to investigate the power of this method in discriminating among various genetic models, and concluded that the test is quite powerful. They have also proposed a new test procedure – called the SOS Criterion Test – which is computationally less heavy because it does not entail computation of $(\hat{\theta}^*_k, \hat{w}^*_k)$ for each simulated set of pedigrees. Details on generation of simulated pedigrees are given in Boehnke et al. (1988) and Lange and Matthysse (1989).

INVESTIGATING MODE OF INHERITANCE WITHOUT
PERFORMING SEGREGATION ANALYSIS

In a seminal paper Penrose (1953) proposed computing relative risks for various pairs of relatives to obtain an idea of the genetical background of common diseases. The parameter Penrose (1953) suggested is not the standard relative-risk used in epidemiology, but was defined as :

$$\lambda = \frac{\text{Prob } \{I_2 \text{ is affected } | I_1 \text{ is affected } \}}{\text{Prob } \{I_2 \text{ is affected } \}},$$

where I_1 and I_2 are two related individuals. The standard definition of relative-risk used in epidemiology is :

$$\rho = \frac{\text{Prob } \{I_2 \text{ is affected } | I_1 \text{ is affected } \}}{\text{Prob } \{I_2 \text{ is affected } | I_1 \text{ is normal } \}},$$

Penrose (1953) argued that the recurrence risks (numerator of λ) should be contrasted against the population prevalence in a genetic context. The risk parameters λ and ρ are of course functionally related as:

$$\rho = \frac{(1-\delta)\lambda}{1-\delta\lambda},$$

where δ denotes the prevalence of the disease in the population (Majumder et al., 1983). Since data on bilineal relatives are easier to collect than unilineal relatives, we (Weiss et al., 1982; Majumder et al., 1983) had investigated the behavior of λ for various types of bilineal relatives (sib pairs, uncle/aunt-nephew/niece pairs, first-cousin pairs, etc.) and found that the statistic does not behave well even for simple genetic models (single-locus dominant or recessive, with or without incomplete penetrance and sporadic cases). However, in a recent paper, Risch (1990) discovered some interesting and useful properties of λ for pairs of unilineal relatives for both single-locus and multi-locus models. Risch's (1990) computations are based on a study by James (1971) who showed that :

$$K_R = \text{Prob } \{I_2 \text{ is affected } - I_1 \text{ is affected } \}$$
$$= \delta + (1/\delta)Cov(X_1, X_2),$$

where I_1 and I_2 are relatives of "degree" R, $X_i = 1$ if I_i is affected; $= 0$, otherwise $(i = 1, 2)$. James (1971) also derived $Cov(X_1, X_2)$ in terms of the additive and dominance variances. Using these formulas, Risch (1990) showed that for a single-locus model, the values of $\delta^2(\lambda - 1)$ for various unilineal relative pairs are :

$$
\begin{aligned}
\text{Parent – Offspring} &= \tfrac{1}{2}V_A \\
\text{Grandparent – Grandchild} &= \tfrac{1}{4}V_A \\
\text{Great grandparent – Great grandchild} &= \tfrac{1}{8}V_A, \text{ and so on.}
\end{aligned}
$$

In other words, the parameter $\lambda - 1$ decreased by a factor of 2 for each decreasing degree of unilineal relationship. When multiple loci are involved, these formulas still hold if penetrances of the disease-causing alleles at individual loci additively determine the penetrances for multi-locus genotypes. [For details, see Risch (1990).] Even so, the values of $(\lambda - 1)$ for a given degree R are, in practical situations, quite different under single-locus and multi-locus models. Therefore, λ can be used for the purpose of discriminating among models or even in deciding whether there is a strong genetic contribution in the determination of the disease phenotype. While this approach is not as powerful as segregation analysis in discriminating among models, it is still useful in practice for the study of complex disorders such as schizophrenia (Risch, 1990).

ACKNOWLEDGMENTS

Thanks are due to Mr. R.N. Das for some computational help and to Mr. S.K. Nath for useful discussion.

REFERENCES

Besag, J., 1974, Spatial interaction and the statistical analysis of lattice systems, *J. Roy. Stat. Soc.*, Ser. B, 36 : 192.

Boehnke, M., Young, M.R., and Moll, P.P., 1988, Comparison of sequential and fixed-structure sampling of pedigrees in complex segregation analysis of a quantitative trait, *Am. J. Hum. Genet.* 43 : 336.

Cannings, C., Skolnick, M.H., De Nevers, K., and Sridharan, R., 1976, Calculation of risk factors and likelihoods for familial diseases, *Comp. Biomed. Res.* 9 : 393.

Cannings, C., Thompson, E.A., and Skolnick, M.H., 1978, Probability functions on complex pedigrees, *Adv. Appl. Prob.* 8 : 622.

Cassella, G., and George, E.I., 1992, Explaining the Gibbs sampler, *Amer. Stat.* 46 : 167.

Dempster, A.P., Laird, N.M., and Rubin, D.B., 1977, Maximum-likelihood from incomplete data via the EM algorithm, *J. Roy. Stat. Soc.*, Ser, B. *39 : 1.

Efron, B., 1979, Bootstrap methods : Another look at the jackknife, *Ann. Stat.* 7 : 1.

Elston, R.C., and Stewart, J., 1971, A general model for the genetic analysis of pedigree data, *Hum. Hered.* 21 : 523.

Gelfand, A. E., Hills, S.E., Racine-Poon, A., and Smith, A.F.M., 1990, Illustration of Bayesian inference in normal data models using Gibbs sampling, *J. Am. Stat. Assocn.* 85 : 972.

Gelfand, A.E., and Smith, A.F.M., 1990, Sampling-based approaches to calculating marginal densities, *J. Am. Stat. Assocn.* 85 : 398.

Gelman, A., and Rubin, D., 1991, An overview and approach to inference from iterative simulation, Tech. Rep., Dept. of Statistics, Univ. of California - Berkeley.

Geman, S., and Geman, D., 1984, Stochastic relaxations, Gibbs distributions and Bayesian restoration of images, *IEEE Trans. Pattern Anal. Machine Intelligence* 6 : 721.

Ghosh, J.K., and Sen, P.K., 1985, On the asymptotic performance of the log likelihood ratio statistic for the mixture model and related results, *in :* "Proceedings of the Berkeley Conference in Honor of Jerzy Neyman and Jack Kiefer, Vol.2," L.M. LeCam and R.A. Olshen, ed., Wadsworth, Monterey, p. 789.

Guo, S.W., and Thompson, E.A., 1992, Monte-Carlo estimation of mixed models for large complex pedigrees, Tech. Rep. No. 229, Dept. of Statistics, Univ. of Washington, Seattle.

Hartigan, J.A., 1985, A failure of likelihood asymptotics for normal mixtures, *in :* "Proceedings of the Berkeley Conference in Honor of Jerzy Neyman and Jack Kiefer, Vol. 2," L.M. LeCam and R.A. Olshen, ed., Wadsworth, Monterey, p. 807.

Hasstedt, S.J., 1982, A mixed-model likelihood approximation on large pedigrees, *Comp. Biomed. Res.* 15 : 295.

Hasstedt, S.J., and Cartwright, P., 1970, "PAP : Pedigree Analysis Package," Tech. Rep. No. 13, Dept. of Medical Biophysics and Computing, Univ. of Utah, Salt Lake City.

Hastings, W.K., 1979, Monte Carlo sampling methods using Markov chain and their applications, *Biometrika* 87 : 97.

James, J.W., 1971, Frequency in relatives for an all-or-none trait, *Am. J. Hum. Genet.* 35 : 47.

Kimberling, W.J., Smith, S.D., Ing, P.S., and Tinley, S., 1989, A comment on the analysis of families with prelingual deafness, *Am. J. Hum. Genet.* 45 : 157.

Lange, K., and Matthysse, S., 1989, Simulation of pedigree genotypes by random walks, *Am. J. Hum. Genet.* 45 : 959.

Li, C.C., 1953, Some general properties of recessive inheritance, *Am. J. Hum. Genet.* 5 : 269.

Li, C.C., 1987, A genetical model for emergenesis, *Am. J. Hum. Genet.* 41 : 517.

Lykken, D., 1982, Research with twins : the concept of emergenesis, *Psychophysiology* 19 : 361.

Majumder, P.P., Chakraborty, R., and Weiss, K.M., 1983, Relative risks of diseases in the presence of incomplete penetrance and sporadics, *Stat. Med.* 2 : 13.

Majumder, P.P., Das, S.K., and Li, C.C., 1988, A genetical model for vitiligo, *Am. J. Hum. Genet.* 43 : 119.

Majumder, P.P., Ramesh, A., and Chinnappan, D., 1989, On the genetics of prelingual deafness, *Am. J. Hum. Genet.* 44 : 86.

Majumder, P.P., and Nath, S.K., 1992, Statistical analysis of family data on complex disorders in man, *J. Genet.* 71 : 89.

Marazita, M.L., Nance, W.E., and Arnos, K.S., 1989, Further comments on the genetics of prelingual deafness, *Am. J. Hum. Genet.* 45 : 637.

Metropolis, N., Rosenbluth, A.W., Rosenbluth, M.N., Teller, A.H., and Teller, E., 1953, Equation of state calculations by fast computing machines, *J. Chem. Phys.* 21 : 1087.

Morton, N.E., 1982, "Outline of Genetic Epidemiology," S. Karger, Basel.

Morton, N.E., 1991, Genetic epidemiology of hearing impairment, *Ann. NY Acad. Sci.* 630 : 16.

Morton, N.E., and MacLean, C.J., 1974, Analysis of family resemblance. III. Complex segregation analysis of quantitative traits, *Am. J. Hum. Genet.* 26 : 489.

Ott, J., 1979, Maximum likelihood estimation by counting methods under polygenic and mixed models in human pedigrees, *Am. J. Hum. Genet.* 31 : 161.

Risch, N., 1990, Linkage strategies for genetically complex traits. I. Multilocus models, *Am. J. Hum. Genet.* 46 : 222.

Ritter, C., and Tanner, M.A., 1992, Facilitating the Gibbs sampler : The Gibbs stopper and the Griddy-Gibbs sampler, *Jour. Amer. Stat. Assocn.* 87 : 861.

Schork, N., and Schork, M.A., 1989, Testing separate families of segregation hypotheses : Bootstrap methods, *Am. J. Hum. Genet.* 45 : 803.

Tanner, M.A., 1991, "Tools for Statistical Inference," Springer- Verlag, New York.

Thompson, E.A., and Guo, S.W., 1991, Evaluation of likelihood ratios for complex genetic models, *IMA J. Math. Appl. Med. Biol.* 8 : 149.

van der Genugten, B.B., 1977, Identification in statistical inference, *Statist. Neerlandica* 31 : 69.

Wald, A., 1941, Asymptotically most powerful tests of statistical hypotheses, *Ann. Math. Stat.* 12 : 1.

Weiss, K.M., Chakraborty, R., Majumder, P.P., and Smouse, P.E., 1982, Problems in the assessment of relative risk of chronic disease among biological relatives of affected individuals, *J. Chron. Dis.* 35 : 539.

Williams, D.A., 1970, Discussion of "A method for discriminating between models," by A.C. Atkinson. *J. Roy. Stat. Soc.* 32 (B) : 350.

HALDANE AND THE ANALYSIS OF LINKAGE

J.H. Edwards

University of Oxford
South Parks Road, Oxford
England

INTRODUCTION

Haldane described the first clear linkage in mammals and estimated the first linkage in man. He also developed the theory of mapping functions and introduced the terms the morgan and centimorgan, and the approach to linkage, based on the relative likelihoods of trial values of recombination fractions, which is now established. These contributions are discussed against the background of the mendelian controversy and the recent foreground of shotgun linkage.

THE BACKGROUND TO MENDELISM

Haldane's life (1892 - 1964) spanned the whole period of modern genetics from the revival of Mendelism to the decipherment of the genetic code. The appreciation of Mendel's work followed publications by de Vries in French and German and by Correns in German and, later, by Bateson who arranged a translation into English in the Transactions of the Royal Horticultural Society, to which he appended footnotes. His book, 'Mendel's Principles of Heredity, A Defence', was published two years later (Bateson, 1902). Punnett, his colleague and successor at Cambridge, wrote a small and lucid book 'Mendelism' which was to undergo many editions (Punnett, 1905). In the twenty years following the rediscovery of Mendelism in 1900 this basic concept of genetical atomism became established. That is, the first part of 'Mendel's law of the pea', which dealt with allelic independence and random gametic allocation, now termed segregation, or Mendel's first law, when related to a single locus, applied to many characters in many species and had the makings of a universal law. The extraordinary opposition to this 'first law', as it is now known, led Bateson, Punnett and others to give priority to establishing the near universality of this law against the united opposition of Karl Pearson and the biometricians, who founded their own journal, *Biometrika*, for the protection and nurturing of evolution by gradualism. In this they had the support of the editor of *Nature*, an astronomer, who ensured mendelian papers were not published. Hardy's short paper of 1908 on the stability of allelic proportions was published in the USA in *Science*. It was against this background that the interest in what is now known as 'Mendel's second law', that pairs of alleles from different loci were assorted at random, and the interpretation of exceptions to it which constitute linkage, took second place.

Human Population Genetics, Edited by P.P. Majumder
Plenum Press, New York, 1993

Advances in the understanding of linkage were further blocked in England by the aversion of both Punnett and Bateson to the chromosomal model: they were influenced by Boveri and considered cut-and-join events could not take place regularly without disrupting these presumably delicate structures of unknown function: the first chiasmata were not seen until 1909. This prolonged and often bitter controversy on the first law, without which the study of exceptions to his second law could hardly flourish, was finally resolved in 1918 by Fisher (Fisher, 1918) who demonstrated that the atomism of mendel sufficed for both gradual and, as Bateson termed it, saltatory, evolution.

This was published by the Royal Society of Edinburgh and financed by Leonard Darwin, who paid for the type to be set while awaiting formal acceptance, after rejection by the Royal Society of London whose referees were Punnett and Pearson. It introduced the concepts of variance, of its analysis, and of proportional representation as a measure of heritability. This controversy on the applicability of Mendelism, by now mainly restricted to a numerate rearguard, was largely resolved for those who could follow Fisher's mathematics.

Against this background of organised, and even institutionalised, scientific disagreement on the relevance of 'one law for the pea' to other plants and animals the apparently leisured advance of linkage in England, as opposed to America, is hardly surprising. In this same twenty years the chromosomal theory of inheritance was advanced (Sutton, 1902, 1903), the stability of allelic frequencies clarified (Hardy, 1908), the linear model of co-transmission of loci confirmed in the fruit-fly on substantial data (Sturtevant, 1913), the first clear case of linkage in mammals described (Haldane et al., 1915) and the relationship of recombination rate to chromosomal position analysed and the terms morgan and centimorgan defined (Haldane, 1919). Shortly afterwards Fisher published the procedure for the most likely position of loci given their order and pairwise recombination fractions (Fisher, 1922).

These two decades closed with the founding of the Genetical Society at a meeting in the Royal Society in June 1919 followed shortly by its first meeting in Cambridge with 34 members. In 1924 there were 108 members, including one 'involved in medical research'. The third decade opened with its second meeting and a symposium on the 'Heredity of sex in plants and animals' when Haldane described what is now known as his rule. The minutes recorded 'if one sex is absent or infertile in a species cross, it is the one which from other evidence may be inferred to be heterozygous in sex' (Lewis, 1969).

By 1922 the basic procedures for the detection of linkage and its estimation under the conditions of field, farm and laboratory had been defined. A satisfactory, if erroneous, model of two-strand cross-over was widely accepted, and a nomenclature with such words as linkage, interference, recombination fraction and centimorgan established. Morgan, his flies, and some of his group, attended the fourth meeting of the Genetical Society in 1922. Morgan spoke on 'Mutants in *Drosophila*', and Sturtevant on the comparison of *melanogaster* and *simulans*. The first six meetings were chaired by Bateson. 'At about 1924 chromosomes became a feature of the meetings' (Lewis, 1969). Throughout this time Bateson, who had documented disturbed cosegregation in numerous species, dominated the development of mendelism, and introduced the words *genetics, allelomorph, homozygote* and *heterozygote*, continued to doubt the simple model implicit in the chromosome theory of inheritance. Punnett, who spent part of the the war years 1914-1918 improving poultry production by developing genetic markers for sexing chicks at birth, had accepted the chromosomal theory earlier.

Bateson, like Darwin, was one of the great innumerates whose deep insights were unobscured by formal reasoning based on abstract models, but, whose 'profound knowledge of the history of science led him to doubt the validity of long chains of reasoning, however convincing' (Haldane, 1930). In the same essay Haldane described how he could 'remember the time when Mendelism was considered grossly heretical at Oxford'.

It should be appreciated that Mendel only claimed to have a 'law for the pea', and he had to test many peas before he found a strain sufficiently law-abiding. An associated assumption, implicit in his terms dominant and recessive, and later embalmed in Bateson's nomenclature of allelomorph, now contracted to allele, was that genes only had two forms. This seemed inconsistent with other species. Mice, which Mendel had bred, produced colours of fur and eye which did not fit so simple a model. Various plants were equally lawless. Mendel kept bees, and although his notes are lost it seems likely he was confused by their odd segregation due to drones being haploid.

In 1923, three years before Bateson died, E B Ford described his conversion. (Ford, 1979). He wrote:
'One day I happened to be in London and chanced to meet Leonard Darwin, son of Charles Darwin, in Piccadilly. He said 'Bateson is back from America, I have just seen him in the Athenaeum.' 'What did he say about *Drosophila* researches at Columbia?' 'He was reading and I did not disturb him' said Darwin, 'but it would be interesting to know. Let us go back and get his views.' After greeting him and asking him if he believed in the chromosome theory of heredity he replied 'Yes I do and all my life's work has gone for nothing'. He retained his 'elasticity of mind' until his death' (Haldane, 1930). Twenty years earlier he and Punnett had observed the cosegregation which Sutton had predicted (Sutton, 1902,1903).

In 1958 I was invited by Dr Howard Levene to visit the Genetics Department, the successor to Wilson's department, in Columbia University in New York. It was there that Sutton had seen synapses at meiosis just before Morgan arrived in 1903. I set off in pursuit of the original 'fly room' now in another department. It was remarkably small with walls and tables covered with eyes - a postgraduate in art history, who did not know of the room's past, explained that she was working on the eye in Picasso and had cut out eyes from photographs of all his works and was measuring them. This was the room in which Morgan's group had first seen the white-eyed fly almost exactly ten years after Bateson had read Mendel. It was their first mutant after many hundreds of thousands of flies had been bred in milk-bottles filched from doorsteps and reared on spoiled bananas. This fly, with its many million progeny, transformed our views of genetics, aroused Haldane's interest in linkage analysis and eventually converted Bateson to the chromosomal theory. Bateson was a connoisseur of art, though not modern art, and obsessively opposed to the misapplication of mathematics. Later Haldane was coauthor of a paper on a white-eyed mutant in a mosquito, in this case linked to the heterogametic sex chromosome.

HALDANE AND LINKAGE THEORY

In 1917 Haldane was sent to convalesce in Simla after being wounded in Mesopotamia. He was later stationed at Mhow in central India, where he instructed on explosives, and then at New Delhi. He was aged 26 and while in India read the papers from Morgan's 'fly-group'; his first papers on the mathematics of linkage, presumably written mainly in India, were published after his return to Oxford (Haldane, 1919a, 1919b).

In 1922 Haldane moved to Cambridge to become Reader in Biochemistry under Gowland Hopkins. Bateson had moved from Cambridge in 1912 to direct the John Innes Horticultural Institution at Merton near London but Punnett remained as his successor. Darlington joined Bateson in 1922, initially unpaid, and started his great work on chromosomes, eventually graduating to a salary. Looking with one eye and drawing with the other, through a second-hand microscope - even then of antique design - purchased for him he transformed, and partly created, the science of cytogenetics. This was the first microscope at the John Innes (Lewis, 1983). Morgan's laboratory defined their variants with a jeweller's eyeglass, their most powerful instrument. The giant chromosomes of some insect larvae remained unrecognised for over fifty years to be rediscovered in 1933 (Balbiani, 1881; Painter, 1933).

Darlington was joined, on a part-time basis, by Haldane in 1927 after he had gained some notoriety by first being defeated by, and then defeating, the 'sex vires', the six men entrusted with the morality of the Fellows of Trinity   It was in this extraordinary concentration of talent in both Merton and Cambridge that Haldane's abilities to describe in words and to elaborate in numbers, with his compulsion to combine extensive algebraic manipulations of a sometimes superfluous exuberance with approximations bounded by common sense, developed. Of Haldane's great contemporaries in mathematical genetics Wright was a more profound naturalist and Fisher a deeper mathematician, but Haldane had a wider, if secondary, knowledge of facts and an extraordinary range of general knowledge which he could integrate with clarity of thought and word. It is doubtful if anyone has read the printed version of his many algebraic elaborations of numerous terms in infinite series except the printer. He was later to pun in Greek on their contributions as Tectonic and Halieutic (Haldane, 1963).

It was against this background of exposure to Darlington, who combined microscopy, draughtsmanship and clear prose, and Bateson, who, in what Haldane has termed a 'lapidary phrase', practised his scientific directive 'treasure your exceptions' that Haldane developed his continuing, if subsidiary, interest in relating genes to chromosomes. Haldane, like Bateson, concentrated on 'Mendel's first law' and developed most of his evolutionary models on the assumption that Mendel's second law held - that is that the segregation of alleles at pairs of loci was independent as Mendel had observed in pursuing his seven loci in the edible pea, now known to have seven pairs of chromosomes.

He supported this priority in his last major paper, first delivered in Orissa, his 'Defence of Beanbag Genetics' (Haldane, 1963), which is basically the genetics of a mass of genes unrestrained by both the constraints of chromosomes and the confusion of sex.

Haldane's first paper on linkage (Haldane, Sprunt & Haldane, 1915), described the first clear example in mammals, extending earlier work (Darbishire, 1904), which Weldon had directed in hopeful anticipation of disproving Mendelism. This included colour variants in mice at the albino and pink-eye dilution loci, which in mice, though not in man, are linked.   The third author was his sister Naomi, later Naomi Mitchison.   The first paragraph includes the ominous phrase 'one of us has already been killed in France'; Haldane was then an officer in the Black Watch, a famous Highland regiment, where enthusiasm for solitary nocturnal raids across the lines earned him a personal tribute from Marshal Foch.   Their title was 'Reduplication in mice' following the conventional explanation that unequal re-combination, a word then hyphenated, was due to post-meiotic duplication. The nature of albinism is now finally resolved at the DNA level (Spritz et al., 1990) while pink-eyed dilution was only resolved just in time for his centenary (Lyon et al.,1992), and, even more recently, shown to be homologous with tyrosinase-positive albinism. These loci are involved in the two common, and not easily distinguished, forms of albinism in man.

Haldane's father, J S Haldane FRS, with whom he collaborated from childhood, was also a Fellow of New College. The warden, Spooner, was an albino, and well known for the spoonerisms he always denied having made.

In 1919 Haldane wrote his paper on 'The combination of linkage values, and the calculation of distances between the loci of linked factors' and on efficient estimates. (Haldane, 1919a,1919b). Following the results in the fruit-fly (Morgan et al., 1915; Morgan and Bridges, 1916) he assumed a simple cut-and-join two-strand model of meiosis, and considered the imaginary gametic output of an individual heterozygous for three linked loci. The alleles, A ,B,C and a,b,c at loci ABC were defined by upper and lower case following Mendel, but the case referred to parental origin rather than allelic nature.

He argued that if the distribution of cross-over events were random, then the Poissonian distribution of events between two loci x units apart could be split into odd and even terms, and only the former would lead to an evident recombinant.

The odd numbers form the sequence of proportions

$$\exp(-x)(x/1! + x^3/3! + x^5/5! \dots)$$

which he noted was $\exp(-x)\sinh(x)$, so that the probability of an odd number of cross-overs, which is the recombination fraction (y) in the two-strand model, is related to the physical distance (x) by

$$y = (1 - e^{-2x})/2$$

and

$$x = -\ln(1 - 2y)/2 \quad \text{which expands to} \quad ((2y)/1 + (2y)^3/3 + (2y)^5/5 \dots)/2$$

He suggested that a suitable measure of distance would be the 'morgan', 'on the analogy of the ohm, volt etc'. which was defined in terms of centimorgan. In Haldane's words 'Morgan's unit of distance is therefore a centimorgan'.

He tabulated and analysed 36 triads of loci published by Morgan and Bridges (Morgan and Bridges, 1916), and showed that the relationship between structural and functional distance was not consistent with a random distribution, the value relating predicted to observed recombination fractions lying between the predicted value and the value of a single randomly distributed event when $x = y$. This he interpreted in terms of 'rigidity' of the chromosome discouraging neighbouring crossovers, and introduced an empirical compromise function of

$$x = 0.7y + 0.3(\ln(1 - 2y)/2)$$

Kosambi (Kosambi, 1944), a mathematician from Pune (Poona) with wide interests, provided a more elegant but equally empirical solution by omitting half the terms $((2y)^3/3, (2y)^7/7, \dots)$ in the expansion above, which is equivalent to

$$x = (\tanh^{-1}(2y))/2$$

This had little cytological justification but the irrelevant support of being equivalent to the addition rule for velocities in relativity and the practical advantage of being identical to Fisher's normalising transformation of the correlation coefficient (Fisher, 1921) which was tabulated. Later Carter and Falconer (Carter and Falconer, 1951) modified it further to conform better to data on the mouse by omitting another half of the expansion, their result having terms in $1, 7, 11$ etc., which is equivalent to

$$x = (\tan^{-1}(2y) - \tanh^{-1}(2y))/4$$

which is not easily inverted or related to mechanism.

It is now clear that recombination involves four strands, that every tetrad has at least one recombinant event, that their distribution differs along the chromosome, and also differs by sex, and probably by age, as Haldane was later to show in poultry (Haldane and Crew, 1925). Attempts to impose biological reality into mapping functions, either through following Haldane and weighting several of these functions (Rao et al., 1977) or devising other functions, are not simple (Sturt, 1976). As the following table shows there is little point in using any but the simplest and there are advantages in retaining Haldane's mapping function for general use.

Distance in centimorgans (w) of various values of the recombination percentage, theta

Theta	H_w	$w-H_w$	K_w	$w-K_w$
5	5.27	0.27	5.02	0.02
10	11.16	1.16	10.14	0.14
15	17.83	2.83	15.48	0.48
20	25.54	5.54	21.18	1.18
25	34.66	9.66	27.47	2.47
30	45.81	15.81	34.66	4.66
35	60.20	25.20	43.37	8.37
40	80.47	40.47	54.93	14.93
45	115.13	70.13	73.61	28.61

w, Hw and Kw refer to the direct function (w = y) and the functions of Haldane and Kosambi respectively.

Over distances of less than 10 centimorgans or so the even simpler function theta = w is sufficient. It appears, at any rate in man and mouse, that as Haldane had surmised on his analogy of rigidity, interference is a short-range effect but the various elaborations of his mapping function only relate to segments exceeding about ten centimorgans in length.

In human data the errors in assuming the possibility of the rare event of neighbouring chiasmata, whose extreme rarity is now well documented cytogenetically in the male in man, mouse and fly and consistent with recent data from the female *Mus spretus* hybrid, probably exceed those from assuming this happens without restraint. (Kosambi's function, which is usually used, only appreciably restrains long-range interference, while Haldane's imposes no restraint.) This has the advantage that, in linkage involving codominant loci, or rare dominants in segments of less than 10 centimorgans or so, it is possible to replace unreliable inference by uncertain deduction and work directly through the examination of haplotypes. This also bypasses the problem of assuming an identity of gene frequencies within and beyond the tested families, an assumption now known to be capable of errors of inference of two orders of magnitude in forensic applications. This assumption is liable to even greater error if imposed on close linkage, where the ancillary assumptions of allelic independence and the unrelatedness of the mutant alleles in all the families ascertained by disease, and the further assumption, now grossly at variance with recent data, that all alleles have the same mutation rate, impose grave problems of assessing the closeness of the approximations involved.

While errors of this magnitude may be trivial in forensic use when odds ratios of a thousand-million to one can be elaborated by simple erroneous inference, such errors can lead to problems in linkage analysis as they have in tuberose sclerosis (Edwards, 1991) and appear to be doing in Huntington's chorea.

Haldane's purely theoretical studies included the consequences of inbreeding in linkage, accepting the challenge of Jennings who stated the he would 'find it a relief if someone else would deal thoroughly with the laborious problem' (Jennings, 1917). This was one of his more exuberant algebraic papers (Haldane and Waddington, 1931).

The intellectual challenge of linkage in polyploids also attracted him, and later Fisher. He included some reference to linkage among his numerous papers applying mathematics to inheritance. While some of these, often almost asides, were largely mathematical extravaganza which at least made clear that the problems were not amenable to simple analytical solutions, and that many were beyond the formal limits of analysis, some laid the foundations for later approaches.

HALDANE AND HUMAN LINKAGE

Two papers on linkage in humans, both on colour blindness and haemophilia, had a profound significance, while a third, one of his last, initiated a subject which I believe to be of major importance although a victim of both neglect and misunderstanding.

In 1937, Julia Bell and Haldane made the first estimate of linkage in man (Bell and Haldane, 1937). The chromosome was not in doubt, but it is now known that the X chromosome is of sufficient length for most pairs of loci to be outside the range of detectable cosegregation. In fact the loci they studied are now known to be very close and the recombination they observed was almost certainly due to the recombinant family not having the classical haemophilia, now known as haemophilia A, (factor VIII deficiency), but Christmas disease or haemophilia B, (factor IX deficiency), a clinically indistinguishable condition about a fifth as common. They analysed six families.

However, the rigour of their analysis - even the explicit mention of assuming non-mosaicism in a potential founder member - and the novel directness of the approach later extended by Morton (1955), as well as their advancing the possibility of using linkage to predict genetic status before disease was manifest, give this paper profound significance.

In their analysis they defined a model, with explicit assumptions on germinal mosaicism (assumed absent, although 'if the mutation only occurs at meiosis ' ... ' if it arises at a somatic division, mosaics may exist'), segregation (even), gene frequency (known approximately for colour blindness, and for the haemophilia gene assumed to be very rare in normal males and in females not related to a haemophiliac through an all-female line), mutation rates (too low for two distinct mutants in one family to be likely). They omitted the possibility of serially related enzymes having indistinguishable consequences. This is surprising, as Haldane had major expertise in enzymes and made many contributions to their dynamics, including a substantial book. Even now, over fifty years later, it is common to assume the null hypothesis of locus homogeneity in testing sets of data, rather than to assume heterogeneity and estimate its extent. A few months ago the assertion of a single locus in a form of neurofibromatosis was made on the basis of three informative families all showing similar linkages.

Having fixed the model, excluded unlikely possibilities, and defined the task 'to find out whether its (the recombination fraction's) estimated value differed significantly from 1/2, its value in the absence of linkage, and on the other hand to obtain the best estimate of its true value.' They proceed to define the various likelihood functions for these families, using the simplification of, in his terms x for recombination and y for cosegregation $(x + y = 1)$, but even this led to products of up to the power of 14. Having fixed the biological model they then defined the *a priori probabilities* assuming a length of the human X chromosome of a morgan - following their only guide - the fly - on which they assembled data on 3828 locus pairs. The human X is now known to be over two morgans long. After this combined feat of logic and algebra, with the good documentation of their data, the authors concluded:

'The linkage here investigated is so close that on quite a small amount of material it is possible to demonstrate its existence without leaving grounds for reasonable doubt. If such close linkage is typical of the species, the search for further cases will be an easier task than appeared probable at first sight, and its results will be of considerable practical value.

The present case has no prognostic value, since haemophilia can be diagnosed before colour blindness. If, however, to take a possible example, an equally close value were found between the genes determining blood group membership and that determining Huntington's chorea, we should be able, in many cases, to predict which children of an affected person would develop the disease, and to advise on the desirability or otherwise of their marriage.'

It is surprising that he did not appreciate the prognostic value of advising women who were potential carriers of haemophilia. This is now one of the major, and more reliable, uses of linkage analysis to predict carriers of mendelian disease.

Ten years later Haldane, with C A B Smith, who succeeded him as Professor of Biometry when he went to India, returned to this problem. They clarified the logical status of their bayeian approach, documenting their treatment with a clear diagram, and advancing the use of a range based on likelihoods without reference to any underlying assumption of a prior distribution (Haldane and Smith, 1947).

Later his Ph.D student Renwick, in analysing the data which he and Lawler had collected on the nail-patella syndrome and the ABO locus, these approaches were used to demonstrate the feasibility of iterative procedures based on tailor-made likelihood functions fitted to each pedigree (Renwick and Lawler, 1955). Later Renwick was able to compute likelihoods for any defined set of parameters with a precision limited only by time. So far as I know no tailor-made approach has been made for any set of data of this magnitude, and, since Ott's generalisation (Ott, 1974), most problems can now be accommodated with little loss of information. When electronic computers became available Renwick was using the Martin Aircraft Company's computer in Baltimore and his program required the simultaneous use of up to 12 tape decks.

HALDANE AND FALSE CLUES TO LINKAGE

In 1955 Haldane and Carter drew attention to the need to allow for the number of markers used to detect a linkage in relation to the probability of a correct inference (Haldane and Carter, 1955). They introduced the useful term 'false clues', and gave as a worked example the use of 25 markers and a level of significance defined by two standard error, or about 2.5% for the one tail appropriate for linkage detection. They wrote 'The probability of a false clue, that is to say, the probability that a new mutant will appear to be linked with any particular marker is therefore 0.2275. The probability that at least one of the twenty five markers will give a false clue, is $1 - 0.97725^{25}$ or 42.7 %. However, this large risk is justified because if 500 progeny are scored, the probability of discovering a linkage is over 50%'.

Notwithstanding the apparently simple nature of the issues involved, their exact treatment by Thompson (1984), and the likelihood analogue of Haldane and Carter's approach by Kidd and Ott (1984), the need to allow for this was ignored by Botstein et al. (1980) in their recipe for future action and denied by Lander (1988) in his recommendations for the generalised shotgun approach and its extension to non-Mendelian disorders. Various other descriptions of shotgun genetics, a term in no way intended to denigrate this powerful approach to problems beyond the reach of direct studies through lack of guidance of mechanism, or access to relevant tissues, have followed in ignoring this problem. However its resolution is necessary if the logic of the single-shot, with a likelihood ratio of 1000:1, the conventional 'rule of three' for lods favouring linkage advanced when there is an *a priori* expectation against linkage of 50:1, advanced by Morton (1955) in the context of sequential analysis. While this is no longer a part of the usual explicit formal procedures the basic argument holds. In fact most of us are closet sequential analysts as 'stopping rules' in practice depend on results: the resulting biases, leading to underestimation of genetic distance, are greatly amplified by an editorial preference for positive results.

However, we are not concerned with the minor adjustments which exact treatment could offer, but with matters of orders of magnitude now that many shotgun approaches involving hundreds of shot are in use and, if correctly analysed, the information available since the number of loci related to a defined phenotype increases with the number of loci while the number of locus pairs increases as almost the square of this number: the cost of collection and analysis per locus decrease steadily with the number of loci studied. It has even been suggested (Lander, 1988) that, if there appears to be a hit, a narrower cone of shot should be fired at this region by typing for a set of closely linked neighbours. Obviously, if the target is the same, this should not be used as a basis for an independent measure of credulity. This is an adequate explanation of at least one 'false clue' in the analysis of the Amish manic-depressives. Even if the algebra comes and goes, a claim of obviously remarkable cosegregation, even if an inappropriate analysis offered a false clue, should in no way be used to undervalue the importance of so remarkable a family if studied by appropriate methods.

Ott (1991) does not accept this logical distinction between single-shot and shotgun inferences, but the matter is at least worthy of discussion, as is anything Haldane expressed in words with or without a simple worked example. If Carter and Haldane are right this is not a minor matter. Upon its clarification depends whether the various and numerous recipes for action now being executed on both mendelian and, with greater prospect of erroneous inference, on other disorders, are founded on bedrock or sand. Sand may be all we have, but at least it is possible either to invest in solid foundations or in small and simple structures.

Ott states, after mentioning an earlier simple approach (Kidd and Ott, 1984), 'An important aspect of multiple comparisons which has thus so far been neglected is that an increase in the number of markers increases the prior probability of linkage. This effect and the one discussed in the previous paragraph work in opposite directions and approximately cancel each other out ...' (Ott, 1991). I find a logical error in this. The prior expectation of a linkage with one of the markers tested will obviously increase with their number. However, the prior expectation of any particular marker being linked will not. Ott assumes, if I understand his argument, that in a set of markers the one with the highest likelihood of linkage will be the one which is linked. In fact, it will merely be the one most likely to be linked. The matter is not new but the resources in blood, gold and talent now being invested on the assumption that Carter and Haldane were in error are large and are increasing at the cost of other approaches.

While this small aside of Haldane and Carter may be important, the paper in which it appeared was of major significance. It introduced the concept of the 'radius' of a marker in relation to some event - in this case a recessive lethal - whose presence might be manifest through a relative lack of some phenotypes defined by this marker. Later they stated 'we must reckon with the possibility that some lethal genes may increase fitness in the heterozygote and will therefore spread in outbred populations.' We are an outbred population, and a good scattering of recessive lethals, provided they acted very early in fetal life, would both reduce heterozygosity in those surviving birth, and also the detrimental effect of recessives mild enough to reach birth but detrimental after birth. The shortage of cousin marriages in some disorders, including ataxia telangiectasia and spinal muscular atrophy, is certainly suggestive of at least two 'black loci', or groups of such loci, within the range or 'radius' of these disease loci. Smith, in an earlier paper, had stressed the advantages which the children of cousin marriages can offer in studies of linkage (Smith 1953).

HALDANE'S INFLUENCE ON LINKAGE ANALYSIS

The major contribution of Haldane, who just missed the opportunity to use computers with their ability to undertake in a second what he could achieve in a morning, but could hardly be expected to tolerate the boredom of doing repeatedly even if he had time, was to set the foundations for linkage analysis with a realistic example and extend this approach ten years later. Fisher, with Finney, Mather, Bailey and others had attempted to resolve the problems in the intervening decade by attempting to maximize likelihood algebraically assuming that the implicit large-number approximations allowed the use of weighted estimators to be applied to human sibships without excessive error after summations involving many sibships. Without electronic aid this was the only practical approach but the approximations are considerable and now unnecessary. It needed the empirical likelihood approach of Haldane and Smith for this to be clarified by Morton.

Haldane's contribution to the use of likelihoods in linkage is not simple. He used a bayesian approach throughout except in his use of confidence intervals when, if the criterion of a likelihood, as opposed to a bayesian, interpretation is that the measure of confidence in choosing between the alternative values of parameters is invariant to any transformation of their values, then Haldane and Smith have some claims to priority.

By avoiding analytical maximum likelihood approaches and using likelihoods, with the option of empirically finding a maximum, Morton (1955) was able to classify the majority of informative situations in human sibships and use the early models of computer to tabulate 'z scores' and provide a method of analysis which, compared to the labour of collecting the data, was not onerous. In its original formulation it was a sequential analysis based on prior assumptions of the nature of chromosomes and recombination frequencies. He (Morton,1956) extended this approach, demonstrating the first heterogeneity in linkage, although this had almost certainly been found unknowingly by Bell and Haldane twenty years earlier.

Morton's tables were responsible for the bulk of the linkage analyses produced in the twenty years between their publication and the widespread use of Ott's generalisation to pedigrees of arbitrary complexity. With thought and care Morton's tables allowed information to be transferred between parts of straggling or incomplete families, and in the hands of Cook at the Galton Laboratory and of Race and Sanger in their analysis of an X-linked blood-group locus, most families could be analysed efficiently. Haldane left England shortly after Morton's papers and, so far as I am aware, had little opportunity to acquire linkage data in India, although, with Jayaker (Haldane and Jayaker, 1962) he wrote on the symbolic representation of human relationships.

We need another paper of substance to either consolidate the foundations of much recent work, or, if faulty, to rebuild the foundations or lighten the superstructure. As Fisher pointed out, linkage is so difficult a subject that we must constantly check the mathematics involved against reality. We now have a discrepancy of the order of two between the length of much of the human genome based on microscopy at meiosis and the length inferred from various methods of multilocus mapping and automated ordering.

EPILOGUE

The first papers on formally considering the efficiency and precision of estimates of linkage, and on the problem of mapping points defined by mutual relationship onto a map defined by a few firm landmarks - initially only the ends of the chromosomes - were conceived and, almost certainly largely written, in New Delhi in 1918 (Haldane, 1919a, 1919b). Now, 75 years later, these are the fundamental problems in mapping the human genome, in so far as linkage data are to be integrated with physical mapping. With little solid and original contribution to this difficult problem since Morton's treatment of the general problem in 1955 and its generalisation by Ott in 1974, using an approach advanced by Elston and Stewart in 1976, there has been little advance beyond extension to include multiple parameters and numerous markers allowing several moderately close loci with limited allelic variability to provide joint information, a situation now becoming rarer with the increasing numbers of multiallelic markers. However the logical foundations sufficient to support the simpler initial models are not necessarily firm enough to support recent developments and the study of foundations after they have been built on is notoriously difficult. Nor are we always on firm ground in using such procedures to predict the probability of individuals being carriers of alleles which are not expressed, an application first suggested by Bell and Haldane.

The extent to which Haldane was responsible for the application of likelihood methods to linkage analysis is confused by the casualness of recent authors, including myself, in the precise use of such terms. The position is clearly stated in a review of a definitve work: 'For here, within the compass of a single well-studied biological problem, one finds likelihood methods, Bayesian methods, significance tests, maximum-likelihood estimation, and confidence intervals all muddled up topgether' (Edwards, 1987). The matter is further confused as Haldane and Smith considered only a single chromosome and Morton initiated his approach within the framework of sequential analysis, redefining lods to the base of common logarithms and using the term 'score' which had an established and distinct usage in maximum-likelihood estimation. Morton's approach was advanced within a non-sequential framework by Smith (1953), and is now rarely used as an explicitly sequential procedure. It is at least clear that Haldane and Smith (1947) used the term maximum likelihood and considered a confidence range based on a likelihood diagram, with likelihood on the vertical axis, in a way which was invariant to any transformation of the recombination fraction which defined the horizontal axis.

Haldane wrote that the mathematics of evolution was 'too mathematical to interest most biologists and not sufficiently mathematical to interest most mathematicians. It is reasonable to suppose that in the next half century it will be developed into a responsible branch of mathematics.' That was over fifty years ago. Unfortunately, in so far as linkage is of importance in both accelerating evolution and providing data for its decipherment, in this prediction he seems to have been in error in spite of, or possibly because of, the ease with which computers can be both used and misused. The development of computerised procedures which are simple enough to use without the need to understand them, and the complete lack of any repository of raw data relevant to linkage in man, in distinction to the mouse, and, very soon, the pig and the sheep, whose assured paternity and documented grandparentage does not impose such difficult problems as human data, does not give confidence for the future. When Haldane has his bicentenary most of these problems, which were introduced by Haldane's early papers, and were probably written in India, may have been solved and the insoluble subset defined. But they will not have been solved unless they are recognised as existing, and of sufficient interest and importance to attract mathematicians of the stature of Haldane. There is plenty of room at the top. Firm foundations can be built without elaborate equipment, and can be tested initially on imaginary data which small computers are so well equipped to produce, until such time as raw data on linkage in man are readily accessible. As Haldane might have pointed out, in this field India already has all the necessary resources except data.

ACKNOWLEDGEMENTS

I am particularly indebted to the several books about Haldane and his work by Dronamraju, to the Royal Society obituary (Pirie, 1966) and to Dr A G Searle for discussions

REFERENCES

Bailey, N.T.J., 1961, "Introduction to the Mathematical Theory of Linkage."
  Oxford: Clarendon.
Balbiani, E.G., 1881, Sur la structure du noyau des cellules salivaires les larves de
  Chironomus., *Zool Anz* IV:637.
Bateson, W, 1902,. "Mendel's Principles of Heredity." A Defence. Cambridge:
  Cambridge University Press.
Correns, C., 1900, G. Mendel's Regel ueber das erhalten der Nachkommenschaft der
  Rassenbastarde., *Ber. Deutsch. Bot. Gesellsch.* xvii:158.

Edwards, A.W.F., Review of Analysis of Human Genetic Linkage by Ott, J, *Biometrics,* 43:254.

Edwards, J.H., 1990, The linkage detection problem, *Ann. Hum. Genet.*54:253.

Elston, R.C., and Stewart, J., 1976, A general model for the analysis of pedigree data, *Hum. Hered.* 21:523.

Fisher, R.A., 1918, The corelation between relatives on the supposition of Mendelian inheritance, *Trans. Roy. Soc. Edinburgh* 52:399.

Fisher, R.A., 1921, On the probable error of a coefficient of correlation deduced from a small sample, *Metron* 1:1.

Fisher, R.A., 1922, The systematic location of genes by means of crossover observations, *American Naturalist* 56:406.

Ford, E.B., 1979, "Understanding Genetics", Faber and Faber. London.

Haldane, J.B.S., 1919a, The combination of linkage values and the calculation of distances between linked factors , *J. Genet.* 8:299.

Haldane, J.B.S., 1919b, The probable errors of calculated linkage values, and the most accurate methods of determining gametic from certain zygotic series, *J. Genet.* 8:291.

Haldane, J.B.S., and Jayakar,S.D., 1962, An enumeration of some human relationships, *J. Genet.* 58:81.

Haldane, J.B.S., Sprunt, A.D., and Haldane, N.M., 1915, Reduplication in mice, *J. Genet.* 5: 133.

Haldane, J.B.S., 1930, "Possible Worlds and other essays". Chatto and Windus, London.

Haldane, J.B.S., 1934, Methods for the detection of autosomal linkage in man, *Ann. Eugen.* 6:26.

Haldane, J.B.S., 1934, The relative efficiency of two methods of measuring human linkage, *American Naturalist* 68:286.

Haldane, J.B.S., 1963, A defence of beanbag genetics, *Perspectives in Biology and Medicine* 7:343.

Haldane, J.B.S., and Smith, C.A.B., 1947, A new estimate of the linkage between the genes for colour-blindness and haemophilia in man, Ann. Eugen. 46:85.

Mendel, 1866, Versuche ueber Pflanzen-Hybriden, Verh. Naturforsch. *Ver. Bruenn* 4:3.

Morgan, T.H., and Bridges, C., 1916, Sex linked inheritance in Drosophila, *Carnegie Institute of Washington* 21.

Morton, N.E., 1955, Sequential tests for the detection of linkage, *Amer. J. Hum. Genet.* 7:277.

Morton, N.E., 1956, The detection and estimation of linkage between the genes for elliptocytosis and the Rh blood group, *Am.J.Hum.Genet,* 8:80.

Ott, J., 1974, Estimation of the recombination fraction in human pedigrees, *Am.J.Hum.Gen.* 26:588.

Ott, J., 1991, "Analysis of Human Genetic Linkage". Baltimore: Johns Hopkins University Press.

Painter, T.S., 1933, A new method for the study of chromosome rearrangements and the plotting of chromosome maps , *Science* 78:585.

Haldane J.B.S. and Jayaker.S.D. 1962, An enumeration of some human relationships, *J.Genet.*58:81.

Smith, C.A.B., 1953, The detection of linkage in human pedigrees, *J. Roy. Stat. Soc.* 15B:153.

Sturt, E., 1975, A mapping function for human chromosomes, *Ann. Hum. Gen.* 39:266.

Sturtevant, A.H., 1913, The linear arrangement of six sex-linked factors in Drosophila, *J. Exp. Zool.* 14:43.

Thompson, E.A., 1984, Information gain in joint linkage analysis, *IMA J. Math. Appl. Med. Biol.* 1:31.

RECENT DEVELOPMENTS IN THE THEORETICAL ASPECTS OF LINKAGE ANALYSIS

Jurg Ott[1]

[1]Department of Genetics and Development
Columbia University, Unit 58
722 West 168th Street
New York, NY 10032

INTRODUCTION

The phenomenal pace of development of the genome project in recent years has been most beneficial to the field of linkage analysis. Many exciting theoretical and practical developments would not have been possible without the advances in molecular genetics. Presumably, the prospect that a complete human linkage map of approximately 2 cM resolution will soon be available will also have far reaching effects on society. In the not too distant future much of the genetic make-up of an individual may be recognized at birth. It is timely and demonstrates Haldane's foresight that he envisioned this situation a long time ago. In an article entitled *The future of biology*, written before 1927, he wrote of what are now called genetic markers and predicted that

> "we could use them, by a method worked out on flies by Morgan of New York and his associates, as landmarks for the study of such characters as musical ability, obesity, and bad temper. When a baby arrived we should have a physical examination and a blood analysis done on him, and say something like this: 'He has got ... from his father, so it's twenty to one that he will get the main gene that determined his father's mathematical powers; but he's got ... from his mother ... so it looks as if her father's inability to keep away from alcohol would crop up in him again; you must look out for that.'"

Haldane seems to have seen predominantly positive sides of these prospects for he goes on to say that "when that day comes intelligent people will certainly consider their future spouses' hereditary make-up, and the possibility of bringing off a really brilliant combination in one of their future children..."

Since this article is not on matters of philosophy or ethics, I will not dwell on these aspects of modern genetics. But I thought it enlightening that Haldane considered these aspects of genetics more than 65 years ago when less than a handful of genetic markers were available.

Below, I would like to outline a review of recent developments in linkage analysis. In particular, I will be focussing on five areas in which progress has been most notable: Likelihood formulation (a somewhat technical look at ways of calculating pedigree likelihoods), Simulation methods, Data errors, Approximate analysis methods, and Models for complex traits.

LIKELIHOOD FORMULATION

Haldane is probably most well-known for his pioneering achievements in theoretical population genetics. However, he has also done remarkable work in linkage analysis. To my knowledge, the first authors presenting multigenerational pedigree likelihoods with respect to gene frequencies, recombination fractions, and mutation rates were Bell and Haldane (1937) in their analysis of hemophilia versus color blindness (for a historical review of linkage analysis, see Ott, 1991). These authors were ahead of their time. It was only much later, sixteen years after Morton (1955) introduced the concept of lod scores for the routine analysis of nuclear families, that a general formulation of pedigree likelihoods became available (Elston and Stewart, 1971; see also Elston et al., 1992). That work paved the way for the development of generally available computer programs for linkage analysis. The LIPED (Ott, 1974) and LINKAGE programs (Lathrop et al., 1984) are both based on the Elston-Stewart algorithm and are the programs used most by researchers in human linkage analysis.

While LIPED is restricted to analyzing two loci at a time, the LINKAGE programs can in principle handle any number of loci. In practice, due to limited computer memory and execution speed, no more than five or six loci can be currently by analyzed jointly. In human gene mapping, one generally analyses a large number of codominant loci in specific two-generation or three-generation families, the so-called CEPH (for Centre d'Etude du Polymorphisme Humain) reference families. For this situation, a clever method of factorizing the multilocus likelihood was devised (Lathrop et al., 1986), which permits consideration of many more loci jointly than with an unmodified Elston-Stewart algorithm. This factorization technique has been implemented in a specialized version of the LINKAGE programs (CILINK, CMAP, etc.) and works very well.

Another route for representing and calculating pedigree likelihoods was taken by Lander et al. (1987). Instead of recursively calculating the likelihood over individuals as is done in the Elston-Stewart algorithm, these authors carry out recursion over loci (for a review, see Weeks, 1991) and implemented this procedure in the MAPMAKER computer program. A similar algorithm has been used in the CRIMAP program (Green, personal communication). In the likelihood representation of Lander et al. (1987), the number of individuals in a pedigree is a limiting factor while in the Elston-

Stewart algorithm, the limiting quantity is the number of loci. Therefore, the MAP-MAKER program is limited to analyzing codominant loci in CEPH families. Also, the MAPMAKER and CRIMAP programs do not generally calculate pedigree likelihoods exactly but take certain shortcuts, which is part of the reason for their high execution speed.

In the LIPED and LINKAGE programs, a pedigree loop (eg. due to consanguinity) is handled by a technique originally proposed by Lange and Elston (1975). A newer program for likelihood calculation in general pedigrees, MENDEL (Lange et al., 1988), handles loops differently. It carries out Elston-Stewart type likelihood calculations as long as possible, that is, it works on one individual at a time until it reaches the members forming a loop for which the required calculations are then carried out simultaneously. For the user, the MENDEL and LINKAGE programs are comparable in their capabilities. MENDEL is more flexible in the types of analyses it can carry out but it tends to require more computer time and memory than LINKAGE.

In recent years, researchers have turned to formulating pedigree likelihoods in such a way that they are suitable for implementation on parallel computers. An example is the report by Goradia et al. (1992) discussing the statistical issues involved in vectorizing likelihood calculations. If parallel computers really prove to be become successful and much faster than other computers, many linkage programs will have to be adapted to this hardware.

Exact calculations of multipoint pedigree likelihoods have until recently always been carried out under the assumption of no interference. While this assumption is generally considered to be unrealistic, it was for technical reasons that absence of interference has been assumed — gamete probabilities for an ordered set of loci are then simply the product of the (marginal) recombination probabilities in each interval between loci (Ott, 1991). Many map functions incorporate specific assumptions on interference; however, most of these functions are not "multilocus-feasible" (Liberman and Karlin, 1984), that is, they may predict negative gamete probabilities. With a suitable map function, a transformation permits calculation of gamete probabilities from recombination fractions under interference (Liberman and Karlin, 1984). This method has been implemented in the CILINK program of the LINKAGE package and is suitable for analyzing codominant genetic loci for gene mapping under interference in CEPH families (Weeks et al., 1993). When this method is applied to six DNA marker loci on chromosome 10, Weeks et al. (1993) find significant evidence in favor of positive interference as modelled by the Sturt map function.

SIMULATION METHODS

Simulation (Monte Carlo) methods are computer-based techniques to emulate a particularly complicated phenomenon or process such that its properties may be studied under "laboratory" conditions. In human genetics, a process often simulated is the flow of genes in a family from parents to offspring. One may then simulate on the computer the observation of phenotypes and study the properties of a particular analysis technique or how well an anticipated linkage can be detected on the basis of a given number and structure of families. Reliable and efficient simulation methods have become important tools in statistical genetics, particularly in the analysis of complex traits for which the mode of inheritance is largely unknown. Below, simulation methods are discussed, which focus on generating genotypes at a marker locus linked with a disease locus, given known disease phenotypes.

Two such methods of computer simulation have been introduced in recent years. The first method, developed by Boehnke (1986), originally assumed that the disease genotype for all individuals must be known. The relevant conditional probabilities of marker genotypes given disease genotypes are then easy to derive and have been implemented in a computer program called SIMLINK. An improved version of this method works by first generating disease genotypes given disease phenotypes and then, in a second phase, generating marker genotypes given the generated disease genotypes (Ploughman and Boehnke, 1989).

The second simulation method is based on "risk" calculations for the marker genotypes of an individual, given known disease phenotypes, perhaps known marker genotypes on some of the family members, and the already generated marker genotypes (Ott, 1989). This method, implemented in the SLINK program (Weeks et al., 1990), is more general than the one implemented in SIMLINK in that it allows for partial marker typing at the locus to be simulated, which is not possible with SIMLINK. On the other hand, SIMLINK executes faster than SLINK for a given problem.

The simulation methods implemented in SIMLINK and SLINK are typically applied to determine the usefulness for linkage of a set of m pedigrees collected before marker typing is begun. For the simulation, one must assume a situation of linkage one is likely to encounter in the future, for example, a marker locus with four equally frequent alleles linked to the trait locus with a known (true) recombination fraction of 2%. One then generates a number n of replicates, that is, marker genotypes given the observed disease phenotypes. These replicates are analyzed as if each of the n sets of m families had been observed in reality. Three statistics are generally obtained from the n replicates to assess informativeness for linkage: 1) The ELOD, which is the expected lod score at the true recombination fraction under which the data were generated; it has the nice property that it is additive over families. 2) The power, which estimated as the proportion of replicates with a maximum lod score exceeding some constant such as 3; it has a direct probabilistic interpretation. 3) The expected maximum lod score, that is, the average over replicates of the maximum lod score in each replicate, is also often used; it has neither a direct probabilistic interpretation nor is it additive over families but it is always as large or larger than the ELOD, which be one of the reasons for its popularity with researchers.

Simulating marker genotypes under no linkage is much easier than under linkage because, with a true recombination fraction of 50%, the marker genotypes are independent of disease phenotypes. This type of simulation is used to estimate the p-value, that is, the empirical significance level associated with an observed maximum lod score. Traditionally, the maximum lod score achieved in a linkage analysis is used directly as a measure for evidence for linkage and linkage is declared significant when the maximum lod score is equal to or exceeds the value of 3. However, in nontraditional approaches to linkage analysis, researchers sometimes carry out various analyses on the same data, each time varying some genetic parameters or the disease classification such that they effectively maximize the lod score over model parameters, which inflates the lod score and makes its interpretation difficult (Weeks et al. 1990). It is predominantly in this situation that computer simulation under no linkage but under the maximization technique applied in reality provides a reliable significance level where a p-value of 0.001 is equivalent to a maximum lod score of 3 in fully informative data.

The p-value is estimated as the proportion of replicates with a maximum lod score as large or larger than the one observed in reality. Because p may be quite small, a large number (several thousand) of replicates may be required for a reasonably accurate estimate of the p-value. Importance sampling, a technique well known in sampling theory (Hammersley and Handscomb, 1983) but, until recently, largely unknown in human genetics, can then be applied (Ott and Terwilliger, 1992; Kong et al., 1992). It works by simulating under linkage such that, ideally, about half of all replicates show a maximum lod score exceeding the desired cutoff point; to compensate for simulating under linkage rather than under no linkage, a correction factor is applied to the probability of occurrence of each replicate.

One of the problems with the current methods of computer simulation is that they are often extremely slow, particularly when consanguinity loops and untyped individuals are present in a family and/or several marker loci should be generated. A new method of computer simulation, chromosome-based simulation, is currently being developed, which will be able to rapidly generate a large number of marker loci for given phenotypes at a disease locus and can also take into account interference (Speer et al., 1992; Terwilliger et al., 1993). It works by generating crossover points along a chromosome and then randomly assigning marker genotypes given the crossover points and the disease phenotypes. This method will be useful in the context of locating a disease gene on a known map of the human genome. For example, to track down a disease gene, one initially carries out two-point analyses until a marker with an interesting maximum lod score is found. Then the neighborhood of that marker is investigated to see whether the initial lod score increases or decreases. One of the questions in this context is: What constitutes an interesting lod score? What critical limit should alarm a gene mapper to investigate the neighborhood of a marker? This is the kind of questions that can be addressed with chromosome-based computer simulation.

In addition to the approaches discussed above, other simulation methods have been developed, with the main purpose of carrying out fast approximate rather than slow exact linkage analyses. These methods will be referred to below, under Approximate Methods.

DATA ERRORS

For many years, linkage analyses were carried out without much regard to errors in identifying phenotypes at trait or marker loci; very few authors considered the possible effects of errors and the effects of neglecting errors in the analysis. In an early paper, errors were treated in the framework of misclassifying recombinants as nonrecombinants and vice versa and it was shown that disregarding in the analysis the misclassification errors leads to a consistent overestimation of the recombination fraction (Ott, 1977).

On the basis of marker typing on a total of close to 2,000 individuals in 48 pedigrees from a South Pacific island population, Lathrop et al. (1983) carried out a comprehensive investigation of various sources of pedigree error. From the observed distribution of genetic marker inconsistencies between parents and offspring, they estimated that the error of a paternal link was 4%, that of a maternal link was zero, and the overall marker typing error was 1%.

A few years later, a brief abstract pointed out that in mapping a trait locus to a map of markers, assuming wrong lengths of map intervals leads to inconsistent estimates of the map position of the trait, where too small a map interval can lead to wrong locus orders while too large a map interval tends to cause smaller inconsistencies (Ott and Lathrop, 1987).

It has only been for a very short time that more attention is being paid to the effects of data errors, particularly in genetic marker typing and data transcription. Even a small percentage of errors can have severe effects, that is, reduced informativeness of the data and, thus, power to order loci, and overestimation of map lengths (Terwilliger et al., 1990; Buetow, 1991).

Errors leading to mendelian inconsistencies are, of course, easy to recognize because they lead to a zero likelihood of the pedigree data. They are, however, not necessarily easy to pinpoint, that is, it may be difficult to see which individual in a pedigree is responsible for the inconsistency. A method has recently been developed to pinpoint inconsistencies in pedigrees (Ott, 1992b). In this method one calculates, for each individual with marker observations in a pedigree, the sum of squared differences, SS, between two arrays of predictions (conditional probabilities), 1) the genotype probabilities, $P(g \mid x)$, given that individual's phenotype, x, and 2) the genotype probabilities given all phenotypic information, F, in the pedigree. Because of the mendelian inconsistencies, these conditional probabilities must be computed under incomplete penetrance, for example, with a penetrance structure as shown in table I (see below). In the absence of pedigree errors, one expects only random deviations between the two predictions and, thus, a smooth unimodal distribution of SS = $\Sigma_g[P(g \mid x)\text{-}P(g \mid F)]^2$. Any outlying, unusually high SS values are indicative of errors. In a real case of a large pedigree, the method pointed out the correct individual as the "culprit", later identified by molecular genetics methods as a case of nonpaternity (Ott, 1992b).

After obvious errors are eliminated, laboratory errors in genetic marker determination not detectable as mendelian inconsistencies still seem to occur at an approximate frequency of around 1% or higher. Therefore, statistical methods have been and are being developed for error detection. The most straightforward approach is to look for double recombinants within a close distance. The rationale for this technique is that genotype errors often show up as apparent double recombinants while, because of positive interference, real double recombinants in close proximity of each other are extremely unlikely. The CRIMAP program (P. Green, unpublished) identifies recombinants along chromosomes and is therefore often used to identify possible genotype errors. Such events are then scrutinized in the laboratory and often turn out to be due to marker errors. A possible disadvantage of this method is that more attention is being paid to recombinants than nonrecombinants, which tends to bias recombination fraction estimates towards smaller values. However, this disadvantage is certainly more than offset by the possibly substantial gain in accuracy due to elimination of genotype errors. In practice, this technique has proven to be most useful.

After a reasonable effort has been made to detect as many errors as possible, in most pedigree data, a low error rate presumably persists. One may then simply ignore the possibility of errors and proceed with data analysis assuming absence of errors, which is the usual approach taken by researchers. Another possibility is to allow for errors via incomplete penetrance. In codominant inheritance, with full penetrance, the genotypes, $g = 1/1$, $1/2$, etc., lead to the associated phenotypes, $x = 1/1$, $1/2$, etc., with probabilities (penetrances) $P(x \mid g) = 1$ for $x=g$ and $P(x \mid g) = 0$ for $x \neq g$. A simple error structure may now be postulated (Keats et al., 1990) by assuming penetrances $P(x \mid g) = 1-\epsilon$ for $x=g$, and $P(x \mid g) = \epsilon/(m-1)$, where m is the number of genotypes at the locus under consideration and the value of ϵ is either estimated from the data or fixed at a reasonable value (Ott 1992b). Table I shows an example of such a penetrance structure for a locus with three alleles ($m = 6$ genotypes) and $\epsilon=0.05$. It turns out, however, that calculating pedigree likelihoods under incomplete penetrance is many orders or magnitude more time consuming than with full penetrance. The reason for this is that under incomplete penetrance, formerly codominant markers are no longer codominant and the special techniques for rapid calculation of pedigree likelihoods mentioned above are no longer applicable. In practice, therefore, this method is rarely applied but it has been used in specialized situations (Lincoln and Lander, 1992).

For highly polymorphic marker loci, population allele frequencies are usually estimated on the basis of a random sample of unrelated individuals. Because of problems of identifiability of alleles between families, allele frequencies are often taken to be equal to $1/n$, where n is the number of alleles. In

Table I Example of a penetrance structure allowing for errors at a codominant locus

Genotype	Phenotype					
	1/1	1/2	1/3	2/2	2/3	3/3
1/1	.95	.01	.01	.01	.01	.01
1/2	.01	.95	.01	.01	.01	.01
1/3	.01	.01	.95	.01	.01	.01
2/2	.01	.01	.01	.95	.01	.01
2/3	.01	.01	.01	.01	.95	.01
3/3	.01	.01	.01	.01	.01	.95

linkage analyses for marker mapping, using possibly wrong allele frequencies does not seem to pose any particular problems. However, in linkage analyses between disease and marker loci, with incomplete penetrance at the disease locus and absence of marker typing for some of the pedigree members, falsely assuming marker allele frequencies of $1/n$ tends to provide strong false positive evidence for linkage or, if there is linkage, tends to inflate the maximum lod score (Ott, 1992c). Consequently, special efforts must be taken to ensure that in highly polymorphic markers, specific alleles can be recognized across families (Knowles et al., 1992).

After a linkage analysis between a disease and some markers has been carried out one might be interested in finding those individuals in a pedigree whose disease phenotype is crucial for the lod score obtained. This question has been addressed by Xie and Ott (1990) who developed a computer program, which flips affection status for one individual after another n a pedigree. Each time, the effect of the change in affection status is measured by the associated change in maximum lod score. Those individuals whose change in affection status causes a strong change in the lod score are then particularly scrutinized. Such investigations are now known as sensitivity analysis (Hodge and Greenberg, 1992) and are recommended particularly after linkage analyses of complex traits in which changes of affection status of some key individuals have dramatically altered the results of previously published linkage studies.

APPROXIMATE ANALYSIS METHODS

For many chromosomes, progress in molecular genetics and efficient linkage strategies has furnished dense maps of genetic markers, and it is only a question of perhaps a year until 1-2 cM maps for all chromosomes will be available. For disease loci, progress in genetic mapping methods has not quite kept up with this pace or, to put things another way, the problem of disease gene mapping is much more complicated than that of marker mapping. Typically, in a pedigree of medium size without loops, a maximum of 4 to 6 marker loci can be analyzed jointly with a disease locus. The limiting factor is either computer memory or the time it takes to calculate likelihoods.

The introduction of highly polymorphic markers has exacerbated this problem. On the one hand, a large number of marker alleles is desirable as it tends to make more parents heterozygous and thus potentially informative for linkage. On the other hand, it is less computer intensive to handle markers with a small number of alleles. Clearly, in the presence of completely polymorphic markers flanking a trait locus, there is no need to look at additional markers further away from the trait locus because they do not contribute additional linkage information. Therefore, with highly polymorphic markers, a smaller number of flanking loci is required than when markers have low polymorphism.

Given the technical problems of carrying out extensive multipoint analyses of trait loci, the use of approximate analysis methods, at least at the initial stages of a project, may be more economical than always carrying out full likelihood analyses (Ott, 1992a). Disregarding complicating factors such as marker errors or uncertainties in marker map positions may be seen as approximations implicit in many of today's analyses. Here, however, methods specifically designed to lead to simplified and approximate analyses shall be discussed.

One of the oldest approximate methods consists of carrying out a multipoint analysis (trait and many marker loci) by appropriately combining the results of twopoint analyses (trait locus versus one marker locus at a time). A modernized version of this approach is that of Morton and Andrews (1989). Statistical properties of this approach are largely unknown except that it seems to be less affected by marker errors than are other methods (Buetow, 1991). On the other hand, with low levels of marker errors, map distances estimated by this method presumably are too short. A novel procedure for combining twopoint lod scores into an approximate multipoint lod score has recently been described (Curtis and Gurling, 1993). For the case of assigning a new locus to a map of marker loci, the procedure depends on calculating a lod score from each marker or pair of markers, beginning with those closest to the new locus, but in each case utilizing only that fraction of meioses for which other, closer markers may be expected to be uninformative. The method is designed such that adding the resulting lod scores will not introduce any systematic bias.

The most recent approaches to approximate analyses have applied computer simulation. Lange and Matthysse (1989) and Lange and Sobel (1991) approximate lod scores using a Metropolis algorithm, which allows the use of a large number of diallelic markers in addition to a disease locus. Thomas and Cortessis (1992) apply Gibbs sampling in a Bayesian framework, that is, they assumed prior distributions of parameters and "update" these to obtain posterior distributions, given observed data, where the updating is carried out via Gibbs sampling of genotypes of family members. These methods are restricted to trait loci governed by a single disease gene — the common assumption in linkage analysis. For quantitative traits, a computer simulation method for estimating segregation and linkage parameters has been proposed by Guo and Thompson (1992). These authors employ Gibbs sampling for jointly estimating the recombination fraction and other parameters such as the variance due to additive polygenic effects. All current Monte Carlo methods work properly only when at most two alleles are present at each locus investigated.

An important aspect of Monte Carlo approximation methods is that in addition to the "normal" variability of the parameter estimates, one also has variability due to the randomness of the method. In other words, estimates of parameters such as the recombination fraction deviate from the true parameter value, where the likely deviation between the two can be assessed by the estimated variance of the estimate. In simulation methods, one also estimates (approximates) the estimate, and the likely discrepancy between true and approximated estimates are assessed by yet another layer of variability. In the few applications thus far reported, however, the variability due to the approximation method was ignored, with the approximated estimates being taken as the true estimates, perhaps because the Monte Carlo methods in current use do not easily furnish estimates of their variability.

Approximate analysis methods may serve as a substitute for a full analysis when a thorough analysis is just not computationally feasible. But even when a full analysis is technically feasible, approximate techniques may be applied in the form of preliminary investigations, for example, in the genomic search of a disease gene location. When promising areas have been identified, these can then be followed up by complete analyses. One nice aspect of approximate Monte Carlo methods is that the accuracy of the results can be varied such that optimal strategies can be planned.

COMPLEX TRAITS

Many genes responsible for well-defined single-gene diseases have been mapped, eg. that for cystic fibrosis, various genes for retinitis pigmentosa (Humphries et al., 1992), etc. The great challenge remaining for genetic analysis is now posed by the so-called complex traits, that is, diseases with unclear mode of inheritance and unclear disease definition. Examples are schizophrenia and other psychiatric traits.

In linkage analyses of complex traits, one usually assumes a single major disease locus, but joint action of a small number of disease loci is generally thought to be more realistic (Risch, 1990). In linkage analysis between disease and a single marker, falsely assuming a single-locus model when in fact the disease is due to more than one locus tends to retain much of the linkage information, but the recombination fraction estimate can be expected to be strongly biased towards larger values (Risch et al., 1989); it is only in special cases of using a wrong model that linkage may be mimicked (recombination fraction underestimated) when in fact no linkage exists (Terwilliger et al., 1991). Thus, the practice of using single-locus models for complex traits has been vindicated by theoretical investigations for various situations, but this is generally true only for twopoint analysis, not for multipoint analysis (Risch et al., 1989). In multipoint analysis, the tendency for inflated recombination fractions can be expected to lead to false negative results.

Many pertinent methodological problems in the linkage analysis of complex traits have been investigated by a MacArthur Foundation task force (Spence et al., 1993), for example, secular trends, assortative mating, and bilineal pedigrees (dominant disease entering a pedigree from two sides). The general conclusion was that these factors can reduce the power of linkage analyses but do not necessarily invalidate it.

Instead of assuming a single disease gene, a two-locus disease model may be adopted in which two (generally unlinked) loci are jointly responsible for causing disease. The two loci may then act in such a way that both are required for disease (epistasis), or disease may result from one *or* the other locus (heterogeneity). A method for permitting epistatic interaction of two loci had earlier been presented by Ott and Falk (1982). It postulated a disease locus and a modifying locus and required knowledge of the phenotype at both the disease and modifier locus.

Generally, two-locus models refer to the situation that only one phenotype exists, which is under the control of two different loci. Special versions of the LINKAGE and MENDEL programs have recently been developed that allow analysis under such models. The possibility of analyzing simulated family data under single-locus or two-locus models has allowed to study the relative merits of the two analysis models. It turns out that for family data generated under two-locus models, if only one of the disease loci is linked to a genetic marker, analysis under a single-gene model is almost as efficient as analysis under the correct two-locus model (Vieland et al., 1992; Dürner et al., 1992). However, when each of the two disease loci (unlinked with each other) are linked with a marker locus, two-locus analysis can provide substantially more linkage information as measured by the expected maximum lod score than standard one-trait-locus, one-marker-locus methods. On the other hand, when the two disease loci each are dominant and act in a heterogeneity-type fashion, one-locus models that allow for heterogeneity extract essentially as much information as the two-trait-locus methods (Schork et al., 1993). For such "dominant or dominant" models, one also finds that bilineal pedigrees provide sufficient linkage information to warrant their inclusion in a given linkage study.

Neuman and Rice (1992) studied theoretical properties of nine specific two-locus models. They developed formulas for the recurrence risk to five classes of relatives of the

Table II Five epistatic two-locus models (f = penetrance; A, a = alleles at locus 1; B, b = alleles at locus 2) considered by Neuman and Rice (1992)

	Ep-1 BB	Bb	bb	Ep-2 BB	Bb	bb	Ep-3 BB	Bb	bb	Ep-4 BB	Bb	bb	Ep-5 BB	Bb	bb
AA	f	f	0	f	f	0	f	0	0	f	f	0	f	f	0
Aa	f	f	0	0	0	0	0	0	0	f	0	0	f	0	0
aa	0	0	0	0	0	0	0	0	0	f	0	0	0	0	0

proband. For various rates of population prevalence and risk to sibs, they derived the possible range of penetrances and gene frequencies in each model. Such theoretical investigations are prerequisites for a good understanding of the statistical properties of two-locus models and, thus, for a meaningful practical application of them. Table II shows five two-locus epistatic models investigated by Neuman and Rice (1992). For example, each of the two loci in the Ep-1 model act in a dominant manner. As pointed out by these authors, it is interesting to note that for several parental matings, none of the parents is at risk but they may produce offspring with at-risk genotypes. An example is the mating AAbb×aaBB and the offspring genotype AaBb.

Two-locus disease models were developed with the hope that they are closer to the true state of nature for complex traits than single-locus models and that these more realistic models should make it easier to localize disease loci by linkage analysis. Therefore, these models should more or less accurately predict population parameters such population prevalence, K_p, and risk to a sibling, K_{sib}. It turns out that some fixed values of K_p and K_{sib} are compatible with only a very narrow range of penetrances. Table III shows some examples.

Considering several population parameters in addition to the two shown in table III, Neuman and Rice (1992) conclude that Ep-2, Ep-3, and Ep-5 are suitable mod-

Table III Possible ranges of the penetrance, f (in %), for given population prevalence, K_p, and risk to sib, K_{sib} (Neuman and Rice, 1992)

K_p	K_{sib}	Ep-1	Ep-2	Ep-3	Ep-4	Ep-5
.001	.10	20–35	20–55	40–80	20–45	20–55
.01	.10	20–25	20–35	30–40	20–30	20–35

els for schizophrenia for which K_p is somewhat less than 0.01 and $K_{sib} \approx 0.10$. Thus far, only very few linkage analyses have been carried under two-locus models. One of the technical problems with analyses under these models is that they require much more computer time than under single-locus models.

In diseases with a known mode of inheritance, negative linkage results are useful in the sense that they can exclude regions of the genome as possible sites of the disease locus. In complex traits, such exclusions depend on the mode of inheritance assumed for the trait; for a given set of family data, the exclusion region may vary strongly with the assumed mode of inheritance. Therefore, for complex traits, the practice of "exclusion mapping" for complex traits is generally unreliable, and only positive linkage results are meaningful.

Multiple testing between disease and a number of marker loci is not a problem in mendelian diseases, because the disease gene will sooner or later be found if only one searches long enough. The prior probability of linkage increases with an increasing number of markers tested, which more than offsets the increased type I error (Ott, 1991). For complex traits, however, these considerations do not apply because presence of a major gene responsible for the trait is generally not known but only hypothesized. Therefore, allowance must be made for the number of different markers investigated for linkage with the trait. A simple formula is to increase the critical limit of 3 to a new critical limit, $3 + \log_{10}(m)$, where m is the number of markers tested (Kidd and Ott, 1984; Ott, 1991). For example, after 100 markers have been tested, the critical lod score for significant linkage should be 5 rather than 3.

A few cases of successful linkage investigations involving complex traits are known. The most prominent examples are breast cancer for which in some families a gene on chromosome 17 has been identified (Hall et al., 1990), a form of diabetes (MODY) for which a gene on chromosome 7 was shown to be responsible (Vionnet et al., 1992), and melanoma for which a gene was assigned to chromosome 9 (Cannon-Albright et al., 1992). In psychiatric genetics, on the other hand, several initial reports of high lod scores eventually turned out to be false positive findings and at this point there is no clear evidence for a single gene influencing psychiatric traits such as schizophrenia or affective disorder. It is disturbing that lod scores of magnitudes between 5 and 10 can turn out to be false positives. A few factors have been identified, which can inflate lod scores in the absence of linkage, notably searching for linkage evidence by applying multiple diagnostic classifications in the same data, and using wrong gene frequencies of genetic markers in the presence of untyped individuals. However, these effects are hardly sufficient to fully explain the high lod scores found. Another possible reason might be preferential selection of families or individuals, which are particularly indicative of linkage. Such selection may well occur without any intent to falsify data on the part of the responsible investigator. For this reason, it is of utmost importance that disease diagnosis and genetic marker determination be done completely independently. Rao (1989) discusses several cases of selectively ascertained data and presents methods of cross-examination of data, which should be applied more often than is currently done.

OUTLOOK

A discussion of recent theoretical developments in linkage analysis would be incomplete without trying to predict what the current trends might lead to in the future. Clearly, faster computers will solve many problems, which currently plague linkage analysts and force them to take short-cuts and apply approximate analyses. Improvements in algorithms are also expected to lead to an increase in efficiency of likelihood calculations. For example, the way pedigree loops are currently handled in the LINKAGE programs is somewhat inefficient. For each multilocus genotype of one of the individuals in a loop, a complete pedigree likelihood calculation is carried out. Instead, one might first "peel" branches of the pedigrees as close to the loop as possible and only afterwards handle the loops in the known fashion; this way, peeling of pedigree branches would have to be done only once. Several opportunities for such improvements in algorithms are relatively easy to spot, but it is time consuming to implement them in linkage programs. One of the problems is that even highly qualified programmers do not generally have sufficient training in statistical genetics to be able to make such program modifications. This also explains why programs for linkage analysis are generally not very user-friendly.

I also expect that approximate methods will become common knowledge such that screens for possible genomic locations of disease loci can be carried out rapidly. For complex traits, such screens might identify several loci jointly acting to produce disease, and a follow-up with powerful two-locus models will then verify or refute these tentatively identified locations. One of the limiting factors in finding loci for complex traits will presumably be the number of families available for study (Suarez et al., 1991).

ACKNOWLEDGMENTS

This work has been supported by grant HG00008 from the National Center for Human Genome Research, and NIH grant MH44292.

REFERENCES

Bell, J., and Haldane, J.B.S., 1937, The linkage between the genes for colour-blindness and haemophilia in man, *Proc. Roy. Soc. Ser. B* 123:119.

Boehnke, M., 1986, Estimating the power of a proposed linkage study: a practical computer simulation approach, *Am. J. Hum. Genet.* 39:513.

Buetow, K.H., 1991, Influence of aberrant observations on high-resolution linkage analysis outcomes. *Am. J. Hum. Genet.* 49:985.

Cannon-Albright, L.A., Goldgar, D.E., Meyer, L.J., Lewis, C.M., Anderson, D.E., Fountain, J.W., Hegi, M.E., Wiseman, R.W., Petty, E.M., Bale, A.E., Olopade, O.I., Diaz, M.O., Kwiatkowski, D.J., Piepkorn, M.W., Zone, J.J., and Skolnick, M.H., 1992, Assignment of a locus for familial melanoma, MLM, to chromosome 9p13-p22, *Science* 258:1148.

Curtis, D., and Gurling, H., 1993, A procedure for combining two-point into a summary multipoint map, *Hum. Hered.* (in press).

Dürner, M., Greenberg, D.A., and Hodge, S.E., Inter- and intrafamilial heterogeneity: effective sampling strategies and comparison of analysis methods, *Am. J. Hum. Genet.* 51:859.

Elston, R.C., and Stewart, J., 1971, A general model for the analysis of pedigree data, *Hum. Hered.* 21:523.

Elston, R.C., George, V.T., and Severtson, F., 1992, The Elston-Stewart algorithm for continuous genotypes and environmental factors, *Hum. Hered.* 42:9.

Goradia, T.M., Lange, K., Miller, P.L., and Nadkarni, P.M., 1992, Fast computation of genetic likelihoods on human pedigree data, *Hum. Hered.* 42:42.

Guo, S.W., and Thompson, E.A., 1992, A Monte Carlo method for combined segregation and linkage analysis, *Am. J. Hum. Genet.* 51:1111.

Haldane, J.B.S., 1987, The future of biology, *in*: "On Being The Right Size," J. Maynard Smith, ed. Oxford University Press, Oxford.

Hall, J.M., Lee, M.K., Newman, B., Morrow, J.E., Anderson, L.A., Huey, B., and King, M.C., 1990, Linkage of early-onset familial breast cancer to chromosome 17q21, *Science* 250: 1684.

Hammersley, J.M., and Handscomb, D.C., 1983, "Monte Carlo Methods," Chapman and Hall, New York.

Hodge, S.E., and Greenberg, D.A., 1992, Sensitivity of lod scores to changes in diagnostic status, *Am. J. Hum. Genet.* 50:1053.

Humphries, P., Kenna, P., and Farrar, G.J., 1992, On the molecular genetics of retinitis pigmentosa, *Science* 256:804.

Keats, B.J.B., Sherman, S.L., and Ott, J., 1990, Human Gene Mapping 10.5 - Report of the committee on linkage and gene order, *Cytogenet. Cell Genet.* 55:387.

Kidd, K.K., and Ott, J., 1984, Power and sample size in linkage studies, *Cytogenet. Cell Genet.* 37:510.

Knowles, J.A., Vieland, V.J., and Gilliam, T.C., 1992, Perils of gene mapping with microsatellite markers, *Am. J. Hum. Genet.* 51:905.

Kong, A., Frigge, M., Irwin, M., and Cox, N., 1992, Importance sampling. I. Computing multimodel p-values in linkage analysis, *Am. J. Hum. Genet.* 51:1413.

Lander, E.S., Green, P., Abrahamson, J., Barlow, A., Daly, M.J., Lincoln, S.E., and Newburg, L., 1987, MAPMAKER: an interactive computer package for constructing primary genetic linkage maps of experimental and natural populations, *Genomics* 1:174.

Lange, K., Weeks, D., and Boehnke, M., 1988, Programs for pedigree analysis: MENDEL, FISHER, and dGENE, *Genet. Epidemiol.* 5:471.

Lange, K., and Matthysse, S., 1989, Simulation of pedigree genotypes by random walks, *Am. J. Hum. Genet.* 45:959.

Lange, K., and Sobel, E., 1991, A random walk method for computing genetic location scores, *Am. J. Hum. Genet.* 49:1320.

Lathrop, G.M., Lalouel, J.M., Julier, C., and Ott, J., 1984, Strategies for multilocus linkage analysis in humans, *Proc. Natl. Acad. Sci. USA* 81:3443.

Lathrop, G.M., Lalouel, J.M., and White, R.L., 1986, Calculation of human linkage maps: likelihood calculations for multilocus linkage analysis, *Genet. Epidemiol.* 3:39.

Liberman, U., and Karlin, S., 1984, Theoretical models of genetic map functions, *Theor. Popul. Biol.* 25:331.

Lincoln, S.E., and Lander, E.S., 1992, Systematic detection of errors in genetic linkage data, *Genomics* 14:604.

Morton, N.E., 1955, Sequential tests for the detection of linkage, *Am. J. Hum. Genet.* 7:277.

Morton, N.E., and Andrews, V., 1989, MAP — an expert system for multiple pairwise linkage analysis, *Ann. Hum. Genet.* 53:263.

Neuman, R.J., and Rice, J.P., Two-locus models of disease, *Genet. Epidemiol.* 9:347.

Ott, J., 1974, Estimation of the recombination fraction in human pedigrees: efficient computation of the likelihood for human linkage studies, *Am. J. Hum. Genet.* 26:588.

Ott, J., 1977, Linkage analysis with misclassification at one locus, *Clin. Genet.* 12:110.

Ott, J., and Falk, C.T., 1982, Epistatic association and linkage analysis in human families, *Hum. Genet.* 62:296.

Ott, J., 1989, Computer-simulation methods in human linkage analysis, *Proc. Natl. Acad. Sci. USA* 86:4175.

Ott, J., 1991, "Analysis of Human Genetic Linkage," Johns Hopkins University, Baltimore.

Ott, J., 1992a, The future of multilocus linkage analysis, *Ann. Med.* 24 (in press).

Ott, J., 1992b, Detecting marker inconsistencies in human gene mapping, *Hum. Hered.* (in press).

Ott, J., 1992c, Strategies for characterizing highly polymorphic markers in human gene mapping, *Am. J. Hum. Genet.* 51:283.

Ott, J., and Lathrop, M.G., 1987, Estimating the position of a locus on a known map of loci, *Cytogenet. Cell Genet.* 46:674 (abstr).

Ott, J., and Terwilliger, J.D., 1992, Assessing the evidence for linkage in psychiatric genetics, *in*: "Genetic Research in Psychiatry," J. Mendlewicz and H. Hippius, eds. Springer, New York.

Ploughman, L.M., and Boehnke, M., 1989, Estimating the power of a proposed linkage study for a complex genetic trait, *Am. J. Hum. Genet.* 44:543.

Rao, C.R., 1989, "Statistics and truth," Council of Scientific and Industrial Research, New Delhi.

Risch, N., 1990, Linkage strategies for genetically complex traits. II. The power of affected relative pairs, *Am. J. Hum. Genet.* 46:229.

Risch, N., Claus, E., and Giuffra, L., 1989, Linkage and mode of inheritance in complex traits, *in*: "Multipoint Mapping and Linkage Based Upon Affected Pedigree Members: Genetic Analysis Workshop 6," R.C. Elston, M.A. Spence, S.E. Hodge, and J.W. MacCluer, eds. Liss, New York.

Schork, N.J., Boehnke, M., Terwilliger, J.D., and Ott, J., 1993, Two-locus linkage analysis: an efficient strategy for mapping complex genetic traits, (in preparation).

Speer, M., Terwilliger, J.D., and Ott, J., 1992, A chromosome-based method for rapid computer simulation in human genetic linkage analysis, *Am. J. Hum. Genet.* 51 (suppl):A202 (abstr).

Spence, M.A., Bishop, D.T., Boehnke, M., Elston, R.C., Falk, C., Hodge, S.E., Ott, J., Rice, J., Merikangas, K., and Kupfer, D., 1993, Methodological issues in linkage analyses for psychiatric disorders: secular trends, assortative mating, bilineal pedigrees (report of the MacArthur Foundation Network I Task Force on Methodological Issues), *Hum. Hered.* (in press).

Suarez, B.K., Van Eerdewegh, P., and Hampe, C.L., Detecting loci for oligogenic traits by linkage analysis, *Am. J. Hum. Genet.* 49 (suppl):14 (abstr).

Terwilliger, J.D., Weeks, D.E., and Ott, J., 1990, Laboratory errors in the reading of marker alleles cause massive reductions in lod score and lead to gross overestimates of the recombination fraction, *Am. J. Hum. Genet.* 47:A201 (abstr).

Terwilliger, J.D., Lehner, T., and Ott, J., 1991, Differential sex dependent penetrance of autosomal dominant diseases mimic linkage to the boundary of the pseudoautosomal region, *Am. J. Hum. Genet.* 49 (suppl):362 (abstr).

Terwilliger, J.D., Speer, M., and Ott, J., A chromosome-based method for rapid computer simulation in human genetic linkage analysis, *Genet. Epidemiol.* (in press).

Thomas, D.C., and Cortessis, V., 1992, A Gibbs sampling approach to linkage analysis, *Hum. Hered.* 42:63.

Vieland, V.J., Hodge, S.E., and Greenberg, D.A., 1992, Adequacy of single-locus approximations for linkage analyses of oligogenic traits, *Genet. Epidemiol.* 9:45.

Vionnet, N., Stoffel, M., Takeda, J., Yasuda, K., Bell, G.I., Zouali, H., Lesage, S., Velho, G., Iris, F., Passa, Ph., Froguel, Ph., and Cohen, D., 1992, Nonsense mutation in the glucokinase gene causes early-onset non-insulin-dependent diabetes mellitus, *Nature* 356:721.

Weeks, D.E., Ott, J., and Lathrop, G.M., 1990, SLINK: a general simulation program for linkage analysis, *Am. J. Hum. Genet.* 47:A204 (abstr).

Weeks, D.E., 1991, Human linkage analysis: strategies for locus ordering, *in*: "Advanced Techniques in Chromosome Research," K.W. Adolph, ed. Marcel Dekker, New York.

Weeks, D.E., Lathrop, G.M., and Ott, J., 1993, Multipoint mapping under genetic interference, *Hum. Hered.* (in press).

Xie, X., and Ott, J., 1990 Determining the effect of a change in affection status on the lod score. *Am. J. Hum. Genet.* 47:A205 (abstr).

HALDANE AND POPULATION STRUCTURE

D. F. Roberts

Department of Human Genetics
University of Newcastle upon Tyne
19 Claremont Place
Newcastle upon Tyne, NE2 4AA
U.K.

INTRODUCTION

I did not know Haldane well, met him two or three times only and heard him speak on a similar number of occasions and on none did he mention population structure or its importance in human genetics.   Nor did I have any immediate recollection of its occurrence in his lectures and published works.   This review therefore entailed re-reading as many as possible of Haldane's writings in a search for any statement that he made on the subject.   This task has provided much pleasure, for his writings were full of reminders of some of the great classics of genetic work that we now take for granted.   But especially this re-reading brought back memories of when I first discovered Haldane and particularly some of his earlier works.

I was intrigued by these of course for the wrong reasons.   He was an excellent writer.   He was fun to read.   He gave glimpses of mind-boggling future possibilities. He made one think.   Thus in his essay "The origin of life" published in 1929 I first came across the verse from Augustus de Morgan's *A budget of paradoxes* "Big fleas have little fleas...".   This stanza Haldane used in his argument that the link between living and dead matter is somewhere between a cell and an atom. Then in his essay "The argument from design" published in 1944, he reminded us that whereas in the past "cut-throat competition, both between species and to a lesser extent within them, was a necessary condition of evolution, it is so no longer".   For man can "control the evolution of animals and make unprecedented creatures, such as the Jersey cow and the Angora rabbit, for our needs".   The processes of so doing were simple genetic procedures of hybridisation and selection.   What would Haldane have said at the modern advances of biotechnology, whereby we can make in the laboratory synthetic biological substances, and achieve many other miracles! Among his later works, in the essay (1957) "Aunt Jobisca, the Bellman and the Hermit" he discusses the nature of scientific statement.   Then the essay "A passage to India" (1958) explained what to many was inexplicable, namely why he retired to India, for there he recounts the myth of the end of the Pandavas and how Yudhisthira came to be admitted to Svarga with his dog.   For that essay, showing loyalty and concern for others, also explains many things about him - why despite his strong social conscience and writings he did not align himself with the Eugenics movement in Britain (unlike Fisher), why when invited to lecture in Gröningen in

Human Population Genetics, Edited by P.P. Majumder
Plenum Press, New York, 1993

March 1940 he refused to discuss racial differences "in the event of a German invasion I did not want to increase my hosts' chances of being murdered, by criticizing the official Nazi theories on this topic".

But apart from the pleasure of rediscovering old delights, it was salutary to be reminded how much we owe to Haldane. I am not referring here to his writings as an almost exhaustible source of quotations for examination questions e.g. "Mutagenesis is a biochemical rather than a biophysical process. Discuss". No, we are indebted to him, not only for his solid contributions to population genetics, but also for the amazing predictions that he made, the farsighted ideas that were later formalised by others. For instance it was Haldane who was responsible in 1920 for the idea that "A gene makes a particular chemical species of enzyme or antigen" and acknowledged (Haldane 1954, pages 15 and 17) that he had developed it from Cuenon writing in 1903. Similarly in 1933 (Haldane, 1933) in explaining graft rejection he predicted the HLA system, for there he envisaged that each of a fairly large number of genes in an individual makes a specific antigen, and a transplant containing any antigen foreign to the host is destroyed.

DEVELOPMENTS IN POPULATION GENETICS

In his concern with the laboratory and experimental procedures of physiology, biochemistry and similarly, it is perhaps surprising that he was able to contribute so much to the acceptance of neoDarwinism based on the principles of genetics. Here he saw his task as to weigh up Darwin's theory of evolution in the light of the new science of genetics. In a series of papers that were later drawn together in "The causes of Evolution" (Haldane, 1932) he successfully integrated Darwin's theory of evolution by natural selection with the theory of particulate inheritance propounded by Mendel.

In order to understand his success it is necessary to consider the work of Fisher (1930) and Wright (1931). Fisher enquired whether natural selection of random mutations could account for observed degrees of adaptation. For this purpose he constructed a systematic theory of evolution by natural selection that could be tested. His theory predicted that, especially in large populations, even slight selective differentials were sufficient to replace a less advantageous gene by a more favoured counterpart.

Wright was interested in the circumstances that would best favour the origin and spread of novel adaptations. But where Fisher assumed that the attainment of adaptation was best examined by testing each mutation one by one, on its average merits, Wright felt that the merits of a gene combination could not be predicted from the average merits of its component genes. He therefore argued that adaptive evolution is favoured in species composed of many small, nearly isolated, subpopulations. In such circumstances, more different gene combinations would arise, primarily by chance, but if one of them were really advantageous it could multiply and spread through the whole species. This theory would account for the fact that so many species are composed of many nearly isolated subpopulations, and that so many differences between subpopulations, which presumably arose by chance, appear non-adaptive.

Haldane's contributions to population genetics according to Kimura (1983) are not as original as Fisher's, but they are more diverse, more open-minded, and often biologically more appropriate. His series of papers starting in 1924 entitled "A mathematical theory of natural and artificial selection" initiated a systematic mathematical attack on the process of change in gene frequencies by natural selection. He was convinced that a satisfactory theory of natural selection must be quantitative, and that the adequacy of the genetical theory of natural selection could only be tested through quantitative investigation. He considered various kinds of selection such as zygotic, gametic, familial and X-linked. For example in considering a situation where in a very large random mating population the dominant genotypes are favoured and the recessive not, he showed that, in general, the number of generations required for a given change is inversely proportional to the intensity of selection. He applied this to a case of industrial melanism in moths that occurred in Manchester and found that the selection coefficient required was at least 0.332 and probably larger. It was not until 30 years later that Kettlewell and others showed that selection intensities as large as this commonly occur in these moths, especially by predation by birds.

This treatment of the change of gene frequencies is deterministic, since no random elements, e.g. random sampling of gametes, are considered. Because of its simplicity this approach is still the most useful. In 1927 Haldane turned to a stochastic problem and investigated the probability of fixation of mutant genes. Whereas Fisher had discussed the probability of survival of a neutral mutation, Haldane extended the discussion to advantageous mutations. He showed for the first time that in a large random mating population a dominant mutant gene having a small selective advantage k in the heterozygote has a probability of about 2k of ultimately becoming established in the population, and for a recessive with selective advantage t in homozygotes the probability is $\sqrt{t}/N$. In another outstanding paper in 1931 he investigated the situation where mutant genes are disadvantageous singly but become advantageous in combination. Using the two locus case where a population genotype (unlinked) AABB is transformed into one of genotype aabb, he termed species liable to such transformations "metastable". He suggested that the process of species formation may be a rupture of such metastable equilibrium, and that such rupture will be especially likely in small isolated communities. Here we have one of the few approaches to Sewall Wright's theories, propounded independently, which describe evolution as a trial and error process in terms of a multidimensional adaptive surface, and the only one to my certain knowledge where Haldane began to consider evolution in a part of a structured population.

Another occasion on which he mentioned briefly the potential importance of the study of subpopulation differentiation occurs in his concluding remarks to the Jerusalem 1961 conference, where he recommended a comparison of frequencies of a range of genes in the diverse Jewish populations converging on Israel. He said "It would show perhaps that some characters may be regarded as having been unchanged for a long time, while others - such as, for example, the genes responsible for resistance to malaria - are of comparatively recent origin". At the Rome 1961 congress Haldane read Ajit Ray's paper for him on gene diffusion in a primitive tribe in India, in which birth place and clan name were included as variables for understanding population structure. The only article to my knowledge, for which I am indebted to Professor Mukherjee, in which Haldane deals in large part with population structure is a semi-popular paper in New Biology. In this, summarising the "structures" seen in different species, he equates population structures with reproductive patterns, of which random mating is one. He lists the several influences which modify random mating that were known to him (i.e. spatial contiguity, inbreeding customs, assortative mating), and examines their genetic consequences. He makes no mention of the importance of kinship between and within populations, for example as adumbrated by Malecot (1948). He accepts the importance of the Sewall Wright effect, particularly in early human history and, a topic of current interest, concludes that it is only in small isolated populations that natural selection would favour the spread of genes making for certain kinds of altruistic behaviour.

But Haldane did not concern himself greatly with population structure generally or with the theories that endeavoured to formalise its effects e.g. of Wright and Malecot. He seems to have been unaware of the latter, but he certainly knew of Wright's work. One criticism that he made (Haldane, 1959) of Wright is that, once an adaptive gene combination is fixed in a subpopulation, it is unlikely to be exported intact to others, since recombination would dissolve these new genotypes when migrants introduced them to other subpopulations. But otherwise his contribution to the genetics of small subpopulations was indirect, a secondary spin-off from his main findings incorporated by others in their models e.g. Crow says that the approach to inbreeding through identity by descent owes a great deal to Haldane.

DEVELOPMENTS IN HUMAN SCIENCES

Concerned as he was with somewhat artificial situations, the investigations of stocks and pure lines, controlled breeding studies in the laboratory, and experimental procedures where it was possible to replicate and deal with large amounts of data, and with his mathematics which assumed infinitely large random-mating populations, it is not surprising that he tended to ignore developments in the empirical population sciences. Important relevant developments lay in the fields of anthropology and demography: the former showed that populations and their infinite variety of subdivisions were universal and the latter concerned their numerical study.

Whether we like it or not man is a social beast. Humanity finds it necessary to live as members of a group. The basic group is the family - mother, father and offspring - at once a biological and social grouping. The family is universal, occurring in all human societies. It almost always forms a component of a larger unit, even in primitive hunting cultures where the group must remain of the smallest. The extended family is of common occurrence. Families, households and kindreds may be combined in a variety of ways into villages, tribes and other communities territorially, economically, and socially defined, which are themselves further combined in more inclusive social entities. Such larger entities may be composed of lineages, moieties, castes and any of a host of the other subdivisions by which individuals are associated and separated and their actions and their interactions ordered.

To define a population is by no means easy. It can be defined in a number of ways. Always it is a group of individuals. It may be the members of the community in which daily life is lived (e.g. the village); the inhabitants of the social entities into which several communities are combined who consider themselves, and on occasion act, as a unit (e.g. the tribe); the inhabitants of still larger groups which never act as a whole but who nevertheless bear a common culture and perhaps speak a common language; the members of the great regional divisions or continental groups of mankind; or all human beings, members of the species *Homo sapiens* - the species population.

A population then, no matter whether defined by economic, territorial, social, or ethnic associations, is an entity consisting of a number of individuals who, although similar in many respects, always vary amongst themselves, often to a considerable extent. Besides its peculiar sociological characteristics, a population possesses a number of biological and statistical properties. Like an organism it occupies at any moment a definite position in space, possibly nucleated possibly diffuse. It is a continuing entity, permanent in relation to the individuals composing it; for the individual is born into the population which exists before his arrival and continues to exist after his death. It has a definite demographic structure (sex composition, age composition, marital composition) which fluctuates with time, but which can be described at any given moment. A population can be characterised statistically, and distinguished from other populations, by the use of parameters, its group attributes (e.g. birth rates and death rates, means and variances of metrical characters, territorial density, frequencies of discrete biological features) which are meaningless relative to any individual. Most important for our purpose, it has an heredity, and it is this, its genetic characteristics, with which we are particularly concerned.

GENETIC CHARACTERISTICS

(a) genetic constitution

Each population has a genetic constitution, in simple terms a pool of genes which are, as it were, held in trust by the individuals making up the population at any one time.   Every individual born receives a sample of the gene pool from his parents, half from his father and half from his mother, and then passes on half of them to each of his children for subsequent transmission in their turn.   For each of the many characters that are under the control of a pair of genes at a particular locus in a chromosome, for example some blood groups or red cell enzyme types, the number of individuals possessing it in the population (or a sample of it) can be counted.   That is to say the phenotype frequency can be obtained.   From this the frequency of a given gene in the population can be calculated.   The population gene pool at any given moment can thus be specified in terms of the gene frequencies, the total gene pool being the array of gene frequencies over all loci.   This then is a definition of the genetic constitution of the population, the total gene pool, the array of gene frequencies over all loci.   This definition can be extended in terms of sequences of DNA other than genes.

There is another way in which the genetic constitution of a population can be defined, that is applicable when the pedigree of the population is known.   Instead of the array of gene frequencies at a given time, it is instead specified in terms of the contributions to it from particular ancestors, that is to say by the Pack method, in terms of the probable ancestral contributions.   If, say, there are six founding ancestors, one has two children and another three, and there are nine offspring in the first filial generation, then the contribution of these two ancestors to the gene pool of that generation is one ninth and one sixth respectively, on account of their differential fertility.   The probability that an autosomal allele present in a grandparent is present in any one of his grandchildren is one quarter, so the probable genetic contribution of any grandparent to the gene pool of his grandchildren is one quarter and his contribution likewise to the total second filial generation will depend on the fertilities of himself and his children.   The genetic constitution of any generation of descendants can therefore be specified in terms of the contributions to it from particular ancestors.   In real human population generations overlap and the one individual can belong to several different generations simultaneously, with respect to the same ancestor, but the same principle can be applied for any group of individuals of known descent.   Thus the genetic constitution of a population at any given time can be specified in terms of the probable ancestral contributions.

These two definitions of genetic constitution can be employed for different purposes.   Using the first definition, the positions of populations in relation to each other can be compared, either by considering the allele frequencies at a given locus one at a time and mapping them or plotting them along a unidimensional scale, or by multivariate comparison by means of a measure of genetic distance.   These frequencies can be employed in assessments of the amount of admixture and, on certain assumptions, its rate.   The second definition is particularly valuable in tracing secular variation within a population by calculating the probable ancestral contributions at a series of points in time.   For this shows how the genetic constitution of the population so defined varies over the period, and can be used to help identify the factors responsible.

(b) genetic structure

Genetic structure is a different concept.   Whereas genetic constitution as defined above is concerned essentially with individual loci, genetic structure concerns the way in which genes are distributed and combined within populations. As such, it is concerned not with gene frequencies, but with measures of gene relationships - correlation between uniting or adjacent gametes (E), coefficients of inbreeding (F), coefficients of kinship (ϕ), the parameters of the decline in kinship with distance (a - local kinship, b - systematic migration pressure, d - distance), heterozygosity and its rate of decay.   For all these, factors are of relevance that do not enter the simple concept of genetic constitution - effective population size, population distribution, population density, clines, assortative mating, migration, marriage systems.

DYNAMICS OF POPULATION STRUCTURE

The idea of genetic structure seems to have developed from the work of Wahlund (1928) who coined the word "isolate" and first explored the consequences of population subdivision, and Dahlberg (1929) who developed the concept.   The word isolate is employed in different contexts and has diverse meanings, ranging from an identifiable conceptual group of persons within a vast mainstream population to a closely demarcated, relatively small, isolated population entity.   Wright (1931) investigated the mathematical theory relating to the genetics of population subdivision.   Few human populations can be regarded as a homogeneous social entity, but within most there exist some boundaries to breeding.   The boundaries may be geographical or social in nature, communities may be hostile to each other, and these barriers either prevent or at least render more difficult mating between persons on either side of them e.g. of different social strata or castes, of different religious persuasions or historical provenance.   Even in large modern cities the choice of a spouse is restricted.   Every individual has his own circle, in which it is probable that he will marry and the population is therefore made up of many such circles, overlapping with each other, minor subpopulations or isolates.   They rarely coincide with actual geographically defined communities except in reproductively sequestered island populations such as the descendants of the Bounty mutineers of Pitcairn island, or in small religious communities such as the Dunkers, but it is to these isolated populations that the word isolate in recent years has come to be particularly applied, and not to segments within larger populations.   These contrasting situations may be viewed as a matter of degree, the effectiveness of isolating mechanisms between those within and those without the isolate, and in practice it is usually clear how the term is being used.

All these subdivisions, and the departures from random mating that they produce, affect the distribution of genes within the population, and the existence of such a population structure is of considerable importance in population genetics. The size of the population, important in the evolutionary process, is considerably smaller than may appear at first sight.   The large population postulated in Haldane's calculations does not exist, but instead it is conceived of as comprising a series of smaller units.   In the absence of differential selection, mutation, and assuming random mating within each isolate, this division would lead merely by chance processes to <u>genetic differentiaion</u> of the isolates, extreme if they are totally cut off from each other, less so according to the degree of intermigration. The random variation in gene frequency from generation to generation, due to the sampling process of zygotic formation, is negligible in a large population, but large in a small one.   The variance due to drift between local isolates of the same size and the same initial gene frequency in one generation is compounded with the passage of generations, and the gene frequencies of the isolates become more and more divergent, even though the gene frequency for the total population as a whole remains unchanged.   When isolation is partial the rate of divergence depends on the amount of gene flow.

With the passage of time the gene frequencies become so divergent that more and more genes become either lost or fixed.   A second consequence then is that since the smaller numbers increase the probability of fixation at a given locus, division into isolates makes for <u>greater genetic uniformity within each isolate</u>.

There are limits, either gene loss or fixation, beyond which random dispersion cannot go, and the rapidity with which these limits are attained in a given isolate depends on its size, initial gene frequency, and the amount of gene flow. Again however the gene frequency in the total population remains unchanged.

A third consequence is that there is a tendency to <u>increased homozygosity</u> within the population as a whole. Within each isolate the genotype frequencies are those expected from the changed gene frequencies; as the isolates drift apart in gene frequencies, their genotype frequencies are correspondingly divergent. As the gene frequencies for each isolate assume extreme values with the approach of fixation, there is a diminution in the proportion of heterozygotes, since heterozygotes are most numerous when gene frequencies are at intermediate values. In the population as a whole, the frequency of heterozygotes is less than expected from the overall gene frequencies by an amount equal to twice the gene frequency variance amongst isolates. In other words, there is a departure of the genotype frequencies of the total population from those expected from Hardy Weinberg relationships. For those loci at which more than a single pair of alleles is available, the greater the number of coexisting alleles at a locus, the more rapid the rate at which some are lost or fixed, so that population subdivision is particularly effective in restricting the number of coexisting alleles in an isolate. Indeed the rate of decrease in genetic variation in a population with many isolates varies with their number.

It is not only the genotype frequencies that are affected by the existence of isolates. There is greater opportunity for another random variable to take effect, namely the establishment of change in linkages and linkage groups, and chromosome structure. The concept of population structure is central to Sewall Wright's (1931) shifting balance theory of evolution and, just as in this the fixation of individual alleles is important for establishing differentiation, so too is the fixation of new linkage groups.

Following the early work and especially that of Wright, the concept of genetic structure has been much developed by Malecot, Morton, Harpending, Jacquard and many others. Today comprehensive description of the genetic structure may utilise the relationships between alleles, loci, chromosomes, gametes, individuals, aggregates of individuals, sibpairs, genealogies, social classes, and geographical and other subpopulations. The many studies that have been made of genetic structure in populations in many parts of the world have greatly enhanced our understanding of human evolution.

CONCLUSION

Why then did Haldane ignore what is to us so obvious, that human breeding populations are not large, are not homogeneous? Perhaps first it was his interest in mathematical prediction. In 1941 he wrote "There is almost always an element of uncertainty in prediction concerning individuals. But when we deal with millions, possibility becomes certainty and conjecture accurate prediction". Secondly there was a difference in perspective. Haldane was concerned with his experimental procedures, his laboratory work in physiology, biochemistry and similarly, his breeding studies, so that his vision was channelled to the small specific rather than to the world at large. Yet he was not unaware of the world at large - witness his numerous essays on social topics. Thirdly, he was a thinker, starting with an idea and developing it. It is typical that his essay on the origin of lactation started from an apparently irrelevant observation made by his wife Helen Spurway on a nesting pair of lapwing in his garden at Bhubaneswar. The whole of the rest of that essay evolved as a result of his thinking around that topic, and speculation on it in the light of what he had read. Finally, in the words of John Maynard Smith his pupil, colleague, and friend, "He was not himself a good observer... but he read avidly and he listened to what people told him, and he had a knack of drawing conclusions which the observer himself had missed". Perhaps if Haldane had listened and read less and had spent more time observing the world around him, he would have come to recognise the genetic importance of the heterogeneity of human populations. He may well have spent more time on its implications and less on his other pursuits. But if he had done so, we should have been the poorer.

REFERENCES

Dahlberg, G., 1929, Inbreeding in man, *Genetics* 14:421.

Fisher, R.A., 1930, "The genetical theory of natural selection", Oxford University Press.

Haldane, J.B.S., 1932, "The causes of evolution", Harper and Row, New York.

Haldane, J.B.S., 1933, The genetics of cancer, *Nature* 132:265.

Haldane, J.B.S., 1954, "Biochemistry of Genetics", Allen & Unwin, London.

Haldane, J.B.S., 1959, Natural Selection, *in*: "Darwin's Biological Work", P.R. Bell, ed., Cambridge University Press.

Haldane, J.B.S., 1963, The concentration of rare recessive genes in the past and in modern times, *in*: "The Genetics of Migrant and Isolate Populations", E. Goldschmidt, ed., The Williams & Wilkins Company, Baltimore.

Kimura, M., 1983, "The neutral theory of molecular evolution", Cambridge University Press.

Malécot, G., 1948, "Les Mathématiques de l'Hérédité", Masson, Paris.

Wahlund, S., 1928, Zusammensetzung von Populationen und Korrelations-erscheinungen vom Standpunkt der Vererbungslehre aus betrachtet, *Hereditas* 11: 65.

Wright, S., 1931, Evolution in Mendelian populations, *Genetics* 16:97,.

ANALYSIS OF GENETIC STRUCTURE OF POPULATIONS: MEANING, METHODS, AND IMPLICATIONS

Ranajit Chakraborty

Genetics Centers, Graduate School of Biomedical Sciences
University of Texas Health Science Center
Houston, Tx 77225, USA

INTRODUCTION

Genetic structure of a population is characterized by the number of subpopulations within it, the frequencies of different genetic variants (alleles) in each subpopulation, and the degree of genetic isolation of the subpopulations. Understanding of genetic structure of populations had been the focus of attention of population geneticists from the inception of this discipline. Like any other topics of population genetics, the study of population structure also did not escape Haldane's attention (see e.g., Haldane and Moshinsky, 1939; Haldane, 1954), but Sewall Wright (1989-1988), another poineering population geneticist, is credited for the first mathematical conceptualization of the study of genetic structure of populations.

The literature on methods of analysing genetic structure of populations has grown both in volume and diversity since Wright's early work (Wright, 1921, 1931), and a comprehensive review of this subject is beyond of the scope of this presentation. In stead, the purpose of this work is three fold. First, I shall briefly mention the meaning of population structure in various contexts that are of current interest to the population biologists. Second, the principal and popular methods of quantifying population structure will be stated, and third, some of the implications population structure will be addressed to indicate that at the present time this subject is not only of academic interest, its significance affects a larger arena. Obviously, there may be a bias in this presentation, but I hope that this perspective will depict the current status of this subject and will encourage further work needed in this area.

MEANING OF POPULATION STRUCTURE

In nature all organisms live in groups of breeding units, within which mating occurs by certain principles of mate choice. Generally, the principle of mate selection

Human Population Genetics, Edited by P.P. Majumder
Plenum Press, New York, 1993

in unknown, although socio-biologists consider this topic important from the point of view of causes of evolution (Wilson, 1975). Loosely, a group within which mating occurs may be defined as a population. For example, the Hindus of India may be defined as a population, and the different caste groups as its subpopulations. European Caucasians would also be a population, with groups within them defined by nationality or religion would be the subpopulations. By the same token, the entire species of Indian carps (*Cyprinus carpio*) may be defined as a population, subdivied by geographic partitions of the network of river systems where they breed. The carps of each river system will then be the subpopulations.

With such a convention, we might call a population substructured when it consists of components (subpopulations) amongst which gene flow is restricted. There could be a complete absence of gene flow between subpopulations because of social, geographic, ecological, or even biological barriers. In such cases, evolutionary changes between subpopulations occur under complete isolation (no gene flow). New mutations arising in certain subpopulations remain "private" (Neel, 1973), and genetic differentiation between subpopulations occur with a speed governed by mutation rate and the breeding size of subpopulations. Gene flow between subpopulations retards the process of genetic differentiation between populations (Nei and Feldman, 1972; Chakraborty and Nei, 1974: Li, 1976; Slatkin, 1985).

With this in mind, it is obvious that the detection of population structure requires investigations beyond conventional genetic studies; understanding of the ecology and environment of the habitat in which the organism lives, demography of the population, as well as social behavior, all interplay in underpinning the exact nature of populations and its constituent subpopulations. Incisive mind would therefore say, no matter how one defines a population, every population is substructured, since restrictions of gene flow always exist with suitable definition of subpopulations within the population.

The above perspective exhibits the elusiveness of formal description of the structure of a population. Elsewhere I proposed a formalization (Chakraborty, 1990), which states that the understanding of the genetic structure of a population consists of knowing (1) how many "breeding units" are within a given population, (2) what their relative sizes are, (3) the extent and pattern of their genetic variation, (4) to what extent they are related, either by virtue of their evolution or by their interrelatedness, (5) what factors determine the observed gene-genotype or gene-phenotype relationship in the population, and (6) what the causal factors are that maintain the observed genetic characteristics of the population. All of these ingredients, even though not stated in these exact terms, exist in any discussion of this subject (see e.g., Morton, 1973). Nevertheless, it is clear that a full understanding of the genetic structure of a population requires knowledge of a wide range of factors that determine the gene-genotype (-phenotype) relationship and their effect on genetic variation. The causal processes that determine the structure of a population and the observed characteristics of the structure are obviously intertwined. Population structure analysis, therefore, serves two purposes: It defines the genetic composition of a population and the pattern of genetic variation in relation to observed genotype (phenotype) frequencies; and it allows us to infer certain evolutionary conclusions regarding the causes of variation.

The study of population structure has practical meaning besides evolutionary investigations. With the global environmental movement, conservation biologists are increasingly becoming aware of preservation of existing genetic variation, and conservation of biodiversity. Since the evolutionary dynamics of genetic variation in a substructured population differs depending upon the number, size, and degree of isolation of subpopulations, it is of interest to know to what extent the genetic diversity is affected when subpopulations are either created or eliminated by changes in the environment. In some Scandinavian countries this has raised considerable interest in the context of preservation of aquatic species (see e.g., Ryman and Utter, 1987). Conservation biologists are also interested in concepts such as the *Minimum Viable Population Size* (MVPS), whose evaluation requires knowledge of the genetic structure of populations (Ewens, 1990).

More recently, and apparently to some extent surprisingly, the issue of population structure has entered into legal proceedings as well. In the context of forensic applications of genetic markers it is of interest to know with what frequency certain genotypes (phenotypes) occur in a population, so that statistical weights may be placed when genetic profile of a biological specimen found in a crime scene matches with that of a suspect. Since a genetic profile may always be translated into a genotype (as in the case of a single-locus typing), or a combination of genotypes (in the case when multiple loci are studied), the basic issue here is the evaluation of genotype probabilities. When substructuring exists and the subpopulations may be potentially be such that they have been separated sufficiently long to permit differentiation, genotypic proportions in the total population cannot be predicted merely from the frequencies of different alleles in the total population. This observation is well known (Nei, 1965: Li, 1969), since the genotype frequencies in a subdivided population are also functions of variances and covariances of allele frequencies over all subpopulations. When each subpopulation is known without ambiguity, these type of evaluations can be made in full details, even though the number of parameters (e.g., the number of subpopulations and frequencies of each revelent allele in every subpopulation) needed is generally large. To what extent our current knowledge of genetic structure of populations ameliorates this difficulty in providing a forensically valid estimate of frequencies of specific genetic profiles remains a debatable issue, and this debate is far from being settled at least in courts of the United States of America. I shall return to the meaning and significance of population structure in this context later, but the two contrasting views are well summarized in the recent papers of Lewontin and Hartl (1991) and Chakraborty and Kidd (1991).

METHODS OF POPULATION STRUCTURE ANALYSIS

Maruyama (1973) quite succinctly summarized that the existing approaches to the study of genetic problems in a structured population may be divided in two contrasting groups. One of these is the study of structure itself, by identifying the constituent subpopulations and exmining their characteristics and differences between them. In my opinion, these are the descriptive studies, although statistics of such studies provide means of quantifying the extent of substructuring. However, from such investigations one might determine the effects of ignoring the structure of populations in relevant applications. While most genetic studies on population structure fall in this category, the other approach is to try to develop ways of bypassing the structure so as to learn something about the genetic processes despite ignorance of the structure. The focus of this approach is to test certain genetic hypotheses without any precise knowledge of the population structure. Apparently, this later type of studies may appear to be gross simplification of reality, because population substructuring is ubiquitous. In fact, they are no so, because several biologically relevant summary measures of genetic variation have the mathematical property of "invariance", or simply stated, they enjoy properties that hold for any form or nature of substructuring. Using diffusion approximations of the dynamics of gene frequency changes in populations, Maruyama (1973) and Yamazaki and Maruyama (1972) have shown that such invariant statistics of variation provide several insights with regard to the relative roles of causal factors of maintenance of genetic variation.

Study and Analysis of Structure

Beginning with Wright (1921, 1931) the analysis of genetic structure of populations has been approached in various ways. The advocates of these different approaches often stressed philosophical as well as conceptual differences of the formulation of the problem. In spite of this, it may be shown that these different approaches often yield estimates that are empirically as well as analytically very similar (Chakraborty, 1988; Chakraborty and Danker-Hopfe, 1991). In some form or other, all methods are designed to estimate three basic composite parameters, generally called Wright's fixation indices (Wright, 1943, 1951), F_{IT}, F_{IS}, and F_{ST}, which together give a quantification of the extent of genetic substructuring in a population.

Wright's Fixation Indices and Their Interpretations. The three fixation indices were interpreted as correlations in Wright's (1943, 1951) work. The first two, F_{IT} and F_{IS} are defined as correlations between uniting alleles to produce individuals relative to the total population and relative to the subpopulations, respectively. In contrast, F_{ST} is the correlation between two alleles drawn at random from each subpopulation. For a population consisting of s subpopulations, at a locus with r segregating alleles (say, $A_1, A_2, ..., A_r$), allele-specific parameters for each of these indices can also be defined, given by (see Chakraborty and Danker-Hopfe, 1991)

$$F_{ISk} = (P_{ikk} - p_{ik}^2)/[p_{ik}(1 - p_{ik})],$$ (1a)

$$F_{ITk} = (\bar{P}_{.kk} - \bar{p}_{.k}^2)/[\bar{p}_{.k}(1 - \bar{p}_{.k})],$$ (1b)

and

$$F_{STk} = (\overline{p_{.k}^2} - \bar{p}_{.k}^2)/(\bar{p}_{.k} - \bar{p}_{.k}^2),$$ (1c)

where P_{ikk} = frequency of the homozygotes $A_k A_k$ in the i-th subpopulation,
$\bar{P}_{.kk} = \sum_{i=1}^{s} w_i P_{ikk}$ is the proportion of $A_k A_k$ in the total population,
$\bar{p}_{.k} = \sum_{i=1}^{s} w_i p_{ik}$ is the frequency of the allele A_k in the total population,
p_{ik} = frequency of allele A_k in the i-th subpopulation,
w_i = size of the i-th subpopulation relative to the size of the total population
$\quad$ (i.e., $\sum w_i = 1$), and
$\overline{p_{.k}^2} = \sum_{i=1}^{s} w_i p_{ik}^2$, is the weighted average of squared allele frequencies.

Three aspects of this formulation are worth noting. First, under the assumption that we are investigating the genetic structure of a fixed population which has s specified subpopulations within it, equations (1a)-(1c) truly defined three sets of composite parameters for each of the k alleles at a locus. Second, the parametric spaces for these parameters are specified by

$$-p_{ik}(1 - p_{ik}) \leq F_{ISk} \leq 1, \tag{2a}$$

$$-\bar{p}_{.k}(1 - \bar{p}_{.k}) \leq F_{ITk} \leq 1, \tag{2b}$$

and

$$0 \leq F_{STk} \leq 1, \tag{2c}$$

for all k = 1, 2, ..., r; even though each have a correlational interpretation. Third, they are interdependent, satisfying the identity

$$(1 - F_{ITk}) = (1 - F_{ISk})(1 - F_{STk}). \tag{3}$$

In any estimation procedure all three features should be accounted for. These definitions can be extended at a locus level by defining

$$F_{IS} = \sum_{i=1}^{s} \sum_{k=1}^{r} w_i (P_{ikk} - p_{ik}^2) / \sum_{i=1}^{s} \sum_{k=1}^{r} w_i [p_{ik}(1 - p_{ik})], \tag{4a}$$

$$F_{IT} = \sum_{k=1}^{r} (\bar{p}_{.kk} - \bar{p}_{.k}) / \sum_{k=1}^{r} [\bar{p}_{.k}(1 - \bar{p}_{.k})], \tag{4b}$$

and

$$F_{ST} = \sum_{k=1}^{r} (\overline{p_{.k}^2} - \bar{p}_{.k}^2) / \sum_{k=1}^{r} (\bar{p}_{.k} - \bar{p}_{.k}^2), \tag{4c}$$

which still satisfy the identity (3), in the parametric space

$$-max_{\{i,k\}} p_{ik}(1 - p_{ik}) \leq F_{IS} \leq 1, \tag{5a}$$

$$-max_{\{i,k\}} \bar{p}_{.k}(1 - \bar{p}_{.k}) \leq F_{IT} \leq 1, \tag{5b}$$

and

$$0 \leq F_{STk} \leq 1. \tag{5c}$$

It should be recognized that in the context of analysing genetic structure of a given fixed population, the underlying parameters that fully describe the structure are the genotype frequencies $P_{ikk'}$ for all $k(k+1)/2$ genotypes in each of the s subpopulations. For an autosomal codominant locus since the allele frequencies are specified by the genotype frequencies (obtained by gene count) without any further assumption, the total number of independent parameters is $r(k-1)(k+2)/2$, since total of genotype frequencies in each subpopulation must add to one. The composite parameters, allele-specific, or locus-specific (equations 1a - 1c, or 4a - 4c), are expressed as ratios of functions of allele and genotype frequencies. Any estimation method requires estimation of these functions.

Nei's gene diversity formulation of Wright's fixation indices. Nei(1973) used the concept of gene diversity to quantify the extent of substructuring, using a probabilistic in stead of correlational interpretation. Gene identity is defined as the probability that two randomly chosen genes are identical in their allelic state, the complement of which is called gene diversity. At a locus level, therefore, Nei (1973) defined gene diversity within the i-th subpopulation

$$H_{Si} = 1 - \sum_{k=1}^{r} p_{ik}^2 = \sum_{k=1}^{r} p_{ik}(1 - p_{ik}), \tag{6}$$

whose weighted average over all subpopulations becomes the gene diversity within subpopulations

$$H_S = \sum_{k=1}^{r} w_i H_{Si} = \sum_{i=1}^{s} \sum_{k=1}^{r} w_i p_{ik}(1 - p_{ik}). \tag{7}$$

The gene diversity in the total population is

$$H_T = 1 - \sum_{k=1}^{r} \bar{p}_{.k}^2 = \sum_{k=1}^{r} \bar{p}_{.k}(1 - \bar{p}_{.k}). \tag{8}$$

To represent the fixation indices in the terms of the gene diversity indices, Nei introduced another quantity, the proportion of individuals whose two alleles are non-identical in their genotypes. This is obtained by taking the complement of total frequency of all homozygotes in the entire population

$$H_O = \sum_{i=1}^{s} w_i (1 - \sum_{k=1}^{r} P_{ikk}) = 1 - \sum_{i=1}^{s} \sum_{k=1}^{r} w_i P_{ikk}. \tag{9}$$

Substituting these in equations (4a)-(4c), Nei (1977) derived

$$F_{IT} = (H_T - H_O)/H_T, \tag{10}$$

$$F_{IS} = (H_S - H_O)/H_S, \tag{11}$$

and

$$F_{ST} = (H_T - H_S)/H_T, \tag{12}$$

so that irrespective of the number of alleles at a locus, Wright's fixation indices can be defined by gene diversity within the subpopulations and that in the total population. Like in equations (4a)- (4c), the composite parameter F_{ST} depends upon only the allele frequencies, not on genotype frequencies, while the other two depends on the genotype frequencies in the subpopulations. Since in some work the quantity H_O is also called the observed gene diversity in the total population, there are claims (Weir and Cockerham, 1984) that Nei's gene diversity interpretation of in terms of observed statistics; they do not represent parameters. Examination of equations (4a)-(4c) and (10)-(12) indicates that this criticism is incorrect, since like F_{IT}, F_{IS} and F_{ST} Nei's gene diversity indices are functions of the parameters P_{ikk}, for a specified population and a locus with r and s known.

Nei (1973) coined a new term for F_{ST}, the co-efficient of gene differentiation (G_{ST}). His decomposition of gene diversity in the total population (H_T) in two components, H_S and $D_{ST} = H_T - H_S$, is mathematically equivalent to the decomposition of Gini-Simpson's measure of diversity (see Rao, 1982a,b; Rao, 1984; Chakraborty and Rao, 1991). Nevertheless, Wright (1965) and Kirby (1975) have shown that Nei's coefficient of gene differentiation, G_{ST} ($= F_{ST}$) defined by the equation (12) can also be expressed as the weighted average of allele-specific F_{STk} parameters. Furthermore, as will be shown later, apart from the descriptive quantification of substructuring, the quantity G_{ST} can be used for examining the evolutionary dynamics of gene differentiation, and it can be extended for multiple loci (simply by taking averages of H_S and H_T over loci), and multiple hierarchical levels of substructuring within a population (Chakraborty, 1974, 1980, 1985; Nei, 1975, Chakraborty et al., 1982). Also note that in this formulation no assumption is made with regard to what causes the departure of genotype frequencies from their expectations under the Hardy-Weinberg rule, nor any assumption is made with regard to what causes the allele or genotype frequency differences between subpopulations. Therefore, these parametric representations for a specified subdivided population hold for any type of mating structure within subpopulations, or even in the presence of local differences of selection pressures in subpopulations. When random mating prevails within each subpopulation, F_{IS} becomes zero, and $F_{IT} = F_{ST}$, so that the population structure is characterized by a single composite parameter F_{ST}.

Variance Component Representation of Fixation Indices. Cockerham (1969, 1973) redefined the fixation indices in terms of intraclass correlations derived from the analysis of variance of allele frequencies. The basic premise of his analysis is that the subpopulations are replicate of each other, so that indicator variables are defined for both alleles of a random individual drawn from a random subpopulation. Using a linear-random effect model, additive effects of between-subpopulations (a), between-individuals within a subpopulation (b), and within- individual (c or w) components of variation may be defined. Allele- specific values of fixation indices $(F_{ITk}, F_{ISk}$ and $F_{STk})$ are related to the variance components (a_k, b_k, and c_k) by

$$F_{ITk} = (a_k + b_k)/(a_k + b_k + c_k), \tag{13}$$

$$F_{ISk} = b_k/(b_k + c_k), \tag{14}$$

and

$$F_{STk} = a_k/(a_k + b_k + c_k), \tag{15}$$

The analysis of variance model is constructed from genotype data with regard to a specific allele (A_k), so that all alleles other than A_k are grouped into a single class $\bar{A}_k$, giving only three genotypes $A_k A_k$, $A_k \bar{A}_k$ and $\bar{A}_k \bar{A}_k$. In terms of the above notations, the three variance-components take the form

$$a_k = \sum_{i=1}^{s} w_i (p_{ik} - \bar{p}_{.k})^2, \tag{16}$$

$$b_k = \sum_{i=1}^{s} w_i F_{ISik} p_{ik} (1 - p_{ik}), \tag{17}$$

and

$$c_k = \sum_{i=1}^{s} w_i(p_{ik} - P_{ikk}). \tag{18}$$

Since $p_{ik} \geq P_{ikk}$ for all i and k, c_k is non-negative, and so is a_k from equation (16). But in situations where one or more F_{ISik} values are negative, b_k can become negative even in its parametric value. While this is acknowledged in Cockerham's (1969) work, negative parametric value of a variance component is not usual in the formulation of the analysis of variance model. Locus-specific values of the fixation indices are defined in Cockerham's formulation by summing a_k, b_k and c_k values over all alleles at a locus, and expressing the fixation indices as respective ratios of sums, analogous to equations (13)-(15). The same algorithm applies for defining the fixation indices for multiple loci (Weir and Cockerham, 1984).

Multivariate Variance Component Representation of Fixation Indices.
Long (1986) and Smouse and Long (1988) noted when the number of segregating alleles at a locus is more than 2, i.e.,$r > 2$, the variance components a_k, b_k and c_k over alleles are not independent parameters. Their formulation proceeds from the premise of a multivariate analysis of variance (MANOVA) analysis, where a genotype is represented as a pair of vector-valued (r-1 dimensional) indicator variables, so that the total dispersion matrix Σ is decomposed into three parts, Σ_a, Σ_b and Σ_c in the analogy of a, b, and c of a bi-allelic locus. With $\Sigma = \Sigma_a + \Sigma_b + \Sigma_c$, the locus-specific fixation indices take the form

$$F_{IT} = (r-1)^{-1}tr[\Sigma^{-1/2}(\Sigma_a+\Sigma_b)\Sigma^{-1/2}], \tag{19}$$

$$F_{IS} = (r-1)^{-1}tr[(\Sigma_b+\Sigma_c)^{-1/2}\Sigma_a(\Sigma_b+\Sigma_c)^{-1/2}], \tag{20}$$

and

$$F_{ST} = (r-1)^{-1}tr[\Sigma^{-1/2}\Sigma_a\Sigma^{-1/2}], \tag{21}$$

where tr denotes the trace of a matrix. In this formulation, again, while Σ_a and Σ_c are positive semi-definite matrices, Σ_b can be negative- definite. Elsewhere it is shown (Chakraborty, 1988; Chakraborty amd Danker-Hopfe, 1991) that for $r = 2$, equations (19)-(21) are mathematically equivalent to equations (13)-(15), but for r > 2, the numerators and denominators of equations (13)-(15) represent the diagonal elements of the respective Σ-matrices. Furthermore, with the factor of $(r-1)^{-1}$ in front, the MANOVA approach, in essence, proposes the average of ratios of variance components, instead of the ratio of averages, as suggested by Weir and Cockerham (1984). This is important, since this difference can sometimes make the multivariate estimate of fixation indices different from that derived from Weir and Cockerham's approach. Some numerical illustrations of such situations are given in Chakraborty and Danker-Hopfe (1991).

Estimation of Fixation Indices and An Illustration

The above formulation indicates that under a uniform of set of notations all of the major approaches of describing Wright's fixation indices can be postulated

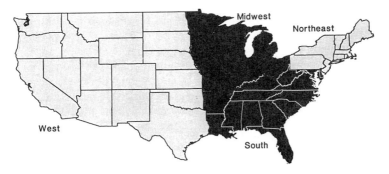

Figure 1. Regional subdivision of the USA used in the National Health Examination Survey(1980).

without any loss of statistical rigor. The point to be emphasized is that in this formulation the population is a specified one, and it contains a fixed number of subpopulations (s) from each of which sampling will be done for estimating the fixation indices. The underlying parameters are the genotype frequencies in each subpopulation $(P_{ikk'})$; but the fixation indices are ratios of composite functions of the genotype and allele frequencies. As in other statistical applications, in estimating fixation indices the problem of bias is relevant, when numerators and denominators are separately estimated to estimate the ratios. Some authors discussed this problem to suggest bias-corrected estimates (see e.g., Nei and Chesser, 1983). However, no estimate of any of the fixation indices is truly unbiased; at best they are consistent estimators in the terminology of the theory of estimation. A full description of various estimators of fixation indices under each of the above formulations are given in Chakraborty and Danker-Hopfe's (1991) review on this subject, and they will not be re-iterated for space limitation. Nevertheless, it has been shown that: (i) when Cockerham's variance component estimate of allele-specific F_{IS} is negative, it is bounded by Nei's bias-corrected and biased estimators; (ii) but when Cockerham's estimate is positive, Nei's bias-corrected estimator is bounded by his biased estimator and that of Cockerham. (iii) Furthermore, the expected difference between these various estimators is of the order of the inverse of the number of genes sampled from a subpopulation. This makes the differences between the estimates generally smaller than their standard errors. (iv) In addition, although the above formulation, e.g., equation (4c) or (12), indicates that Nei's gene diversity estimator of F_{ST} depends only on the allele frequencies, but not on genotype frequencies, while Cockerham's or Long's variance component estimators of F_{ST} are dependent on genotype frequencies as well. (v) Finally, Long's MANOVA estimator of locus-specific estimator of F_{IS} is mathematically identical to Curie-Cohen's (1982) estimator. Analytical proofs of these properties are given in Chakraborty and Danker-Hopfe (1991), along with the demonstration that in spite of the philosophical differences of the premises of these estimators, the numerical values of the estimators as well as their standard errors are generally close, so that the relative efficiencies of these estimators are nearly equal.

This may be illustrated with an example, which is deliberately chosen to make an important point with regard to the implications of population structure in human populations. In 1980, in the context of a National Health Examination Survey, the

US Department of Health, Education and Welfare (National Health Survey, 1980) conducted a stratified random sampling in the entire country, dividing the whole USA in four regions, Northwest, Northeast, South, and West, shown in Figure 1. Within each region, blood samples were collected from individuals of two major racial groups, Whites and Blacks, proportional to their demographic sizes within each state. Phenotype data included blood group loci A_1A_2BO, Rh (C, D, and E), and Secretor (Se), and protein loci Haptoglobin (Hp), Transferrin (Tf) and Group-Specific component (Gc). Since A_1A_2BO, Rh-D, and Se loci involved dominance relationships between alleles, Table 1 shows the estimates of all three fixation indices within the White and Black samples using the genotype data on the remaining five systems (Rh- C, Rh-E, Hp, Tf, and Gc).

Striking similarities of the numerical values of the estimates as well as their standard errors are evident from this table. Furthermore, none of the F_{IT} or F_{IS} values appear significant. The F_{ST} values between geographic samples of Blacks are larger than that in Whites, which is consistent with the observation that within the Afro-Americans of the USA there is a gradation of the amount of Caucasian genes by geopolitical subdivision of populations. In the Southern United States the Blacks have generally less influence of Caucasian admixture than in Northeast or West (Reed, 1969; Chakraborty, 1986). Numerically, the F_{IT} or F_{IS} values within the Blacks are higher than the Whites, but in spite of the fact that the Black populations are admixed, their F_{IS} indices are not significant. This illustrates that the calculation of genotype frequencies from allele frequencies using the Hardy-Weinberg rule is adequate even in the presence of genetic admixture. The smaller insignificant F_{IS} value within the Whites is also indicative of the fact that the gene differentiation between regional classification of US Whites is virtually non-existent.

Table 1. Estimates of fixation indices for geographic subdivisions of two major racial groups within USA (based on five loci)

Estimators	White			Black		
	F_{IS}	F_{ST}	F_{IT}	F_{IS}	F_{ST}	F_{IT}
Nei's method:						
Weighted	-0.0027	0.0002	-0.0025	0.0274	0.0039	0.0312
	±0.0091	±0.0001	±0.0090	±0.0273	±0.0018	±0.0267
Unweighted	-0.0032	0.0002	-0.0030	0.0141	0.0065	0.0205
	±0.0087	±0.0001	±0.0087	±0.0255	±0.0033	±0.0246
Large sample	-0.0035	0.0004	-0.0031	0.0112	0.0087	0.0198
	±0.0088	±0.0001	±0.0087	±0.0256	±0.0034	±0.0247
Cockerham's method:						
Weighted	-0.0027	0.0002	-0.0025	0.0274	0.0059	0.0332
	±0.0091	±0.0001	±0.0090	±0.0274	±0.0027	±0.0264
Large sample	-0.0031	0.0006	-0.0025	0.0253	0.0084	0.0334
	±0.0091	±0.0002	±0.0090	±0.0274	±0.0028	±0.0264

Table 2. Gene Diversity Analysis of Geographic subdivisions of two racial groups within the United States with 8 blood group and protein loci

Locus	Gene diversity (Total)	Between racial groups	Between regions within racial groups	Between individuals within regions
A_1A_2BO	0.502	0.024	0.002	0.498
Rh-C	0.411	0.038	0.001	0.373
Rh-D	0.436	0.009	0.001	0.426
Rh-E	0.238	0.002	0.001	0.235
Se	0.500	0.001	0.002	0.497
Hp	0.500	0.018	0.004	0.478
Tf	0.037	0.000	0.001	0.036
Gc	0.317	0.011	0.001	0.305
Pooled	0.368	0.010	0.002	0.356
	±0.058	±0.005	±0.001	±0.057
(in %)	100.000	2.730	0.370	96.890

Since the different methods yielded nearly equivalent results, the same survey data can now be used to examine at what level of subdivision (racial versus geopolitical) the gene diversity is most prominent. This can be done by Nei's nested gene diversity analysis, following the algorithm given in Chakraborty et al. (1982). Table 2 shows the results of this analysis where allele frequency data for all loci are used, estimating the allele frequencies by gene counting for the five codominant systems, and using the maximum likelihood method for the A_1A_2BO, Se, and Rh-D systems.

As in most earlier studies (Lewontin, 1972: Nei and Roychoudhury, 1972; 1982), this analysis also indicates that the most predominant component of genetic diversity is that between individuals; only 2.7% of the diversity is between individuals of different racial groups, while less than 0.4% of the diversity can be ascribed to between regional populations within each racial group. This is particularly noteworthy, since currently in some applications of genetic markers it is argued that genetic variation between groups within races may be larger than that between races (Lewontin and Hartl, 1991; NRC, 1992). The above findings are evidently in sharp contrast of such assertions.

IMPLICATIONS OF POPULATION STRUCTURE

Having shown that population substructuring is ubiquitous, and its extent can be reasonably well-estimated from genetic markers, one might ask what do they imply? From a purist point of view, the answer is if substantial substructuring is found its immediate implication is that without having direct observations on all subpopulations within the entire population any genetic inference would be incorrect. For example, the predictions of frequencies of specific genotypes (at a single- or multi-locus level) cannot be reliably done from allele frequencies in the total population alone. The next question, then, becomes what value of F_{IS} would be regarded as substantial? In order to answer this, we would need to describe the process through with subdivision occurs and how it is maintained. Wright's (1931) "isolation by distance" is a mechanistic model to seek answers of this type; Morton's work on bioassay of kinship (see e.g., citations in Morton, 1982) provides both theoretical and empirical observations on how isolation by distance generates and maintains substructuring. Other investigators considered the processes of mutation and drift, without any direct reference to the geographic positioning of subpopulations. Both approaches exhibit that the accumulation of genetic differentiation is a monotone function of time of isolation and the speed of accumulation of substructuring effect is dictated by mutation rate as well as breeding size of populations (see e.g., Nei, 1975; 1987). Local differentials of selection pressure would also affect substructuring, although even in absence of selection substructuring may persist. For example, Figure 2 illustrates the speed at which expected coefficient of genetic differentiation between subpopulations ($G_{ST} = F_{ST}$) can accumulate as a function of time of separation (t) measured in units of 2N generations, where N is the effective size of each subpopulations, and the number of subpopulations (s). The two surfaces are drawn, one when each mutation yields a new allele not seen in the population before (infinite allele model, the upper surface), the other (lower surface) being the case where mutations lead to allelic changes of a forward-backward nature (the stepwise mmutation model). The specific surfaces shown are drawn for loci whose average heterozygosity (H_S) within each subpopulation is 90%, to reflect what is to expect for hypervariable DNA loci.

Even though this figure represents a specialized set of computations (see Li, 1976; Chakraborty et al. 1992a for the relevant theory), its results are quite general. First, the extent of substructuring, summarized in the single value, G_{ST}, reaches an asymptote with respect to the number of subpopulations (s) very quickly. Second, even when the subpopulations do not exchange any gene (no gene flow), most of the accumulation of genetic differentiation occurs early, since the time asymptote also reaches quite fast. Therefore, from evolutionary perspective, the present-day substructuring in any human population is not a reflection of the contemporary populations, rather it is an effect of distant past. This is aptly stated by Smouse and Long (1988), in the sense that most of genetic differences within the human species "arose before we became a cosmopolitan and collectivist species".

These computations, however, ignored the effect of gene flow between subpopulations, in the presence of which the accumulation of genetic differentiation as well as the level of its asymptotic value diminish drastically (Nei and Feldman, 1972; Chakraborty and Nei, 1974). If the history of gene flow is not recent, the asymptotic value of G_{ST} is generally small enough to remain undetected (Chakraborty and Jin, 1992; Chakraborty et al., 1992a). Even if the value of F_{ST} or G_{ST} becomes statistically significant, its impact on the total population (seen through the evaluation of F_{IT}) becomes trivial. Table 1 data exhibits this empirically. Gene frequency differences between regional populations within both Blacks and Whites appear significant as seen from the values of F_{ST} (in relation to their standard errors). However, none of the F_{IT} (or F_{IS}) values are significantly different from

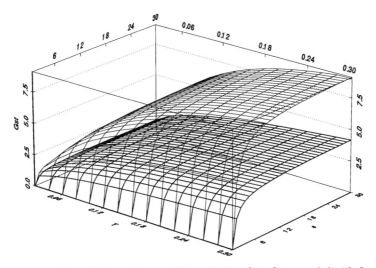

Figure 2. Expected coefficient of gene differentiation (G_{ST}) in a subdivided population as a function of the number of subpopulations (s) and their time of divergence(t) from a common ancestry in units of 2N generations. The upper surface is for the Infinite Allele Model and the lower surface is for the Stepwise Model of mutations.

zero. In other words, treating a substructured population as a single homogeneous one does not seem to lead to any gross deviation, at least in statistical sense. As mentioned before, this observation has a great impact on the issue of the presence of substructuring in forensic applications of genetic markers. I contend that the theory and empirical data on genetic effects of population substructuring, some of which are enumerated here, support the assertion that a conservative estimate of genotype frequencies can be given from allele frequencies of the entire population, even in the presence of population substructure. Morton (1992) also came to this conclusion from other considerations of possible causes of substructuring.

DISCUSSION AND CONCLUSION

This presentation exhibits that the presence of substructuring within any natural population is a biological reality, and there is no ground to deny it. Population structure can be defined and its extent can be evaluated from genetic data without any loss of statistical rigor. Gene diversity- as well as variance-component based estimates of Wright's fixation indices are generally in agreement with each other. While these summary measures provide a description of the extent of substructuring, the causes of substructuring are not specified by their enumeration alone. Even when gene flow between subpopulations does not occur, significant substructuring appears to have a little effect on the genotypic structure in the total population. As a result, detection of substructuring in the absence of identification of its subpopulations, and in the absence of data on each subpopulation, is difficult. No gross error is caused, when the objective is to predict genotype frequencies from allele frequencies, even in a substructured population. Of course, the individual subpopulations might exhibit considerable allele frequency differences, but their evolutionary relatedness (particularly when the time of divergence is as short as the different geopgraphic populations of our species) makes allele frequencies in different subpopulations positively correlated. Alleles that are rare in one population are generally rare in others, while alleles that are common in subpopulations remain common elsewhere too. Mourant et al.'s (1976) allele frequency contour maps illustrate this quite evidently, and this applies to hypervariable DNA polymorphism as well (Chakraborty et al. 1992b).

In many discussions on population structure, particularly in the context of human populations, the heterogeneity of allele frequencies across subpopulations has been exaggerated (e.g., Lewontin and Hartl, 1991). Such arguments rest on the ground that 'exogamy' is not common and even when it exists, it has only recent history. In contrary, it has been shown (Chakraborty and Nei, 1974; Chakraborty and Jin 1992) that even a small amount of gene flow across subpopulations dilute the substructuring affects drastically, and as shown in Figure 2, most of the homogenization process occurs early during the history of gene admixture. Furthermore, recent molecular data (Bowcock et al., 1991) suggests that the history of gene migration across ethnic populations within any of the major racial groups of human may not be as recent as the current demographic data indicates. As a result, the nature of human gene diversity is such that most significant of the genetic variation is between individuals, with very little added from the component of average differences between populations, irrespective of the structure of populations.

While the data shown in this presentation is completely in agreement with the above postulations, one possible criticism of this argument could be how well such theory applies to hypervariable polymorphic loci which are now being used by conservation biologists and forensic geneticists. Are there data to support applications of such theory? While it is true that data on allele frequency differences between ethnic groups within the major racial classification of human at hypervariable DNA loci is far more sparse compared to the classical markers, several surveys have shown that each of the above arguments hold equally well, if not more forcefully, to hypervariable loci as well (see e.g., Deka et al., 1991; Edwards et al., 1992; Chakraborty and Jin, 1992). This is due to at least three characteristics of hypervariable loci. First, as a result of larger number of alleles at each of such loci (compared with the traditional loci), individual allele frequencies are generally much smaller, each of which individually contribute little to the gene diversity. Therefore, absence of any specific allele in one subpopulation does not cause concern, since it is shown that the non- shared alleles rarely exceed any appreciable frequencies within any subpopulation (Chakraborty et al., 1992b). Second, gene diversity within subpopulations (H_S) is higher for almost all hypervariable loci studied. This makes the coefficient of gene differentiation between subpopulations (G_{ST}) smaller (Chakraborty and Jin, 1992) for hypervariable loci. Third, although the exact molecular mechanism of production of new alleles at the hypervariable loci is not known, examination of population data indicates that the mutations are more consistent with a stepwise model, caused by replication slippage of tandemly repeat core sequences (Edwards et al. 1992). This being the case, the computations shown in Figure 2 indicate that the level of gene differentiation for hypervariable loci should be even smaller for hypervariable loci, for which gene diversity within subpopulations (H_S) is larger (say, greater than 90%) and mutations follow a stepwise model.

In view of these, it is fair to conclude that while the study of substructuring in populations is important for evolutionary understanding as well as conservation of species, its impact on providing bounds on single- or multi-locus genotype frequencies is not great enough to raise concern such as the ones expressed by Lewontin and Hartl (1991).

ACKNOWLEDGEMENTS

This work was supported by US Public Health Research Service grants GM 41399 and GM 45861 from the US National Institutes of Health. I thank Li Jin, M.R. Srinivasan, and Yixi Zhong for their help for some of the computations presented here.

REFERENCES

Bowcock, A.M., Kidd, J.R., Mountain, J.L., Herbert, J.M., Carotenuto, L., Kidd, K.K., and Cavalli-Sforza, L.L., 1991, Drift, admixture, and selection in human evolution: A study with DNA polymorphisms. *Proc. Natl. Acad. Sci. U.S.A.* 88:839.

Chakraborty, R., 1974, A note on Nei's measure of gene diversity in a substructured population. *Humangenetik* 21:85.

Chakraborty, R., 1980, Gene-diversity analysis in nested subdivided population. *Genetics* 96:721.

Chakraborty, R., 1985, Genetic distance and gene diversity: some statistical considerations, *in* "Multivariate Analysis - VI," P.R. Krishnaiah, ed., Elsevier, Amsterdam, p. 77.

Chakraborty, R.,, 1986, Gene admixture in human populations: models and predictions. *Yrbk. Phys. Anthrop.* 29:1.

Chakraborty, R., 1988, Analysis of genetic structure of a population and its associated statistical problems. *Sankhya, Ser.B.* 50:327.

Chakraborty, R., 1990, Quantitative traits in relation to population structure: why and how are they used and what do they imply? *Hum. Biol.* 62:147.

Chakraborty, R., and Danker-Hopfe, H., 1991, Analysis of population structure: a comparative study of different estimators of Wright's fixation indices, *in* "Handbook of Statistics 8: Statistical Methods in Biological and Medical Sciences," C.R. Rao, and R. Chakraborty, eds., North-Holland, Amsterdam, p. 203.

Chakraborty, R., Deka, R., Jin, L., and Ferrell, R.E., 1992b, Allele sharing at six VNTR loci and genetic distances among three ethnically defined human population. *Amer. J. Hum. Biol.* 4:387.

Chakraborty, R., Haag, M., Ryman, N., and Ståhl, G., 1982, Hierarchical gene diversity analysis and its application to brown trout population data. *Hereditas* 97:17.

Chakraborty, R., and Jin, L., 1992, Heterozygote deficiency, population substructure and their implications in DNA fingerprinting. *Human genetics* 88:267.

Chakraborty, R., and Kidd, K.K., 1991, The utility of DNA typing in forensic work. *Science* 254:1735.

Chakraborty, R., and Nei, M., 1974, Dynamics of gene differentiation between incompletely isolated populations of unequal sizes. *Theor. Pop. Biol.* 5:460.

Chakraborty, R., and Rao, C.R., 1991, Measurement of genetic variation for evolutionary studies, *in* "Handbook of Statistics 8: Statistical Methods in Biological and Medical Sciences," C. R. Rao, and R. Chakraborty, eds., North-Holland, Amsterdam, p. 271.

Chakraborty, R., Srinivasan, M.R., Jin, L., and de Andrade, M., 1992a, Effects of population subdivision and allele frequency differences on interpretation of DNA typing data for human identification, *in* "Proceedings of 1992 Symposium on Human Identification", Promega Corporation, Madison. p. 205.

Cockerham, C.C., 1969, Variance of gene frequencies. *Evolution* 23:72.

Cockerham, C.C., 1973, Analysis of gene frequencies. *Genetics* 74:679.

Curie-Cohen, M., 1982, Estimates of inbreeding in a natural population: a comparison of sampling properties. *Genetics* 100:339.

Deka, R., Chakraborty, R., and Ferrell, R.E., 1991, A population genetic study of six VNTR loci in three ethnically defined populations. *Genomics* 11:83.

Edwards, A., Hammond, H.A., Jin, L., Caskey, C.T., and Chakraborty, R., 1992, Genetic variation at five trimeric and tetrameric tandem repeat loci in four human population groups. *Genomics* 12:241.

Ewens, W.J., 1990, The minimum viable population size as a genetic and demographic concept, *in* "Convergent Issues in Genetics and Demography," J. Adams, D.A. Lam, A.L. Hermalin, and P.E. Smouse, eds., Oxford University Press, New York, p. 307.

Haldane, J.B.S., 1954, An exact test for randomness of mating. *J. Genet.* 52:631.

Haldane, J.B.S., and Moshinsky, P., 1939, Inbreeding in Mendelian population with special reference to human cousin marriage. *Ann. Eugen.* 9:321.

Kirby, G.C., 1975, Heterozygote frequencies in small populations, *Theor. Pop. Biol.* 8:31.

Lewontin, R.C., 1972, The apportionment of human diversity. *Evol. Biol.* 6:381.

Lewontin, R.C., and Hartl, D.L., 1991, Population genetics in forensic DNA typing. *Science* 254:1745.

Li, C.C., 1969, Population subdivision with respect to multiple alleles. *Ann. Hum. Genet.* 33:23.

Li, W.-H., 1976, Effect of migration on genetic distance. *Amer. Natur.* 110:841.

Long, J.C., 1986, The allelic correlation structure of Gainj- and Kalam- speaking people, I. The estimation and interpretation of Wright's F- statistics. *Genetics* 112:629.

Maruyama, T., 1973, A diffusion model for geographically structured populations, *in* "Genetic Structure of Populations," N.E. Morton, ed., University of Hawaii Press, Honolulu, p. 40.

Morton, N.E., 1973, "Genetic Structure of Populations," University of Hawaii Press, Honolulu.

Morton, N.E., 1982, "Outline of Genetic Epidemiology," Karger, Basel.

Morton, N.E., 1992, Genetic structure of forensic populations. *Proc. Natl. Acad. Sci. USA.* 89:2256.

Mourant, A.E., Kopec, A.C., and Domaniewska-Sobcazk, K., 1976, "The Distribution of the Human Blood Groups and Other Polymorphisms," Oxford University Press, London.

National Health Survey, 1980, "Selected Genetic Markers of Blood and Secretions," U.S. Dept. of Health, Education, and Welfare, Publ. No. (PHS) 80-1664, Washington DC.

National Research Council, 1992, "DNA Technology in Forensic Science," National Academy Press, Washington DC.

Neel, J.V., 1973, "Private" genetic variants and the frequency of mutation among South American Indians. *Proc. Natl. Acad. Sci. USA.* 70:3311.

Nei, M., 1965, Variation and covariation of gene frequencies in subdivided populations. *Evolution* 19:256.

Nei, M., 1973, Analysis of gene diversity in subdivided population, *Proc. Natl. Acad. Sci. USA.* 70:3321.

Nei, M., 1975, "Molecular Population Genetics and Evolution," North- Holland, Amsterdam-New York.

Nei, M., 1977, F-statistics and analysis of gene diversity in subdivided populations. *Ann. Hum. Genet.* 41:225.

Nei, M., 1987, "Molecular Evolutionary Genetics," Columbia University Press, New York.

Nei, M., and Chesser, R.K., 1983, Estimation of fixation indices and gene diversities. *Ann. Hum. Genet.* 47:253.

Nei, M., and Feldman, M., 1972, Identity of genes by descent within and between populations under mutation and migration pressures. *Theor. Pop. Biol.* 3:460.

Nei, M, and Roychoudhury, A.K., 1972, Gene differences between Caucasian, Negro, and Japanese populations. *Science* 177:434.

Nei, M, and Roychoudhury, A.K., 1982, Genetic relationship and evolution of human races. *Evol. Biol.* 14:1.

Rao, C.R., 1982a, Gini-Simpson index of diversity: a characterization, generalization and applications. *Utilitas Mathematica* 21:273.

Rao, C.R., 1982b, Diversity and dissimilarity coefficients: a unified approach. *Theor. Pop. Biol.* 21:24.

Rao, C.R., 1984, Use of diversity and distance measures in the analysis of qualitative data, *in* "Multivariate Statistical Methods in Physical Anthropology," N. Van Vark, and W.H. Howell, eds., Reidal, Dordrecht, p. 49.

Reed, T.E., 1969, Caucasian genes in American Negroes. *Science* 165:762.

Ryman, N., and Utter, F., 1987, "Population Genetics and Fishery Management," University of Washington Press, Seattle.

Slatkin, M., 1985, Rare alleles as indicators of gene flow. *Evolution* 39:53.

Smouse, P.E., and Long, J.C., 1988, A comparative F-statistics analysis of the genetic structure of human populations from lowland South America and Highland New Guinea *in* "Quantitative Genetics," B.Weir, E.J.Eisen, M.M. Goodman, G.Namkoong, eds., Sinauer Associates, Sunderland, p. 32.

Weir, B.S., and Cockerham, C.C., 1984, Estimating F-statistics for the analysis of population structure, *Evolution* 38:1358.

Wilson, E.O., 1975, " Sociobiology," Harvard University Press, Cambridge .

Wright, S., 1921, Systems of Mating. *Genetics* 6:111.

Wright, S., 1931, Evolution in Mendelian populations. *Genetics* 16:97.

Wright, S., 1943, Isolation by distance, *Genetics* 28:114.

Wright, S., 1951, The genetic structure of populations, *Ann. Eugen.* 15:323.

Wright, S., 1965, The interpretation of population structure by F-statistics with special regard systems of mating. *Evolution* 19: 395.

Yamazaki, T., and Maruyama, T., 1972, Evidence for the neutral hypothesis of protein polymorphism. *Science* 178:56.

STRUCTURE OF HUMAN POPULATIONS IN INDIA

Kailash C. Malhotra and T. S. Vasulu

Anthropometry and Human Genetics Unit
Indian Statistical Institute
Calcutta 700 035, India

INTRODUCTION

The interpretation of the temporal and spatial variation of gene frequencies in populations is one of the main aims of population genetics. In addition to systematic evolutionary processes (mutation, selection and admixture), non-systematic factors such as size, density, and the distribution of a population, the nature of mate selection within a population, and differential fertility and viability also influence the genetic composition of a population. These non-systematic factors are the most important components of the structure of human populations. It is now well known that very different patterns of these non-systematic factors exist in human populations.

In this paper an attempt is made to provide an overview of various aspects of the structure of human populations in India which are likely to have some bearing on the micro-evolutionary processes that, at least in part, have led to the genetic composition of the contemporary Indian populations.

PEOPLING OF INDIA

In order to understand and interpret the structure of contemporary populations in India, we briefly provide an account of the available lithic and skeletal evidence.

The lithic evidence: Numerous sites belonging to various prehistoric periods from early stone age (lower paleolithic) to proto-historic period have been discovered from almost all parts of the country. Figure 1 gives the locations of some of the important sites of various periods. From an inspection of Figure 1 and Table 1 which incorporates the chronology of Indian prehistory from Acheulian to Iron Age, the following main points emerge:

(i) man lived in India since Acheulian period dated over 350,000 years B.P; the evidence has come from a number of sites (for details see Mishra, 1992);

Human Population Genetics, Edited by P.P. Majumder
Plenum Press, New York, 1993

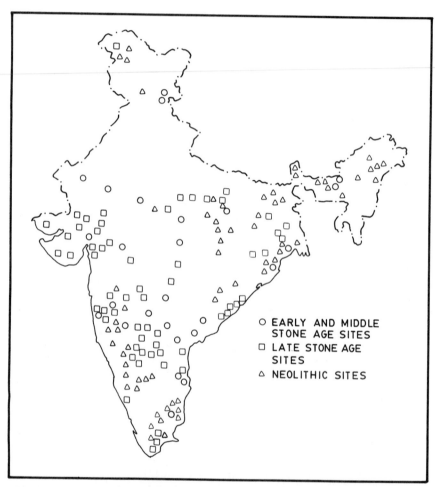

Figure 1. Locations of prehistoric sites discovered from lower paleolithic to proto-historic periods in India

The legend within the figure reads:

○ EARLY AND MIDDLE STONE AGE SITES
□ LATE STONE AGE SITES
△ NEOLITHIC SITES

(ii) by middle paleolithic period man had spread to many parts of the Indian sub-continent (this period is dated 50,000-20,0000 B.P.);

(iii) by Iron Age dated 1000 - 700 B.C. man had spread to almost all parts of the country; and

(iv) many sites/regions in different parts of the country showed uninterrupted human occupation right from early stone age to the Iron Age. (For details see Sankalia 1974; Maloney 1974; Allchin and Allchin 1982).

Table 1. Chronology of prehistoric cultures in India[1]

Culture	Age		
Iron Age	1000	–	700	B.C.
Chalcolithic	2300	–	700	B.C.
Neolithic	2500	–	1000	B.C.
Mesolithic	8000	–	2000	B.P.
Upper Paleolithic	20000	–	10000	B.P.
Middle Paleolithic	50000	–	20000	B.P.
Lower Paleolithic	> 350000		

[1] Source: Sankalia (1974), Agrawal, et al., (1978), and Mishra (1992).

The skeletal evidence: Compared to the lithic evidence, unfortunately the history of skeletal biology of the Indian sub-continent is not so impressive. First, not many sites have yielded human osseous remains, and second, often when such remains are found, the preservation has been rather poor, and materials from several sites still await description and publication. Except for a skull cap of *Homo erectus* found in central India and dated around 200,000 B.P. the remaining evidence pertains to rather younger periods, i.e., from Mesolithic onwards. An examination of the morphological features of the skeletons from various sites of Mesolithic (Lekhahia, Sarai Nihar Rai, Langhnaj), Neolothic (Piklihal, Maski, Brahmagiri, Tekkalakota, T.Narsipur, etc.), Chalcolithic (Harappa, Mohenjodaro, Chandoli, Nevasa, Lothal, Kalibangan, etc.), and Iron Age (Yelleswaram, Brahmagiri, Additanallur, etc.) periods reveal: (i) the presence of proto-Australoids and Caucasoids in all the periods; (ii) the absence of Mongoloids (it may be noted here that no site in the northeast where Mongoloids are present has so far yielded human osseous remains) during all the periods, (iii) considerable variation in the cephalic index of the Caucasoids, from dolicocephals to brachycephals (mostly occurring from Iron Age onwards), and (iv) both proto-Australoids and Caucasoids right from the Mesolithic period to Iron Age show wide geographical spread (southern, central and western India).

Historical evidence shows that a number of migrations brought several people into India. Just to mention a few important ones: Indo-Aryans (2000 - 1400 B.C.), Sakas (2 century B.C.), Kushanas (1 century A.D.), Huns (5-6 century A.D.), Arabs (8 century A.D.), Mongols (12-13 century A.D.), and then of course the Portugese (14 century A.D.) and the British (18 century A.D.). The people of African ancestry were also brought to the west coast beginning from 16th century A.D.

Summarizing archaeological (lithic and skeletal) and historical evidence, it is evident that several waves of immigrants from west and central Asia, and from north and northeast Asia, came to India during at least the last 10000 years. The immigrants primarily came by land, but also used sea routes, and initially mainly settled in the river-valleys and plains. However, in the northeast many colonised the mountain Himalayas.

Numerous anthropometric and genetic studies carried out among the contemporary Indian populations, and on the prehistoric skeletal remains from different periods and geographical areas, suggest the existence or the presence of combinations from the following ethnic elements among the people of India:

(i)   Negritos (Andaman islands);
(ii) Proto-Australoids (west, central and southern regions)
(iii) Caucasoids (all over the country);
(iv) Mongoloids (north-east and sub-Himalayan region; several Tibetan refugees have recently been provided settlement in Orissa, Karnataka and Himachal Pradesh);
(v)   Negroid (west coast and Andhra Pradesh).

There is perhaps no other sub-continent in the world that harbours such a rich array of ethnic elements as India.

In terms of antiquity, both cultural and biological evidence suggests that proto-Australoids on the mainland, and Negritos in the Andaman Islands are probably relic groups from the oldest inhabitants of the country. Europoids came to India in several waves stretching over the last 10,000 to 15,000 years. The antiquity of Mongoloids is not yet fully established, although some populations like Khasis and Garo are likely to have been in the region since neolithic period. The Negroids from Africa were brought in as slaves by the Portugese during historical period beginning from 16th century A.D.

CONTEMPORARY INDIAN SOCIETY

From the previous section, it is evident that the process of peopling of India stretching over indeed a long period brought in not only different biological traits, but also a variety of cultural, religious and technological characteristics. Contemporary India is an agglomeration of thousands of groups displaying a rich diversity in any aspect that one may consider – religion, language, social structure, mode of production, marriage rules, food habits, settlement patterns, etc.

In this section, therefore, we shall try to capture the essence of the enormous diversity but restrict our observations on aspects of population structure that are most likely to have some impact/influence in the microevolutionary processes.

Demographic profile: As per the 1991 census, the population of India is about 850 million, with a population density of 216 persons per square km (range 8 - 655 square km). The annual

210

population growth rate has declined moderately during the 1981-'91 decade, from 2.22 during 1971-'81 to 2.11 during 1981-'91. The sex-ratio was 929. The literacy rate was 52.1 per cent (64% for males and 39 % for females). The latest SRS (Sample Registration System) results show that in 1990, India's birth rate was 29.9. The total fertility rate (TFR) was estimated at 4.1 and the total marital fertility rate (TMFR) around 5.4 births. The death rate in 1990 (SRS) was 9.6/1000 population (10.4 in rural areas and 6.7 in urban areas). Life expectancy at birth was estimated at about 58.3 years for males and 59 years for females. The infant mortality rate (IMR) has been reported at 80/1000 live births (86 in rural areas and 51 in urban areas).

It may be emphasized here that for all the demographic parameters mentioned above, there exist large variability within States, and across different socioeconomic groups within States.

To sum-up, India has a mammoth population with a high population density, relatively high fertility and mortality (both IMR and overall).

Settlement pattern:: Of great importance in understanding micro-evolutionary process, is the pattern in which humans occupy and distribute themselves in space. There are two aspects that are of interest, namely, the distribution of villages/cities of different sizes and the distribution of households within a village.

According to the census of 1981, there were 5,57,137 villages in India; we shall restrict our observations to only rural areas and ignore urban areas, although the description will remain somewhat incomplete. It is interesting to note that the population size of nearly 49% of the villages in India is less than 500, and that nearly three-fourths of these villages harbour less than 1000 persons (Table 2). The villages in the mountainous and forested tracts tend to be usually small in size, compared to those in fertile river tracts.

Table 2. Frequency distribution of villages in India by population size[1]

Village population size	Number of villages	% of total villages
< 200	1,20,073	21.6
200 - 499	1,50,722	27.0
500 - 999	1,35,928	24.4
1000 - 1999	94,486	17.0
2000 - 4999	46,892	8.4
5000 - 9999	7,202	1.3
> 10000	1,834	0.3
Total	5,57,137	100.0

[1] As per Census 1981.

There are essentially three types of patterns discernible in the spatial arrangements of households in a village. They are: (a) nucleated villages - most of the households are located in a small space, often walls being common, surrounded by agricultural fields. Such villages are usually large, (b) Chicken and hen type - these are usually small hamlets with bigger dwellings for parents and small dwellings of married sons. Such villages are usually rather small in population size, represent usually one clan, and are mostly found in hilly and forested tracts, and are associated with tribal populations, and, (c) dispersed houses - in such villages the households (usually large) are scattered each having a courtyard of its own. Such villages are found usually in mountainous tracts of western ghats.

Languages spoken in India: There are four linguistic families with numerous dialects spoken in India. They are: (i) Austro-Asiatic, (ii) Dravidian, (iii) Indo-European, and (iv) Tibeto-Burman.

The Austro-Asiatic languages spoken in India belong to two branches: Mon-Khmer spoken by the Khasis in the north-east and Nicobarese in the Nicobar islands, and Mundari spoken by a number of groups in central and eastern India. It is noteworthy that besides tribal populations, these languages are not spoken by any caste population. Also, this language family lacks script.

The Dravidian family is spoken in southern and central India by both tribal and non-tribal groups. The main branches of this family are Tamil, Telugu, Kannada, Malayalam and Gondi. The only people outside India that speak this language are the Brahui of Baluchistan. (There is of course a strong possibility that some groups in eastern Iran may also be speaking the language.)

In the Indo-European family the languages spoken in India belong to the Indo-Aryan branch. The main languages that constitute this family are Hindi, Bengali, Marathi, Sindhi, Urdu, Bihari, Gujarati, Oriya, Punjabi, Assamese, Rajasthani and Konkani. The other branch of Indo-European is Dardic spoken in Kashmir. Incidentally, the language spoken in Sri Lanka also belongs to this family.

The Tibeto-Burman language family is classified in three sub-families: (a) Tibeto-Himalayan spoken in Tibet, Ladakh and eastern Himalayas; (b) North Assam branch spoken in Assam, and (c) Assam-Burmese branch spoken in several north-eastern States.

As far as the antiquity of these languages is concerned, it appears that Dravidian is the oldest one, followed by Austro-Asiatic. Perhaps Indo-European is the youngest of all the language families spoken in India.

As per 1971 census there were 73.8% speakers of Indo-European languages, 24.2% of Dravidian languages, 0.8% of Tibeto-Burman family and only 1.2% Austro-Asiatic speakers.

Population by religion: There are several religions practiced in the country. These have indeed played a very significant role in shaping the breeding isolates. Till the Muslims arrived in India around 800 A.D., there were broadly speaking only four religions, Hinduism with numerous modified forms, Buddhism, Jainism, and the animistic religions practiced by the tribal

populations. During the Muslim rule spanning three centuries a
large section of the population was converted to Islam. This led
to a religious division within the same caste population. Around
500 years ago, the birth of Sikhism took place, but was confined
largely to the north-western part of India among the Punjabi
speaking population. Christianity arrived in the 16th century
with the Portugese and the British. (Although the Syrian
Christians had arrived earlier around 2 century A.D., they had
remained confined to small pockets in Kerala.) A sizeable
section of the population then became followers of Christianity.
This process can be summarized as fission, the result of which
was that numerous groups emerged with new breeding territories.
It may be noted here that although each of the non-indigenous
religions preached casteless society, none succeeded. Therefore,
the converts be it muslim, christian or sikhs, married among
their own social category.

India is predominantly a Hindu country as returns of 1981
census show: 82.6% Hindus, 11.4% Muslims, 2.4% Christians, 2.0%
Sikhs, 0.7% Budhists, 0.5% Jains, and 0.4% others. It has been
argued by Malhotra (unpublished) that it was primarily due to
ecological reasons that none of the religions could achieve their
goal viz., casteless society.

Population by physical type: As noted earlier in 1.2 above, 5
different physical types can be discerned in the contemporary
Indian population. We attempt here to provide a rough estimate
of the numerical size of these types. We would like to emphasize
that we are not equating the 5 types with the notion of pure
races. As a matter of fact there is a strong evidence of
considerable admixture among these physical types. Malhotra
(1978a), however, showed that, by and large, the 5 types, on the
basis of aggregate of morphological characters can be easily
discerned. The numbers of the 5 types are given in Table 3. It

Table 3. Description of the Indian population by physical type

Physical type	Estimated numerical strength	% of total population
Negrito	306	-
Negroids	15,000	-
Proto-Austroloids	56 million	6.60
Mongoloids	15 million	1.76
Europoids	779 million	91.64
Total	850 million	100.0

is evident that Caucasoids constitute presently the bulk of the
Indian population (around 92%). The other large populations are
Australoids (6.6%), and Mongoloids (1.76%). The Negritos and
Negroids are represented by rather small numbers.

Emerging social organizations: The presence of peopling of India
brought not only different biological elements, but also a wide
variety of cultural and technological traits, and the historical
and social processes led to the evolution of a unique social
organization, the Hindu caste system in the subcontinent.

The people of India are estimated to consist of over 40,000 Mendelian populations. An estimated 37,000 groups are structured in the Hindu caste system. Individually each population in the system is called a *Jati* or a caste (Karve and Malhotra, 1968). Each caste theoretically belongs to one of the five *Varnas*: *Brahmins, Kshatriya, Vaishya, Sudra and Pancham*. The five *varnas* are arranged in a hierarchical order, the *Brahmin varna* is at the top of hierarchy and are followed by the *Kshatriya, Vaishya, Sudra and Pancham varnas*. The social stratification of the society is not only hierarchical, but the hierarchy is rationalized by ritual and religious criteria. The system has been strengthened by an elaborate philosophy. Traditionally, the members of each caste followed a particular profession, practiced endogamy, and had elaborate rules for selecting mates.

Some of the more notable consequences of the Hindu caste system, in terms of evolutionary implications, have been (1) the entire population has been divided into a large number of groups and the system did not permit large-scale inter-caste, inter-religious and inter-ethnic marriages, whereby the gene pool of each caste has evolved over at least 3000 years; (2) because of the nature of occupations, the population sizes of castes show enormous variation; there are several with fewer than 5000 persons, whereas many others number over 50,000 people; (3) several castes are located in rather small territories, often only in a single village, whereas many others are found in several districts; (4) apart from geographical and ecological considerations, the caste system promotes an uneven geographical distribution among many castes because of occupational differences (Malhotra, 1981); (5) the system led to systematic distribution of several castes in a given village (in a typical Indian village in plains one may find 10 - 15 different castes living together, but using the resources to avoid excessive inter-caste competition (Gadgil and Malhotra, 1982); and (6) the system encouraged retention of diversity in many sociocultural practices including those related to the choice of mates among different castes and communities (there are matrilineal and patrilineal castes, polygamous and polyandrous castes, vegetarian and non-vegetarian castes, etc.).

SOME ATTRIBUTES OF POPULATION STRUCTURE AND MICRO-EVOLUTION

As a result of isolating mechanisms that operate because of social, cultural, religious, historical, geographical and spatial factors and restrict the choice of a mate, the Indian population structure is characterized by innumerable subdivided populations of varying size, different degrees of breeding isolation (endogamy) and a great deal of variation in the pattern of mating, all of which invariably results in a wide genetic diversity.

Variation in population size: The population size of the endogamous groups for each caste, subcaste, tribe and other groups vary considerably. Though the actual number of these endogamous groups is not readily available from census figures, however, the records from 1901 census reveal that there were 2378 castes (Hutton, 1963). However, the population sizes of the scheduled castes and tribes are available from the census records, which provide an estimate of variability of sizes and distribution in different regions. For example, according to 1971 census, the tribal population vary from a mere 26 individuals among the Andamanese to about 4 million among the

Gonds and the Bhils in central India. Out of 427 tribes enumerated, there are six major tribes with a size of more than a million each; these are: Bhil, Gond, Santhal, Mina, Munda and Oraon, there are three tribes with a strength of more than 500 thousand, and there are 42 tribes whose size vary between 100 to 800 thousand. And the rest i.e., of about 88.06% of the tribes are of size of less than 100 thousand each (Vidyarthi, 1983). About 55 % of the total tribal population is distributed in the central region comprising Bihar, West Bengal, Orissa and Madhya Pradesh with a total population of more than 20 million (Roy Burman and Harit, 1971). It is inhabited by 125 tribal populations and the extent of variation in their population size is shown in Fig. 2. It is noteworthy that about half of them (48%) are numerically small, (size range from 1 to 25 thousand), whereas 27 tribes (21.6%) have a population size of more than 100, 000.

The total number of endogamous populations in India is not known, but it is estimated to be around 40, 000, comprising about 37,000 castes and subcastes, and about 3000 tribal, religious and other historical migrants populations (Malhotra, 1984). Recently there was an attempt by the Anthropological Survey of India to estimate extent of cultural diversity in each region (Singh, 1993). According to this survey conducted during 1985 to 1991, there are 2753 communities (relatively homogeneous clusters of endogamous groups who share common cultural traits). The regional distribution and extent of variation in size of these communities in the different States are shown in Fig. 3. A wide variation ranging from 7 communities for a population of 0.4 million (1981 census) in Lakshadweep islands to a maximum of 386 communities for a population of 53.55 million in Andhra Pradesh is observed. On an average there are about 145 communities per region/State/union territory (U.T.) (mean 144.84 and s.d. 123.97). Of these, 83% are located within the boundary of a state/U.T., and 13% them are spread over more than one State or adjoining State, whereas only 4% are spread over greater parts of India. At the village level, the number of communities varies from 1 or 2 for a homogeneous population of size of 500 to 1000 persons, e.g. a tribal village, to 8 to 16 communities for a multiethnic village of the same size (Joshi et al., 1993).

Levels of endogamy (Long term effects): Though endogamy is the characteristic feature of subdivided populations within a caste, tribe, and other groups, there is a wide variation in the level of endogamy. This can be generally understood by the following types: (a) Socially sanctioned matings between castes, subcastes of different hierarchy. The caste system is a dynamic institution, that it provides mechanism to retain caste hierarchy and also allows flexibility of socio-cultural practices and laxity in relation to certain types of preferences in choosing a mate from other castes. This is particularly with reference to upward and downward movement of wives from lower social rank to higher rank through marriage, described as *Pratiloma* and *Anuloma* marriages, so that overall stability of caste hierarchy is ensured, though biologically it leads to greater diversity within the community; (b) The second category refers to those marriages other than Anuloma and Pratiloma marriages which are socially not encouraged, but nevertheless occur between caste-clusters, between sub-castes, between a tribe and a caste, and between religious groups, and (c) Apart from this there are undetected chance (random) matings outside the institution of marriage, which are suppressed due to social and other reasons. The extent

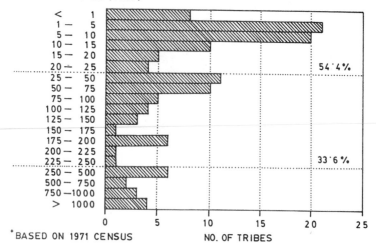

Figure 2. Population size distribution of 125 tribes of Madhya Pradesh, Orissa, Bihar and West Bengal

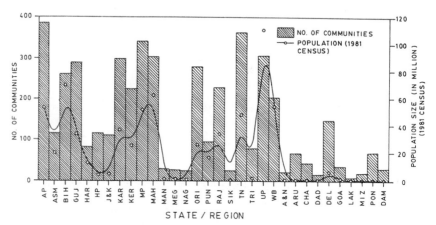

Figure 3. Statewise distribution of number of communities and population size (based on 1981 Census) in India
(Data on number of communities obtained from Joshi et al., 1993.)
(Abbreviations Used: AP=Andhra Pradesh, ASM=Assam, BIH=Bihar, GUJ=Gujarat, HAR=Haryana, HP=Himachal Pradesh, J&K=Jammu and Kashmir, KAR=Karnataka, KER=Kerala, MP=Madhya Pradesh, MAH= Maharashtra, MAN=Manipur, MEG=Meghalaya, NAG=Nagaland, ORI=Orissa PUN=Punjab, RAJ=Rajasthan, SIK=Sikkim, TN=Tamil Nadu, TRI=Tripura, UP=Uttar Pradesh, WB=West Bengal, A&N=Andaman and Nicobar Is., ARU=Arunachal Pradesh, CHA=Chandigarh, DAD=Dadra, DEL=Delhi, GOA=Goa, LAK=Lakshadweep Is., MIZ=Mizoram, PON=Ponoichery, DAM= Daman.)

of such miscegenation is likely to be more than what has been estimated. It is more frequent between the same caste cluster and least frequent between the religious groups. In general, ordering of the extent of endogamy and admixture of the above three types is as follows: between caste-cluster > between adjacent varnas or subcastes > between castes > between caste and tribe > between religious communities. The frequency of the occurrence of these above types of admixture vary quite drastically between regions, between urban and rural populations and between economic classes, because of differences in education, occupational mobility and other economic reasons. Studies indicate that in rural areas, e.g., in a multicaste village, the frequency of admixture is of the order of 2 per ten thousand marriages and in a tribal village, e.g., among the Yanadi, it is about six per thousand marriages (Vasulu, 1989). In the urban areas, the rate of such marriages is higher. For example, in the metropolitan cities of Madras, Poona and Calcutta, it is around three per 100 marriages.

The emerging pattern of social organization with continued maintenance of endogamy and rules of marriage among subdivided populations offers a mosaic picture of the biological composition with wide genetic diversity characterized by gradients of gene frequencies of traits, clinal and regional variations and also uneven distribution of very high or very low frequency of few characters scattered in some populations or a bumpy biological surface. Another important dimension of the Indian populations, especially among small populations, is that it offers potential opportunities for the operation of microevolutionary forces, which bring rapid changes in gene frequency of certain traits.

Several studies have been conducted to: (1) unravel the genetic diversity and the origin of the Indian populations, (2) study the microevolutionary process that are operating in Indian populations and the causes of biological change, and (3) investigate the applicability of theoretical models of population structure to specific subdivided populations. These studies in different parts of the country (geographical, regional, tribal, large scale and small scale) have unequivocally revealed that there is a large genetic diversity in Indian populations. For example, the studies on blood groups and genetic polymorphisms suggest a wide range of gene frequency in all genetic systems and in general there is a great deal of variation (especially for the allele Pl, D, C, E, Se, Fy(a), Gd, PGM2-1, AK, Est.D, PTC etc. and for other genetic systems including RFLPs). Some important findings of these studies are: 1. In general, tribal and non-tribal populations are genetically, and morphologically different from non-tribal populations; (2) the southern tribal populations are different from the central and northeastern tribal populations; (3) geographically contiguous populations (castes and tribes) are genetically more similar than their linguistic affiliations; (4) some of the distinct genetic features of the tribal populations are: absence of A_2, Rh (D) (Bhalla, 1984), and a very high frequency of HbE in northeastern populations, e.g, 85% among Deshi in north West Bengal (Das et al., 1991) and among Sonoval Kachari (Deka et al., 1988), some tribes in Andhra Pradesh (e.g., Koya Dora, Naik Pond etc.) show unique frequency of transferrin variants of D_{Chi}, D_{Gond} and is absent in others, a very high frequency of Rh negative among Saraswat Brahmins in West Coast and complete absence of sickle cell trait in Bihar tribes (Roychoudhury, 1984); (5) the studies with regard to the biological validity of caste hierarchy indicate that (i) there is a large genetic diversity between castes belonging to two different *varnas*, (ii) geographically contiguous castes are genetically and morphologically more close irrespective of caste or social hierarchy, (iii) the component of genetic diversity among castes is usually of the order of 1 to 3%, whereas between castes and tribes it is about 5%; (6) geographical clines are observed in some traits, e.g. in the ABO blood groups, for which B is more frequent in the northern populations, sickle cell trait is concentrated in central, southern and western regions, HbE is predominant in northeast region (Mukherjee and Das, 1990) and G-6-PD deficiency is more prevalent among western India (Walter et al., 1992).

At least in some populations it was possible to identify the causes of the observed genetic differences being due to microevolutionary forces of genetic drift, founder effect, gene flow and to the processes of fusion and fission. For example, among the three religious groups of Gavdas, the observed differences in qualitative and quantitative characters, serological, anthropometrics and dermatoglyphics were attributable to fission and fusion (Malhotra, 1978b). There was an indirect evidence of genetic drift as a plausible cause for the observed high frequencies of attached ear lobe, absence of alleles P_2, Hp^0, low value of allele Hp^1 (0.026) and a high incidence of gene d (15%) and low values of whorls on fingers among the Nandiwalas, a nomadic group of Maharashtra (Malhotra, 1978c). Similar observations of microevolutionary trends were also found among some other small populations, for example, among the tribal populations of Andaman and Nicobar islands and in northeastern region.

Consanguinity and inbreeding:
The two kinship systems: The prevailing practice of preferential marriages among biological relatives in some communities has been in vogue over the last three millenia (Sastri, 1955). The extent of consanguinity varies considerably between populations. The overall pattern can, however, be largely explained by the patterns of marriage and kinship organization and the caste system among the Indo-Aryan speaking communities in North and Dravidian linguistic groups in southern parts and among autochthonous tribal and historical migrants (Karve, 1953; Karve and Malhotra, 1968).

There are three important marriage regulations that specify the levels of breeding isolation or endogamy of the caste in the traditional Hindu society. Each caste constitutes several endogamous divisions (*gotra, kul* etc.). There is a regulation against consanguineous marriages and specifically between descendants of paternal and maternal ancestors within seven and five generations respectively (Sanghvi, 1966). The enforcement of these regulations differ between Dravidian and Indo-Aryan traditions which are followed in the southern and northern regions. The differences between the two systems with respect to pattern of marriage regulations are as follows (Table 4). Apart from the castes, a majority of the tribal and other historical migrants, e.g., in remote hills, and practice consanguineous marriages as part of their traditional culture. Among Muslims, the second largest religious community, the practice of consanguineous marriages is common and is often influenced by the local caste norms, but it differs from other communities in the type and degree of marriage practiced.

Table 4. Kinship system and marriage pattern in India

Dravidian tradition	Indo-Aryan tradition
1. Consanguineous marriages are preferred and often prescribed	The lineage exogamy prevents consanguineous marriages (*sapinda* rule)
2. Village endogamy is practiced	Village exogamy is practiced
3. There is no restriction of marriages with the neighbouring villages.	Marriages with the neighbouring villages is not favoured.
4. Characterized by close marital net-work and less variability or diversity of choice of mates.	Characterized by large marital net-work and greater variability or diversity in choice of mates.

Geographic and social variation: The extent of consanguinity and inbreeding have been investigated in more than 300 populations from different regions (Dronamraju and Meerakhan, 1963; Sanghvi, 1966; Malhotra, 1979; Rao, 1984; Bittles et al., 1988; Ali, 1968; Basu, 1975; Bhalla and Bhatia, 1974; Roychoudhury, 1976a, 1976b). These data provide information on geographic, social, religious and temporal variation.

At the national level, three levels of inbreeding are observed: the first level is less than 0.02 in the four southern states, the second level is between 0.02 to 0.002 in the middle zone in Rajasthan, Madhya Pradesh, northern parts of Maharashtra and Orissa, and the third level is greater than 0.002 in the sub-Himalayan region and Gujarat (Fig. 4). While these are general trends, the extent of inbreeding also varies greatly at the regional or State level. For example, the inbreeding levels are lower in the Telangana region than the other parts of Andhra Pradesh (Sanghvi, 1966; Rao and Saheb, 1982). Similarly in the northern region, the prevalence of consanguineous marriages were also observed among Aroras, Khatris, Gosains and some Brahmin communities. 62.9% of consanguineous marriages with inbreeding levels of the order of 0.028 as high as in southern States is observed among Bhatias of Rajasthan, who migrated from Pakistan, (Bhalla and Bhatia, 1974). Further examples of random inbreeding, especially in small populations, even in the absence of preferences for consanguineous marriages are found among populations of the cis-Himalayan region (Bhalla and Bhatia, 1974).

Maharashtra provides an interesting pattern of distribution of consanguinity due to the influence of both Indo-Aryan and Dravidian traditions. For example, in the Dhangar caste cluster, the district-wise distribution of consanguinity levels shows a north-south gradient; the consanguinity levels are higher in the southern districts (Table 5). The matrilateral type of marriage is present in all the 24 districts and ranges from 8.69% to 46.7%. Interestingly, uncle-niece marriages are most prevalent in the border districts of Andhra Pradesh and Karnataka (Table 5).

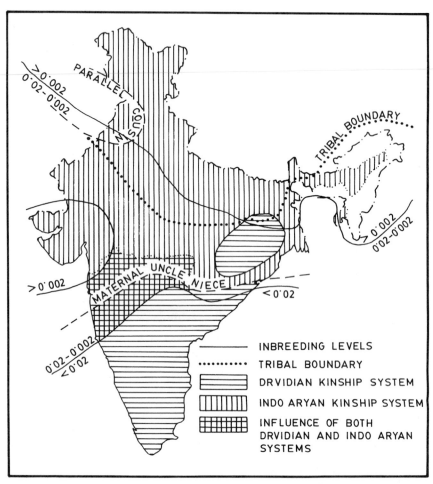

Figure 4. Extent of consanguineous marriages and inbreeding levels in India

The total inbreeding level varies from 0.008 to 0.029, with an
average of 0.016. There are also variations between the four
Dhangar castes irrespective of their spatial contiguity.
Large variations in inbreeding levels between castes is observed
irrespective of social hierarchy in certain parts of Andhra
Pradesh and Tamil Nadu. For example, in Andhra Pradesh the
uncle-niece marriages vary from 25% to 80% and the inbreeding
ranges from 0.027 to 0.071 in some castes and tribes (Table 6).
Some of the interesting features are very high frequency of
uncle-niece marriages in some of the tribal populations. Another
feature is that in some of these populations the proportion of
consanguineous marriages is greater than that of non
consanguineous marriages.   There are also rural and urban
differences in the prevalence of consanguinity. Such differences
have been found in a prospective study in Tamil Nadu (Rao, 1984).
The study has revealed a greater percentage of consanguineous
marriages (47%) in   rural areas than in urban areas (29%).

Table 5.   Inbreeding in districts of Maharashtra in the Dhangar
caste-cluster

District	Consanguineous marriages (%)					F
	N	Matrilateral	Patrilateral	U-N	Total	
Northern region						
Akola	70	10.00	1.43	–	11.43	0.021
Amaravati	90	11.11	2.22	–	13.33	0.008
Buldhana	15	46.67	–	–	46.67	0.029
Jalgaon	44	25.00	–	–	25.00	0.016
Dhulia	137	44.52	–	–	44.53	0.028
Nasik	133	21.80	–	–	21.80	0.013
Aurangabad	94	14.89	–	–	14.89	0.009
Southern region						
Kolaba	48	31.25	4.17	–	35.42	0.022
Ratnagiri	30	20.00	–	–	20.00	0.012
Kolhapur	262	16.03	9.16	4.20	29.39	0.021
Sangli	157	10.19	9.55	7.64	27.39	0.022
Satara	337	12.17	5.64	0.89	18.69	0.012
Poona	331	20.85	3.02	0.30	24.17	0.015
Ahmadnagar	145	28.28	2.76	–	31.03	0.021
Bhir	74	12.16	1.35	–	13.51	0.008
Sholapur	239	15.06	5.44	2.09	22.59	0.015
Osmanabad	69	8.69	1.45	5.80	15.94	0.013
Naded	54	12.96	5.55	5.55	24.07	0.018
Parbhani	70	20.00	2.86	–	22.86	0.014
Eastern region						
Yeotmal	67	14.92	1.49	–	16.42	0.010
Wardha	65	9.23	3.08	–	12.31	0.008
Nagapur	171	16.96	5.85	–	22.81	0.014
Bhandara	55	23.64	7.27	–	30.91	0.019
Chanda	70	20.00	2.86	–	22.86	0.014

Table 6. Extent of consanguineous marriages and inbreeding in different castes and tribes in Andhra Pradesh

Population	Consanguineous marriages			Inbreeding coefficient
	U-N	First Cousin	Total	
Insualr Yanadi	6.7	54.4	72.2	0.044
Plateau Yanadi	12.1	40.5	68.1	0.042
Konda Dora	13.6	49.5	63.1	0.048
Jalary	13.7	33.3	47.0	0.038
Madiga	14.7	13.2	28.4	0.027
Kumaris	15.3	18.0	36.1	0.031
Yadava	17.1	13.4	40.2	0.031
Fisherman	17.9	38.9	56.8	0.047
Golla	20.0	5.0	35.0	0.030
Mukadora	20.8	59.4	80.2	0.061
Vadde	21.9	3.1	25.0	0.029
Jatapu	23.3	51.7	75.0	0.061
Padmasali	24.9	37.8	62.8	0.055
Savara	26.6	34.9	61.5	0.055
Gadaba	33.0	48.1	81.1	0.071

Similarly, uncle-niece marriages are considerably more frequent in rural than in urban areas (Table 7). Among the different religious populations studied, Hindu communities showed relatively greater frequency of consanguineous marriages than Muslims and Christians (Table 8). Similarly, the results indicate that illiteracy is associated with a greater frequency of consanguineous marriages both in rural and urban areas (Table 7).

Among the Muslims though there is no apparent geographical trend in the consanguinity rates (Table 9), but a north-south trend is observed in the frequency of uncle-niece marriages. Also, parallel cousin marriages are almost absent in the south, whereas in the north especially in Jammu and Kashmir and adjacent regions, parallel cousin marriages are practiced (Roychoudhury, 1976a).

Table 7. Inbreeding coefficient in rural and urban areas by literacy levels in Tamil Nadu

	Rural		Urban	
	Wife	Husband	Wife	Husband
Illiterate	0.038	0.038	0.020	0.018
Primary school	0.033	0.036	0.027	0.025
Middle school	0.035	0.037	0.020	0.024
Secondary school and above	0.028	0.032	0.016	0.018

Table 8. Consanguineous marriages among rural and urban and religious populations

Consanguineous marriages	Rural	Urban		Inbreeding Coefficient Rural	Urban
Uncle-Niece	15.8	6.6			
First Cousin	24.8	17.5			
First Cousin once removed	3.4	3.0	Hindus	0.037	0.021
II Cousin	2.3	1.6	Muslim	0.020	0.011
> II Cousin	0.6	0.5	Christian	0.022	0.012
Total	46.9	29.1			
Inbreeding Coefficient	0.0371	0.0204			

Table 9. Extent of consanguineous marriages and inbreeding among Muslims in different states of India

State/region	No. of marriages	Type of marriage			Inbreeding coefficient
		UN	I.C	Total	
Andhra Pradesh	356	3.37	32.87	36.2	0.025
Tamil Nadu	6116	0.88	16.29	19.5	0.012
Kerala	215	-	16.74	22.3	0.012
Maharashtra	2014	-	14.00	20.71	0.010
Madhya Pradesh	351	7.41	19.09	59.3	0.026
Rajasthan	412	-	31.55	41.3	0.022
Uttar Pradesh and Delhi	1000	-	27.7	49.4	0.020
	1483	-	11.4	27.6	0.010
West Bengal	835	-	20.36	22.2	0.013
	471	-	8.28	19.3	0.007

The Parsis and Iranis, migrant groups from west Asia, show the prevalence of all the four types of first cousin marriages with equal intensity; it is about 52.0% for parallel cousins and about 48% for cross cousin marriages.

Temporal variation: Studies on inbreeding for the last four decades have shown no apparent temporal trend in overall consanguinity levels. But such temporal trends were observed among Parsis and Iranis (Sanghvi et al., 1956; Undevia and Balakrishnan, 1978). Among Parsis the percentage of consanguineous marriages declined from 28.7% in 1901 to 8.15% in 1972 (Table 10). The decrease in consanguinity is also seen in

Table 10. Details of consanguineous marriages among the Parsis by year of marriage

| Period | Number of marriages | % of consanguinity | |
		First cousin	Total
1901–1930	101		28.71
1931–1955	358		13.69
1939–1955	259	5.41	9.27
1956–1960	534	5.24	7.12
1961–1965	1158	4.66	7.34
1966–1972	1190	4.79	8.15

the case of first cousin marriages. The inbreeding coefficient also shows a sharp decrease of 0.57 per cent from 0.0422 to 0.0365 during 1939 to 1956, however there is no appreciable change after 1956. But in other Indian populations, such temporal changes are not apparent either from rural or urban areas. For example, in a study in Tamil Nadu, there was no significant decline in the overall consanguinity levels for the past five decades. The inbreeding coefficient for the couples who married during 1919 and for the couples who married during 1960 is about the same 2 per cent (Rao, 1984).

Inbreeding effects: The theoretical expectation that inbreeding elevates the frequency of homozygotes, more particularly so in case of recessive genes, has prompted several studies on the effects of inbreeding in Indian populations for the last four decades. Since the inbreeding effects are quantitatively different in different populations of varying degree of inbreeding levels, accordingly, different populations have been investigated for the purpose: *viz.,* 1. Large populations in which the inbreeding is rare and occasional event. 2. Large populations in which inbreeding is commonly practiced and 3. Small local populations where high levels of inbreeding is favoured. The Indian population structure provide all the three types of populations of varying degree of inbreeding levels in different regions.

(i) Qualitative effects: Inbreeding effects on fertility, mortality and sex ratio have been investigated based on both prospective and retrospective surveys from four different data sources: (a) hospitals, (b) villages, towns, districts, (c) large endogamous populations, and (d) small tribal and migrant populations.

In general, the effects of inbreeding on reproductive outcome show a large heterogeneity and present unclear and sometimes equivocal, trends. In general these can be summarized as follows: (1) The hospital data from Tamil Nadu, Karnataka, Andhra Pradesh show a higher incidence of congenital anomalies among children of consanguineous marriages (Centerwall and Centerwall, 1966; Murthy and Jamil, 1972), (2) Studies among a few endogamous populations (mostly castes) in Andhra Pradesh and among Sayyad Shias and Sunnis show a higher childhood and fetal loss among consanguineous couples. (3) A prospective study in Tamil Nadu, however, did not show appreciable change in reproductive outcome variables (mean number of pregnancies, livebirths, foetal loss) nor in the incidence of congenital malformations for different types of consanguineous marriages. The results were same in rural and urban areas, and also for the various castes and between religious groups (Rao and Inbaraj, 1977). (4) With the improvement of health conditions and medical facilities it is expected that reproductive wastage will be lessened and the frequencies of recessive diseases will be increased.   An evidence of such a trend was recently observed in a study in Karnataka (Bittles et al., 1982, 1988). Screening of a total of 407 infants and children in Bangalore and Mysore during 1982-1984 for inbreeding effects, the study showed a total of 35 genetic diseases in 63 persons (Devi et al., 1987) (Table 11).

(ii) Quantitative effects: Inbreeding effects with respect to quantitative characters were studied in a few populations with regard to anthropometric and dermatoglyphic characters, skin colour, birth weight and menarchial age (Mukherjee 1982, 1990; Rao, 1978,). (1) Inbreeding effects on the physical measurements (stature, head breadth, etc.) showed two significant components (Mukherjee, 1982), (a) linear component of increasing depression with high levels of inbreeding, and (b) a nonlinear deviation in the case of low levels of inbreeding. (2) Inbreeding effects resulting in an increase in variance and a bimodal distribution have also been observed for stature, head length, cephalic indices, menarchial age and ridge-count on fingers among caste populations of Andhra Pradesh (Mukherjee, et al., 1980; Mukherjee, 1982). (3) No systematic change was observed in average birth weight, crown length, head and chest circumference among the new born inbred babies both in urban and rural areas in Tamil Nadu (Rao, 1984).

Another important aspect of the inbreeding effects is the loss of fitness (defined as the genetic load) resulting from selection against deleterious genotypes due to recessive genes which are likely to get exposed with inbreeding. The amount of genetic load, a measure to account loss of fitness due to deleterious genes in a population, is estimated in general from reproductive data by the regression coefficients A and B. However, it is desirable to estimate the genetic load with respect to known recessive disorders. The values of B/A ratio in a population is supposed to indicate the relative importance of genetic load to mutational and segregational loci, and it varies in different populations with varying intensity of inbreeding and levels of health conditions. In a recent review Reddy (1992)

Table 11. Frequencies of genetic disorders diagnosed in the study
population

Type	Frequency	% of total	F
Single gene defects			0.0513
Autosomal dominant	11	2.7	
Autosomal recessive	24	5.9	0.0625
X linked	5	1.2	
Uncertain	4	1.0	
Chromosomal anomalies (Down's Syndrome)	7	1.7	
Polygenic disorders	12	2.9	
Total	63	15.5	

showed that in India, about 40 populations have been studied for
the genetic load based on data from reproductive outcome from
consanguineous marriages in different regions and among rural and
urban categories. At least 10 populations showed a negative
value of B for reproductive mortality, and 4 out of 32 showed B
value significantly different from zero. The overall mean of A
is 0.223 (and it ranges from 0.07 - 0.625) and mean B is 0.0383
(and ranges from -0.9 to -2.37). The number of lethal equivalents
per gamete is 0.6 and ranges from -0.76 to 2.67. The B/A ratio
ranges from -6.5 to 14.3. The average values of the parameters
B, A + B and the average number of lethal equivalents are smaller
when compared to other populations of the world, which suggests
relatively lower magnitude of genetic load for the Indian
populations despite the higher levels of inbreeding.

Gene dispersion: One of microevolutionary consequences of the
marriage patterns that restrict endogamy and practice of
consanguineous marriages is reflected in the marital distance or
the gene dispersion levels among populations. In general, in
such cases the distribution of marriage distance is expected to
be leptokurtic and positively skewed, with low mean and high
variance. More than 100 populations in different regions studied
conform to the expectations (Majumder, 1977) especially among the
tribal and other populations where they practice consanguineous
marriages and village endogamy in a few southern populations.
The northern populations is in agreement with the practice of
village exogamy and avoidance of consanguinity, however, tend to
exhibit multimodal distribution with large means and variances.
In addition to regional variation, the marital distribution also
varies according to social hierarchy: the tribal and lower caste
populations tend to show leptokurtic and positively skewed
distribution, whereas the upper castes who are geographically
more dispersed for occupational and socioeconomic reasons, show
platykurtic and multimodal distributions (Malhotra, 1980).

The patterns of gene dispersion are influenced by
geographical, cultural and occupational restrictions, which can
occur not only among large endogamous populations, but even among

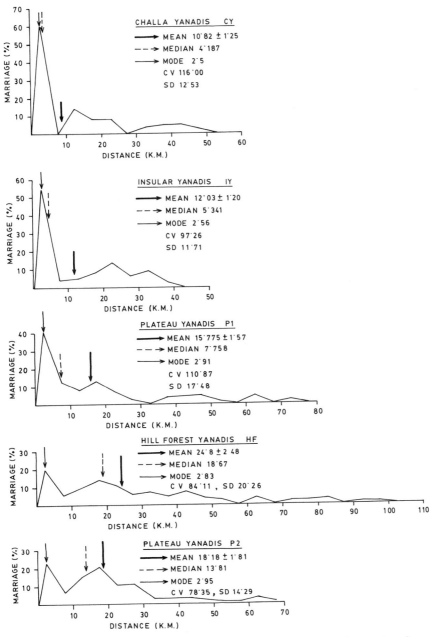

Figure 5. Distributions of matrimonial distance in regional breeding populations of the Yanadi tribe

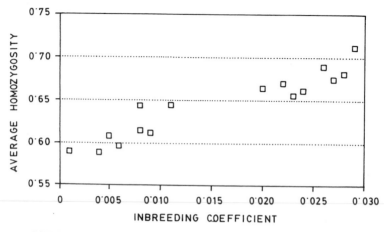

□ 16 TRIBAL POPULATIONS

Figure 6. Relationship between inbreeding coefficient and average homozygosity in 16 tribal populations of Orissa, Madhya Pradesh and Maharashtra (After Das, 1992)

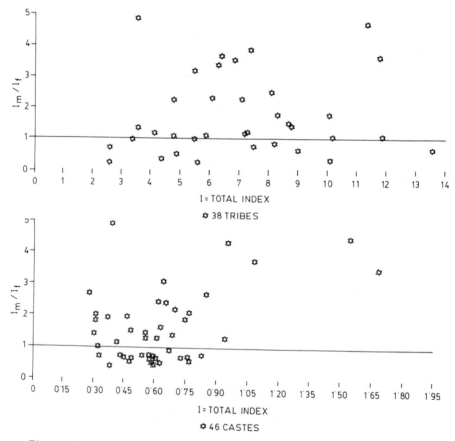

Figure 7. Distribution of Crow's indices of fertility, mortality and total selection intensity in tribal and caste populations of India

subdivided populations. For example, the Yanadi tribe who live in different regions and follow different subsistence patterns form regional breeding populations and differ in their socio-cultural variables. This is reflected in the distribution of marital distance curves (Fig. 5). This clearly shows a changing pattern from a leptokurtic to a platykurtic distribution from hunting-gathering section of the Yanadi CY, IY to settled agricultural section of the tribe HF, P2 populations.

Average Heterozygosity: The extent of genic variation within populations (or average heterozygosities) is more or less of same range in tribal or caste populations, for example, in a study among 4 nontribal and 8 tribal populations the mean average heterozygosity per locus, based on nine loci, is 25.67 with variance 2.12 and the values range from 22.38 to 28.25 (Roychoudhury, 1984). And another important finding is that the genetic relationship between castes and tribes, measured by the ratio of genetic distance to the mean of the average heterozygosities, is of the same order (0.006) as that observed among human races (Roychoudhury, 1984). Among the tribal and other caste populations who practice consanguineous marriages and reflect high levels of inbreeding, it is expected that there should be an increase in homozygosity levels and consequently a decrease in the heterozygosity levels in the genetic characters. Such a relationship was investigated among 16 tribal populations in Orissa, Madhya Pradesh and Maharashtra (Fig. 6) and also among Dhangar caste-cluster in Maharashtra (Das,. 1992; Malhotra et al., 1977). A linear positive relationship between homozygosity (based on 11 and 7 polymorphic systems) and inbreeding levels was observed in both studies.

Opportunity for natural selection:: More than about 100 populations belonging to diverse socioeconomic, linguistic and geographical classes have been studied for variation in Crow's indices of fertility, mortality and total selection (Reddy and Chopra, 1990). The total index varies from 0.258 in the Yanadi tribe in southern Andhra Pradesh to a high value of 2.25 among the Kota tribe in Tamil Nadu. About 67% of the populations show larger index of mortality than fertility component, which is especially pronounced among a majority of the tribal populations in comparison with caste populations. The tribal populations show very high values of the total index due to higher mortality and fertility components, whereas among the caste populations all the indices show lower values indicating a marginal action for the action of natural selection (Fig. 7). There is also a overall decline in I_m and I_t from rural to urban populations that are directly correlated with the influence of socio-economic factors and easy access to public health facilities. However the influence of socioeconomic variables are more pronounced on mortality in rural areas; an opposite trend was observed for urban populations. There is also a north-south trend in total index values, possibly due to the higher consanguinity and inbreeding levels in the southern region.

REFERENCES

Agrawal, D.P., Krishnamurthy, R.V., Sheela Kusumkumar, and Pant, P.K., 1978, Chronology of Indian prehistory from the Mesolithic period to the Iron Age, *J. Hum. Evol.*, 7: 37.

Alchin, B., and Alchin, R., 1982, "The rise of civilization in India and Pakistan," Cambridge Univ. Press, Cambridge.

Ali, S.M., 1968, Inbreeding and endogamy in Kerala (India), *Acta Genet. Stat. Med.*, 18: 369.

Barua, S., 1976, Selection intensity among the consanguineous and nonconsanguineous groups of a Muslim population of 24-Parganas, West Bengal, Abstract, Third Ann. Conf. Ind. Anthrop. Soc.

Basu, S.K., 1975, Effects of consanguinity among North Indian Muslims, *J. Pop. Res.*,2: 57.

Bhalla, V., 1984, Gene diversity in tribal populations of India: Illustrative maps showing the distribution of ABO, MN and Rh polymorphisms, *in* "Proceedings of Indian Statistical Institute of Golden Jubilee Conference on Human Genetics and Adaptation, Vol. II. Human Genetics," Malhotra, K.C. and Basu, A., eds., Indian Statistical Institute, Calcutta.

Bhalla, V., and Bhatia, K., 1974, Analysis of migration effect on the mating systems of a north Indian inbred Hindu community - the Bhatias of Garhi-Dhaneta, *in*: "Human Population Genetics in India", Sanghvi, L.D. et al., eds. Orient Longman Ltd., Bombay.

Bittles, A.H., Devi, A.R.R., and Appaji Rao, N., 1988, Consanguinity, twinning and secondary sex ratio in the populations of Karnataka, South India, *Ann. Hum. Biol.*15:455.

Bittles, A.H., Radha Rama Devi, A., Rao, S.V., and Rao, N.A., 1982, A new-born screening programme for the detection of amino acid disorders in south India, *Biochem. Rev.*, 52:20.

Centerwall, W.R., and Centerwall, S.A., 1966, Consanguinity and congenital anomalies in South India, A pilot study.*Ind. J. Med. Res.* 54:1160.

Das, K., 1992, Relationship between inbreeding coefficient and average homozygosity in 16 tribal groups of India, *Bull. Hum. Genet.*, 17:19.

Das, M.K., Dey, B., and Roy, M., Mukherjee, B.N., 1991, High percentage of Haemoglobin E in three populations of the Malda District, West Bengal, India, *Hum. Hered.*. 41:84.

Deka, R., Reddy, A.P., Mukherjee, B.N., Das, B.M., Banerjee, S., Roy, M., Dey, B., Malhotra, K.C., and Walter, H., 1988, Haemoglobin E distribution in ten endogamous groups of Assam, India, *Hum. Hered.*, 38:261.

Devi, A.R.R., Appaji Rao, N., and Bittles, A.H., 1987, Inbreeding and the incidence of childhood genetic disorders in Karnataka, South India, *J. Med. Genet.*, 24: 362.

Dronamraju, K.R., and MeeraKhan, P., 1963, The frequency and effects of consanguineous marriages in Andhra Pradesh, *J. Genet.*,58:387.

Gadgil, M., and Malhotra, K.C., 1982, Adaptive significance of the Indian caste system: An ecological perspective, *South Asian Study*

Goswami, H.K., 1970, Frequency of consanguineous marriages in Madhya Pradesh. *Acta Genet. Gamellol.*, 19:486.

Haq, F., 1976, Consanguinity and inbreeding among the Muslims of Murshidabad and Birbhum districts of West Bengal, *J. Ind. Anthrop. Soc.*, 11:21.

Hutton, J.H., 1963, "Caste in India," Oxford University Press.

Joshi, N.V., Gadgil, M., and Patil, S., 1993, Exploring cultural diversity of the people of India, *Current Science*, 64:10.

Karve, I., 1953, "Kinship Organization in India," Deccan College, Pune.

Karve, I., and Malhotra, K.C., 1968, Biological comparison of eight endogamous groups of the same rank, *Curr. Anthrop.*, 9: 109.

Majumder, P.P., 1977, Matrimonial migration: A review with special reference to India, *J. Biosoc. Sci.*, 9:381.

Malhotra, K.C., 1978a, Morphological composition of the people of India, *J. Hum. Evol.* 7:45.

Malhotra, K.C., 1978b, Microevolutionary dynamics among the Gavdas of Goa, *in*: "Evolutionary models and studies in Human Diversity", Meier, R.J., Otten, C.M., and Hameed, F.A., eds., Mouton publishers, The Hague.

Malhotra, K.C. 1978c, Founder effect, Gene drift and Natural selection among four nomadic Mendelian isolates, *in*: "Evolutionary models and studies in Human Diversity", Meier, R.J., Otten, C.M., and Hameed, F.A., eds., Mouton Publishers, The Hague.

Malhotra, K.C., 1979, Inbreeding among Dhangar castes of Maharashtra, India, *J. Biosoc. Sci.*, 11: 397.

Malhotra, K.C., 1980, Gene dispersion in Man: The Indian case, *Curr. Anthrop.*, 21: 135.

Malhotra, K.C., 1984, Population structure among the Dhangar caste-cluster of Maharashtra, India. *in*: "The People of South Asia", Lukacs, J.R., ed., Plenum Press, New York.

Malhotra, K.C., 1993, Persistence of Hindu caste system in different religion in India: An ecological analysis. *(Unpublished)*

Malhotra, K.C., Chakraborty, R., and Chakravarti, A., 1977, Gene differentiation among the Dhangar caste cluster of Maharashtra, India, *Hum. Hered.*, 28:26.

Maloney, C., 1974, "People of South Asia", Plenum Press, New York.

Mishra, S., 1992, The age of the Acheulian in India : New Evidence, *Curr. Anthrop.* 33: 325.

Mukherjee, B.N., Das, M.K., 1990, Spatial distribution of two abnormal Haemoglobins - HbE and HbS in Indian subcontinent, *J. Ind. Anthrop. Soc.*, 25: 39.

Mukherjee, D.P., 1982, Inbreeding and genetics of quantitative traits in man, *in*: "Proceedings of Indian Statistical Institute of Golden Jubilee Conference on Human Genetics and Adaptation". Vol. II, Malhotra, K.C. and Basu, A., eds., Indian Statistical Institute, Calcutta. p. 533.

Mukherjee, D.P., 1990, Consanguineous marriages and their genetical consequences in some Indian populations. *in*: "Isolates, Migration and Health". Roberts, D.F., Fujuka, N. and Torizuka, K., eds., Cambridge University Press, Cambridge.

Mukherjee, D.P., Reddy, P.C., Lakshmanudu, M., 1980, Dermatoglyphic effects of inbreeding, *J. Ind. Anthrop. Soc.*, 15: 67.

Murthy, J.S., Jamil, T., 1972, Inbreeding load in the newborns of Hyderabad, *Acta Genet. Med. Gamellol.*, 21:327.

Rao, N.Y.K., Saheb, S.Y., 1983. Inbreeding in different ethnic groups of Indian populations. *Przegald. Anthropol.*, 492: 2.

Rao, P.S.S., 1984, Inbreeding in India, *in*: "The People of South Asia", Lukacs, J.R., ed., Plenum Press, New York.

Rao, P.S.S., and Inbaraj, S.G., 1977, Inbreeding effects on human reproduction in Tamil Nadu, South India, *Ann. Hum. Genet.*,41: 87.

Roy Burman, B.R., and Harit, H.L., 1971, "A Preliminary Appraisal of the Scheduled Tribes of India," Office of the Registrar General of India, New Delhi.

Roychoudhury, A.K., 1976a, Inbreeding in Indian populations. *Trans. Bose Res. Inst.*, 39: 65.

Roychoudhury, A.K., 1976b, Incidence of inbreeding in different States of India. *Demography India*, 1,2: 108.

Roychoudhury, A.K., 1984, Genetic relationship of Indian populations, *in*: "Proceedings of the Indian Statistical Institute Golden Jubilee International Conference on Human Genetics and Adaptation, Vol. 2. Human Genetics," Malhotra, K.C. and Basu, A., eds., Indian Statistical Institute, Calcutta.

Reddy, B.M., 1992, Inbreeding effects on reproductive outcome: A study based on large sample from the endogamous Vadde of Kolleru lake, Andhra Pradesh, India. *Hum. Biol.*, 64:659.

Reddy, B.M., Chopra, V.P., 1990, Opportunity for natural selection among the Indian populations. *Am. J. Phys. Anthrop.*, 83:281.

Sanghvi, L.D., 1966, Inbreeding in India, *Eug. Quartly.*, 13: 291.

Sanghvi, L.D., Verde, D.S., Master, H.R., 1956, Frequency of consanguineous marriages in twelve endogamous groups in Bombay, *Acta Genet. Stat. Med.*, 6:41.

Sankalia, H.D., 1974, "Prehistory and Protohistory of India and Pakistan," Deccan College, Poona.

Sastri, K.A.N., 1955, "A History of South India from Prehistoric Times to the Fall of Vijayanagara Period," Oxford University Press, Madras.

Singh, K.S., 1993, People of India: The profile of national project (1985-92). *Curr. Sci.*, 64:10.

Undevia, J.V., Balakrishnan, V., 1978, Temporal changes in consanguinity among the Parsis and Irani communities of Bombay, *in*: "Medical genetics in India", Vol.2, Verma, I.C. and Puri, R.K., eds., Auroma Enterprises, Pondicherry.

Vasulu, T.S., 1989, Genetic structure of tribal population: 1. Breeding isolation among the Yanadis, *Internat. J. Anthrop.*, 4: 255.

Vidyarthi, L.P., 1983, Tribes of India, *in* "People of India", XV Int Cong Genet, Dec. 11-21, ICMR, Delhi.

Walter, H., Danker-Hopfe, H., Bhasin, M.K., 1992, "Anthropologie Indiens," Gustav Fischer Verlag, Stuttgart.

GENETIC VARIATION AND RELATIONSHIPS AMONG POPULATION GROUPS OF EUROPE

Hubert Walter and Heidi Danker-Hopfe

Department of Human Biology, University of Bremen
D-2800 Bremen 33, Germany

INTRODUCTION

The study of genetic variation in man and its causes is without doubt one of the most interesting and important topics of human biology. Comprehensive presentations of this field of human biological research – the genetics of human populations – have been given among others by Cavalli-Sforza and Bodmer (1971), and more recently by Vogel and Motulsky (1986) and Harrison (1988). It has been shown by numerous population studies in almost every part of the world that the considerable genetic variation in human population is the result of the operation of several factors such as mutation, selection, genetic drift, founder effects, gene flow, geographical, social and cultural isolation, which, however, do not operate independently, but are intricately interrelated. At present it is, however, almost impossible to evaluate the specific contributions of the various factors and their interrelationships to the genetic profile of any human population or group of populations in complete detail.

In this paper genetic variation and relationships among population groups of Europe will be analyzed and discussed considering in particular the linguistic differentiation of the peoples inhabiting this continent. It is worth mentioning that Haldane (1940) was the first who presented a detailed analysis of the distribution of blood group frequencies of European peoples. At that time, of course, data on only ABO blood groups were available. Haldane (1940) considered 75 European populations, but even on the basis of this comparatively small data he pointed out: "Besides the well-known gradient from east to west in which the frequency of B increases and that of O diminishes as we move eastwards, there is a striking variation in the frequencies of A and O among the peripheral populations of western Europe with low frequencies of B, including Scandinavia, Iceland, Britain, Spain, Portugal, and Sardinia" (p. 478). This distribution pattern was corroborated by many subsequent population studies (see e.g., Mourant, 1983; Mourant et al., 1976; Walter 1962).

Presently not only much more ABO blood group data are available for European populations, but also data on other erythrocyte polymorphisms, serum protein and red cell enzyme polymorphisms, the HLA system, etc. are available (see Mourant et al., 1976; Tills et al., 1983; Roychoudhury and Nei, 1988). We are however, far from being acquainted with the detailed distribution pattern of genetic markers in Europe. There is a severe paucity of frequency data not only for many east European populations, where due to several reasons genetic population studies were hardly possible in the past, but also for many populations of

Human Population Genetics, Edited by P.P. Majumder
Plenum Press, New York, 1993

the western parts of the continent, e.g., England, Scotland, Wales, Ireland, parts of Scandinavia, France, etc. This emerges from Tables 1-7, in which the weighted allele and haplotype frequencies, respectively, are shown for the polymorphic systems A1A2B0, MNSs, Rhesus (RH), Haptoglobin (HP), Transferrin (TF), Group Specific Component (GC), and Protease Inhibitor – Alpha$_1$-Antitrypsin – (PI). For the other blood group, serum protein and red cell enzyme polymorphisms the data base is still poor! We therefore had to confine our analyses to the above mentioned seven polymorphic systems of the blood, for which more or less sufficient data were available in the literature. In order to understand the variation of genetic markers among the European populations in detail, representative population samples from almost all European regions should be investigated in the near future, considering as many polymorphic blood group, serum protein and red cell enzyme systems as possible. This could be the objective of a joint international research project.

In the following, the distribution of allele and haplotype frequencies, respectively, of the seven polymorphic systems under study are analyzed with regard to the linguistic differentiation of European populations, irrespective of the comparatively poor data base. Similar attempts have been made by Cavalli-Sforza (1992) on a world-wide scale, by ourselves in India (Walter et al., 1991) and in Europe by Sokal (1991), Sokal et al. (1988, 1989a, 1989b) and Stevenson et al. (1988). The genetic data, on which the present analyses are based, were taken from the comprehensive compilations by Mourant et al. (1976), Tills et al. (1983), and Roychoudhury and Nei (1988). More recent data published in various journals as well as many unpublished data sets collected by us have also been included.

THE LINGUISTIC DIFFERENTIATION OF EUROPE

According to modern linguists – see e.g. Ruhlen (1987) – the majority of the present day spoken European languages belong either to the Indo-European or to the Uralic-Altaic language family, which can be subdivided into various language groups. In addition to these Indo-European and Uralic-Altaic languages spoken in Europe there are two more, which, however, are completely isolated and different from all the other European languages: Basque and Romany.

The following survey shows the present-day linguistic division of Europe. It is based on Ruhlen (1987) and The Cambridge Encyclopedia (1991, p.434):

I. Indo-European Language Family

1. *Germanic Group* (1. Icelandic, 2. Norwegian, 3. Swedish, 4. Danish, 5. German, 6. English, 7. Dutch, 8. Frisian)
2. *Celtic Group* (1. Irish Gaelic (Erse), 2. Scots Gaelic, 3. Manx Gaelic, 4. Welsh, 5. Breton)
3. *Romance Group* (1. Portugese, 2. Spanish (Castilian), 3. Catalan, 4. French, 5. Provencal, 6. Italian, 7. Sardinian, 8. Rhaetian, 9. Rumanian)
4. *Greek Group* (1. Greek)
5. *Albanian Group* (1. Albanian)
6. *Slavic Group* (1. Belorussian, 2. Russian, 3. Ukrainian, 4. Bulgarian, 5. Macedonian, 6. Serbo-Croatian, 7. Slovene, 8. Polish, 9. Czech, 10. Slovak, 11. Sorbian)
7. *Baltic Group* (1. Latvian, 2. Lithuanian)

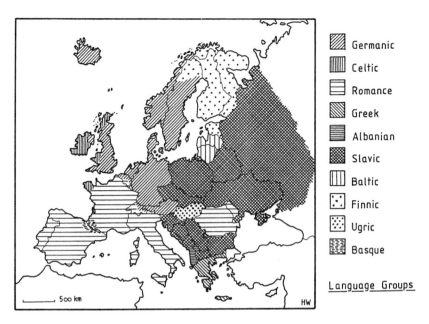

Figure 1. Distribution of language groups in Europe.

II. Uralic-Altaic Language Family

1. *Finnic Group*   (1. Finnish, 2. Lapp (Saami), 3. Estonian)
2. *Ugric Group*   (1. Hungarian)
3. *Turkic Group*   (1. Turkish)
4. *Isolated Languages* (1. Basque, 2. Romany).

Figure 1 shows the distribution of the different language groups in Europe. The Turkic Group (Turkish), though spoken in the southeastern parts of the Balkan Peninsula, was not considered in our analyses; it is mentioned in the above survey only for the sake of completeness.

For a detailed description of all these languages, their history and the number of their present speakers, we refer the reader to Ruhlen (1987). However, some remarks are indispensable in connection with the linguistic grouping of the population samples considered in this study.

In many European countries there is only one native language, e.g. in Iceland, in Denmark or in Portugal, or only one language is used by the overwhelming majority of the inhabitants, such as in Germany, England, Italy or Poland. In these more or less monolingual countries it is therefore generally not difficult to assign a population sample into a specific language group or even language. Greater problems arise in multilingual countries such as Belgium (Dutch and French), Finland (Finnish, Lapp, Swedish) or Switzerland (German, French, Italian, Rhaetian). In samples from these countries it is not always clear to which language group the tested individuals belong. Another difficulty concerns population samples from Scotland, Wales and Ireland. In these regions English is the official language, but many individuals living there are bilingual, that means that they speak in addition to English also a Celtic language. In France the official language is French, but beside it two more languages are still spoken by a certain number of individuals: Breton (a Celtic language) in Brittany (northwest France) and Provencal (a Romance language), which was originally widespread in the southern parts of France. However, almost all Breton and Provencal speakers are now bilingual; they also speak French (Ruhlen, 1987). This bilinguality is found in several parts of the United Kingdom, France, Ireland, as well as in some other European regions. This makes it somewhat difficult to assign a population sample to its actual native language, and therefore the language affiliation should be recorded carefully in all those parts of Europe in which more than one language is spoken. These brief comments underline the complicated linguistic situation in Europe, which is still much more complicated if the various dialects present in all European languages are considered, e.g. in Germany.

Finally some notes should be devoted to the two isolated languages spoken in Europe: Basque and Romany. Basque is the native language of about 500 000 individuals in the northwestern parts of Spain and the neighbouring regions of France. It is not related to any other European language and is thought to be a remnant of the languages spoken in the western parts of Europe before the arrival of the Indo-European speaking populations around 3000 years BC, whose homeland has been placed in the region to the north of the Black Sea (Ammermann and Cavalli-Sforza, 1984). The genetic peculiarities of the Basques, who are characterized, for example, by extremely high frequencies of RH negative phenotype (more than 40% in contrast to approximately 15% in other European populations), have been summarized by Mourant (1983).

Romany, the language of the European gypsies (Roma and Sinti), is of Indo-Aryan origin. Though some dialects of Romany have taken on features of the languages with which these people have come into contact in various parts of Europe, to which they immigrated from the northwestern regions of India since the 14th century (Rishi, 1976), their language is quite different from all the other languages spoken in Europe. The first European country the Roma reached was Greece, where they lived for a long period. From there one branch pushed straight up to the Baltic countries, while the rest dispersed across central Europe and eventually to western Europe; in Germany they first appeared in 1417 (Rishi, 1976). According to Vossen (1983) the number of Roma in Europe is approximately 4 million; most of them live in the eastern and southeastern parts of the continent. The number of Roma, who perished as a consequence of the German "Rassenpolitik" (racial politics) between 1933 and 1945 is estimated to be more than 275 000 individuals (Vossen, 1983). Due largely to the social and cultural isolation of the Roma from the other European populations they show considerable genetic peculiarities (Rishi, 1976; Mourant, 1983), and even among the European Roma some genetic diversification has been observed (Mastana and Papiha, 1992).

VARIATION OF GENETIC MARKERS IN EUROPE

For the analysis of the variation of genetic markers in Europe, seven polymorphic blood group and serum protein systems could be considered as sufficient frequency data are available only for these systems. These seven system are: A1A2B0, MNSs, RH, HP, TF, GC, and PI. In order to evaluate the frequency distribution of the A2 allele of the ABO system only those population studies in which the whole set of phenotypes was typed were considered. The number of such studies is unfortunately much lower than that in which only the four main ABO phenotypes were tested. This applies in particular to east European populations. As for the TF, GC and PI systems it has to be emphasized that only those studies which are based on subtypings were considered.

The population samples were grouped according to their linguistic affiliation. Weighted mean allele and haplotype frequencies, respectively, were calculated for each language and language group, respectively. The homogeneity of allele and haplotype frequency distributions among the ten language groups was tested by means of the nonparametric Wilcoxon test.

A1A2B0 system: Table 1 shows the distribution of A1A2B0 allele frequencies by language and language groups, respectively. Though for some of the language groups at present only very few data are available (e.g., for the Baltic or Basque group) there are clear differences in the distribution of the four alleles of the A1A2B0 blood group system, which turned out to be highly significant (Table 8).

Table 1 shows that there are generally high A1 frequencies in the Slavic, Ugric (Hungarian), Baltic and Romany groups (> 0.22), whereas the lowest frequency of this allele is seen in the Celtic group (0.14). The Slavic, Ugric and Baltic groups are also characterized by high B allele frequencies (> 0.14), which are somewhat lower in the Finnic and Romany groups and considerably lower in all the other language groups, especially in the Basque group (0.03). In contrast to this, the 0 allele frequencies are rather low in these five east European language groups including the Romany group (< 0.57); whereas they are much higher in all others, in particular in the Celtic and Basque groups (> 0.70). As a whole one can point out that the language associated distribution pattern of B and 0 allele frequencies is in good accordance with the well-known east-west gradient of these two frequencies (Haldane, 1940; Mourant, 1983).

This is not the place to go into further details and to analyze the intragroup variations. This has to be left to further studies. But some observations should be annotated briefly. Within the Germanic group the Icelandic speakers are striking with very high 0 allele frequencies (0.76), and the same is seen in the Sardinian (0.74) and Corsian (0.74) speakers in the Romance group. This can be explained without doubt by the geographic isolation of these island populations, whereby the spread of the recessive 0 allele was apparently favoured. While the B allele frequency is rather low within the Romance Group the Rumanian speakers are striking by a comparatively high frequency of this allele (0.12), which can most probably be attributed to gene flow from B allele rich populations coming into Rumania in the course of her history. These few examples illustrate the importance of locally acting factors for the understanding of the genetic variation within the various European language groups.

The A2 frequency finally reveals a distribution pattern, which is of particular interest. The incidence of this allele is by far the highest in the Finnic group (> 0.12), followed by the Romany (0.10) and the Basque groups (0.09), whereas it is obviously less frequent in all the other language groups. Analyzing the Finnic group in detail it is seen that the high A2 allele frequency is essentially caused by the Lapp (Saami) speakers, in which locally this frequency amounts up to even 0.372 (Finnmark in northern Sweden). The reasons for these high A2 allele frequencies in the Lapp speakers as well as in the Finnish speakers and in the Romany and Basque groups are unknown. Genetic drift may have played some role in this connection.

MNSs system: Table 2 shows the distribution of MNSs haplotype frequencies. Remarkable variation is seen only for the haplotypes Ms, NS and Ns. The MS frequency seems to vary. Heterogeneities in frequency distributions turned out to be statistically significant only for the NS and Ns haplotypes (Table 8). However, no clear association between MNSs haplotype frequencies and language groups is discernible except for the Ms haplotype frequencies, which are rather similar in the Baltic, Finnic, Slavic and Ugric (Hungarian) groups and altogether somewhat higher (> 0.33) than in all the other European language groups.

In most of these groups the intragroup variability is quite low. Remarkable deviations from the relevant mean values are seen only for the Icelandic speakers of the Germanic group, the Sardinian and Corsian speakers of the Romance group also for the Lapp speakers of the Finnic group. One can assume that the specific MNSs haplotype frequencies existing in the above mentioned population groups are the result of locally acting genetic differentiation factors such as genetic drift.

The most interesting MNSs haplotype distribution pattern is observed in the Romany

Table 1. Distribution of A1A2B0 allele frequencies by languages

Language **Language Group**	N	n	A1	A2	B	0
Icelandic	9	1 953	0.1240	0.0636	0.0565	0.7559
Norwegian	10	15 351	0.2287	0.0821	0.0680	0.6212
Swedish	28	32 412	0.2164	0.0869	0.0790	0.6177
Danish	5	20 025	0.2035	0.0743	0.0757	0.6465
German	28	148 801	0.2060	0.0665	0.0836	0.6439
English	5	5 142	0.1948	0.0662	0.0613	0.6777
Dutch	7	2 414	0.1973	0.0674	0.0692	0.6661
Germanic Group	**92**	**226 098**	**0.2078**	**0.0711**	**0.0803**	**0.6408**
Irish	5	2 514	0.1333	0.0392	0.0666	0.7609
Gaelic	2	697	0.1544	0.0554	0.1086	0.6816
Celtic Group	**7**	**3 211**	**0.1379**	**0.0427**	**0.0757**	**0.7437**
Portuguese	5	4 314	0.2320	0.0703	0.0620	0.6357
Spanish	11	12 121	0.2338	0.0327	0.0565	0.6770
French	11	12 887	0.2306	0.0632	0.0598	0.6464
Italian	36	39 991	0.2036	0.0445	0.0758	0.6761
Sardian	7	30 583	0.1444	0.0500	0.0662	0.7394
Rumanian	13	10 691	0.2406	0.0514	0.1182	0.5898
Corsian	4	1 937	0.1667	0.0575	0.0394	0.7364
Romance Group	**87**	**112 524**	**0.1978**	**0.0487**	**0.0721**	**0.6814**
Greek	**13**	**5 961**	**0.2090**	**0.0595**	**0.0952**	**0.6363**
Bulgarian						
Serbo-Croatian	4	643	0.2021	0.0616	0.1290	0.6073
Slovakian	2	4 052	0.2752	0.0282	0.1465	0.5501
Czech	5	17 035	0.2488	0.0460	0.1426	0.5626
Polish	4	16 809	0.2282	0.0533	0.1515	0.5670
Russian	1	1 549	0.2150	0.0400	0.1560	0.5890
Slavic Group	**16**	**40 088**	**0.2408**	**0.0473**	**0.1470**	**0.5649**
Latvian	1	444	0.2260	0.0410	0.1820	0.5510
Lithuanian						
Baltic Group	**1**	**444**	**0.2260**	**0.0410**	**0.1820**	**0.5510**
Finnish	15	14 659	0.2142	0.0944	0.1252	0.5662
Estonian	1	97	0.2280	0.0560	0.1100	0.6060
Lapp	12	3 013	0.1373	0.2656	0.0914	0.5057
Finnic Group	**28**	**17 769**	**0.2012**	**0.1232**	**0.1194**	**0.5562**
Ugric Group	**4**	**9 412**	**0.2305**	**0.0674**	**0.1421**	**0.5600**
Basque	**2**	**189**	**0.1809**	**0.0858**	**0.0268**	**0.7065**
Romany	**4**	**687**	**0.2211**	**0.0992**	**0.1121**	**0.5676**
Total	**254**	**416 383**	**0.2080**	**0.0646**	**0.0879**	**0.6395**

Table 2. Distribution of MNSs haplotype frequencies by languages

Language **Language Group**	N	n	MS	Ms	NS	Ns
Icelandic	9	1 953	0.1640	0.4317	0.0645	0.3398
Norwegian	4	957	0.2343	0.3212	0.0664	0.3781
Swedish	3	5 038	0.2409	0.3214	0.0858	0.3519
Danish	2	2 444	0.2240	0.3190	0.0630	0.3940
German	24	35 560	0.2484	0.2959	0.0712	0.3845
English	11	5 710	0.2393	0.3077	0.0623	0.3907
Dutch	5	1 548	0.2453	0.2900	0.0616	0.4031
Germanic Group	**58**	**53 210**	**0.2422**	**0.3059**	**0.0706**	**0.3813**
Irish	3	2 097	0.2708	0.3074	0.0605	0.3613
Gaelic	6	1 542	0.2578	0.3114	0.0401	0.3907
Celtic Group	**9**	**3 639**	**0.2653**	**0.3091**	**0.0519**	**0.3737**
Portuguese	1	302	0.2550	0.3000	0.0710	0.3740
Spanish	11	6 047	0.2446	0.2798	0.0882	0.3874
French	2	217	0.2542	0.2810	0.0964	0.3684
Italian	26	20 142	0.2412	0.3117	0.0895	0.3576
Sardian	6	2 962	0.3291	0.3728	0.0687	0.2294
Rumanian	1	34	0.3000	0.2890	0.1040	0.3070
Corsian	1	132	0.2730	0.2430	0.0790	0.4050
Romance Group	**48**	**29 836**	**0.2510**	**0.3106**	**0.0870**	**0.3514**
Greek	**15**	**3 946**	**0.2716**	**0.2919**	**0.1227**	**0.3138**
Bulgarian Serbo-Croatian	3	184	0.2052	0.3843	0.0689	0.3416
Slovakian	2	1 546	0.2482	0.3195	0.1045	0.3278
Czech	2	733	0.2299	0.3220	0.0754	0.3727
Polish	3	451	0.2341	0.3427	0.1005	0.3227
Russian	1	563	0.2530	0.3430	0.0830	0.3210
Slavic Group	**11**	**3 477**	**0.2410**	**0.3303**	**0.0925**	**0.3362**
Latvian	1	444	0.2660	0.3660	0.0620	0.3060
Lithuanian **Baltic Group**	**1**	**444**	**0.2660**	**0.3660**	**0.0620**	**0.3060**
Finnish	13	12 946	0.2470	0.3950	0.0783	0.2797
Estonian Lapp	5	1 252	0.2673	0.2524	0.1592	0.3211
Finnic Group	**18**	**14 198**	**0.2488**	**0.3825**	**0.0854**	**0.2833**
Ugric Group	**9**	**2 702**	**0.2439**	**0.3307**	**0.1072**	**0.3182**
Basque	**6**	**1 241**	**0.2249**	**0.3219**	**0.1037**	**0.3495**
Romany	**1**	**56**	**0.2110**	**0.5390**	**0.1010**	**0.1490**
Total	**176**	**112 749**	**0.2470**	**0.3183**	**0.0799**	**0.3548**

group: extremely high MS and extremely low Ns haplotype frequencies (0.54 and 0.15, respectively). However, so far only one small Hungarian Roma sample has been typed for the MNSs system (Tauszik and Hummel, 1991). More studies are necessary before general conclusions can be drawn. Considering only the M and N allele frequencies one obtains: M = 0.75, N = 0.25. They are in good agreement with the results of MN studies in various Roma populations (Mourant, 1983; Rishi 1976).

RH system: RH haplotype frequencies are summarized in Table 3. A marked heterogeneity in frequencies is observed, which is statistically significant in all cases (Table 8). Except for the Greek group (0.02) the frequency of the CDE haplotype is very low everywhere. It varies from 0.00 in the Baltic group to < 0.01 in the Celtic group, though locally these frequencies sometimes reach values up to 0.015, e.g. in a Gaelic speaking sample from Scotland. Rather low haplotype frequencies are also found for Cde (from < 0.01 in the Celtic group to ~ 0.03 in the Basque and Romany groups) and for cdE (from < 0.01 in the Finnic group to 0.01 in the Greek group). In several populations these frequencies can reach somewhat conspicuous values, which may be due to locally acting genetic differentiation factors such as genetic drift. For all these RH haplotypes no clear language associated distribution patterns are discernible.

In contrast to this the other RH haplotypes exhibit marked distribution differences, which are more or less associated with the linguistic differentiation of the European populations. Thus the CDe haplotype shows obviously high frequencies in the Romany (0.57), the Romance (0.53) and the Greek (0.50), whereas it is less frequent in all the other language groups, especially in the Basque Group (0.40). The cDE haplotype reveals high frequencies only in the Baltic and Finnic groups (0.18). Its incidence is much lower in the other groups, especially in the Romany (0.08) and the Basque (0.07). High frequencies of the cDe haplotypes are seen only in the Greek and Finnic group (0.04), though locally they can reach rather high also in other groups values, e.g. 0.08 in the Corsian one. The cde haplotype frequency finally is extremely high in the Basque (0.47) and Celtic groups (0.40), but high frequencies are also found in the Germanic (0.39), Slavic (0.38) and Ugric (Hungarian) groups (0.38). The lowest frequencies are recorded for the Romance (0.32) and in particular for the Greek (0.28). In some of the language groups, however, there is a considerable variation of cde haplotype frequencies, e.g. in the Romance, where the Sardinian speakers are striking by very low frequencies of this haplotype (0.22) the Spanish, French and Portuguese speakers are characterized by high frequencies of it (~0.41). In the Finnic group, the Lapp speakers also have by extremely low values (0.19) of the cde haplotype.

Overall the CDe haplotype is rather frequent among the south European language groups (Romance and Greek), and less frequent among the Germanic and Celtic groups of northern Europe. The cDE haplotype shows a reverse distribution pattern, and the cde frequency is seen to be the lowest in the Lapp speakers, but is also low in most of the Romance and Greek speaking populations of southern Europe.

HP system: Table 4 shows the distribution of HP allele frequencies in various European language groups. The variation is rather low as far as the Indo-European and Uralic-Altaic language families are concerned. Obviously differing from these two groups are on the one side the Basque speakers with comparatively high HP*1 and low HP*2 allele frequencies and on the other side the Romany speakers with very low HP*1 and high HP*2 allele frequencies. It is worth mentioning that the Romany group shows a HP allele distribution which is characteristic for all populations of India (Walter et al., 1991). The statistically significant heterogeneity in the distribution of HP allele frequencies (Table 8) is most likely due to the extremely differing frequencies in the Basque as well as in the Romany group.

TF system: The distribution of TF allele frequencies in the various European language groups is shown in Table 5. As can be seen from this table the number of TF subtyped

Table 3. Distribution of Rhesus haplotype frequencies by languages

Language Language Group	N	n	CDE	CDe	Cde	cDE	cDe	cdE	cde
Icelandic	11	4 212	0.0000	0.4406	0.0083	0.1680	0.0169	0.0078	0.3584
Norwegian	13	16 162	0.0002	0.4396	0.0037	0.1589	0.0152	0.0055	0.3769
Swedish	17	34 386	0.0004	0.4167	0.0097	0.1669	0.0188	0.0042	0.3833
Danish	6	21 173	0.0006	0.4198	0.0138	0.1539	0.0175	0.0066	0.3878
German	28	41 449	0.0028	0.4204	0.0133	0.1370	0.0208	0.0071	0.3986
English	7	4 348	0.0025	0.4076	0.0074	0.1507	0.0214	0.0074	0.4030
Dutch	11	4 412	0.0000	0.4385	0.0103	0.1378	0.0179	0.0080	0.3875
Germanic Group	**93**	**126 142**	**0.0012**	**0.4226**	**0.0107**	**0.1523**	**0.0188**	**0.0061**	**0.3883**
Irish	4	2 398	0.0019	0.3942	0.0082	0.1529	0.0292	0.0055	0.4081
Gaelic	5	1 188	0.0152	0.4141	0.0093	0.1670	0.0142	0.0062	0.3740
Celtic Group	**9**	**3 586**	**0.0063**	**0.4007**	**0.0086**	**0.1576**	**0.0242**	**0.0058**	**0.3968**
Portuguese	2	1 416	0.0009	0.4060	0.0028	0.1114	0.0553	0.0046	0.4190
Spanish	13	8 839	0.0043	0.4378	0.0140	0.1005	0.0347	0.0030	0.4057
French	21	16 922	0.0016	0.4284	0.0103	0.1173	0.0238	0.0060	0.4126
Italian	50	27 005	0.0050	0.4605	0.0212	0.1181	0.0342	0.0066	0.3544
Sardian	14	37 973	0.0072	0.6531	0.0113	0.0811	0.0163	0.0069	0.2241
Rumanian	2	935	0.0000	0.4475	0.0169	0.1239	0.0216	0.0165	0.3736
Corsian	5	1 801	0.0000	0.4783	0.0072	0.0968	0.0791	0.0082	0.3304
Romance Group	**107**	**94 891**	**0.0050**	**0.5290**	**0.0141**	**0.1011**	**0.0263**	**0.0064**	**0.3181**
Greek	**17**	**8 420**	**0.0207**	**0.4972**	**0.0230**	**0.1264**	**0.0429**	**0.0125**	**0.2773**
Bulgarian Serbo-Croatian	9	36 338	0.0010	0.4434	0.0149	0.1329	0.0200	0.0031	0.3847
Slovakian	5	3 670	0.0048	0.4043	0.0269	0.1189	0.0495	0.0056	0.3900
Czech	3	1 156	0.0037	0.4665	0.0075	0.1335	0.0436	0.0017	0.3435
Polish	8	16 690	0.0054	0.4038	0.0161	0.1350	0.0391	0.0026	0.3980
Russian	7	5 650	0.0019	0.4232	0.0239	0.1623	0.0276	0.0030	0.3581
Slavic Group	**32**	**63 504**	**0.0025**	**0.4293**	**0.0166**	**0.1352**	**0.0279**	**0.0031**	**0.3854**
Latvian Lithuanian	1	172	0.0000	0.4250	0.0230	0.1840	0.0070	0.0080	0.3530
Baltic Group	**1**	**172**	**0.0000**	**0.4250**	**0.0230**	**0.1840**	**0.0070**	**0.0080**	**0.3530**
Finnish	11	12 092	0.0002	0.4285	0.0123	0.1829	0.0355	0.0017	0.3389
Estonian	2	156	0.0000	0.3586	0.0565	0.1251	0.0292	0.0178	0.4128
Lapp	7	2 114	0.0000	0.5941	0.0023	0.1628	0.0477	0.0000	0.1931
Finnic Group	**20**	**14 362**	**0.0002**	**0.4520**	**0.0113**	**0.1793**	**0.0373**	**0.0016**	**0.3183**
Ugric Group	**13**	**7 733**	**0.0023**	**0.4222**	**0.0158**	**0.1442**	**0.0261**	**0.0055**	**0.3839**
Basque	**8**	**2 006**	**0.0049**	**0.3954**	**0.0294**	**0.0720**	**0.0291**	**0.0034**	**0.4658**
Romany	**6**	**1 223**	**0.0014**	**0.5669**	**0.0290**	**0.0818**	**0.0058**	**0.0025**	**0.3126**
Total	**306**	**322 039**	**0.0032**	**0.4586**	**0.0135**	**0.1335**	**0.0245**	**0.0055**	**0.3612**

Table 4. Distribution of HP allele frequencies by languages

Language Language Group	N	n	HP*1	HP*2
Icelandic	19	2 946	0.4107	0.5893
Norwegian	10	20 381	0.3794	0.6206
Swedish	9	26 174	0.3780	0.6220
Danish	4	4 948	0.3995	0.6005
German	41	81 397	0.3956	0.6044
English	6	5 223	0.3908	0.6092
Dutch	5	2 028	0.4171	0.5829
Germanic Group	94	143 097	0.3907	0.6093
Irish	3	2 221	0.3745	0.6255
Gaelic	7	2 232	0.3972	0.6028
Celtic Group	10	4 453	0.3858	0.6142
Portuguese	4	2 322	0.3941	0.6059
Spanish	7	5 491	0.4042	0.5958
French	11	9 851	0.4024	0.5974
Italian	58	16 956	0.3627	0.6373
Sardian	11	4 196	0.3993	0.6007
Rumanian	13	10 690	0.3362	0.6638
Corsian				
Romance Group	104	49 506	0.3741	0.6259
Greek	26	8 967	0.3484	0.6516
Bulgarian	9	5 255	0.3445	0.6555
Serbo-Croatian	9	1 804	0.3801	0.6199
Slovakian	9	2 985	0.3664	0.6336
Czech	3	1 611	0.3781	0.6219
Polish	10	59 556	0.3650	0.6350
Russian	14	4 429	0.4127	0.5873
Slavic Group	54	75 640	0.3670	0.6330
Latvian				
Lithuanian	3	4 379	0.3621	0.6379
Baltic Group	3	4 379	0.3621	0.6379
Finnish	15	14 543	0.3787	0.6213
Estonian				
Lapp	8	1 504	0.3504	0.6496
Finnic Group	23	16 047	0.3760	0.6240
Ugric Group	21	19 926	0.3623	0.6377
Basque	11	3 043	0.4453	0.5547
Romany	8	1 375	0.2012	0.7988
Total	354	326 433	0.3783	0.6217

populations is still rather small in most of these groups, and for some of them (Celtic, Baltic, Basque) no data are yet available. The discussion of the TF allele distribution pattern certainly suffers because of lack of data.

Nevertheless some interesting observations can be made. Rather high TF*C1 allele frequencies (0.76 – 0.79) are seen in almost all the Indo-European and Uralic-Altaic language speaking populations, whereas the Romany group shows the lowest frequency of this allele (0.75). In contrast to this the TF*C2 allele frequency is the highest in this group (0.21). The TF*C3 allele frequency is rather high only in the Slavic (0.07) and Finnic groups (0.07), and the latter is also characterized by comparatively high TF*B (0.06) and TF*D (0.02) allele frequencies, which are much less frequent in all the other groups. As it is seen from Table 5 some of these TF alleles show remarkable intragroup variation, e.g., TF*B in the Romance group (French speakers!) or in the Finnic group, where the Finnish speakers reveal obviously higher TF*C3, and TF*D frequencies than the Lapp speakers. Rare TF alleles (combined as TF*VAR) are found only in the Germanic, Romance and Slavic groups. As seen from Table 8 the heterogeneity in the distribution of TF allele frequencies is statistically significant only for TF*C1, TF*C2 and TF*D. The data do not show clear associations between the distribution of TF allele frequencies and the linguistic differentiation of Europe.

GC system: Table 6 shows the distribution of GC allele frequencies by language groups in Europe. It has to be emphasized that for many European populations GC subtype frequency data are still insufficient or even completely lacking. Concerning the GC*1F allele it is seen that a rather high frequency (0.19) is characteristic only for the Finnic group – caused by the extremely high frequency of this allele in the Lapp speakers, whereas the prevalence of this allele is much lower in all the other groups, especially in the Basque group (0.08). No marked variation is seen the GC*1S allele frequencies. These frequencies are comparatively high in the Basque (0.36), Ugric (Hungarian, 0.29) and Slavic groups (0.28), whereas they are lower in most of the other groups, especially in the Finnic group (0.21). From Table 8 it is seen that only the distribution heterogeneity of GC*1F and GC*2 allele frequencies proved to be statistically significant. Rare GC alleles – combined as GC*VAR – were reported only for the Germanic, Romance, Greek and Slavic groups. Their prevalence, however, is very low, and and the heterogeneity of frequencies is not statistically significant (Table 8). It may be mentioned, however, that some of the language groups, e.g. the Romance, Slavic and Finnic groups, exhibit a marked intragroup variation in the GC allele frequencies.

Overall the distribution of GC alleles shows some associations with the linguistic differentiation of Europe, and it is the Lapp and the Basque speakers, who differ clearly from all the other European language groups.

PI system: Table 7 shows the distribution of PI allele frequencies by the European language groups. For three of them — Celtic, Greek and Baltic — unfortunately no data have been reported so far, and for many populations of the other groups only limited data are available at present. All PI alleles show a more or less marked heterogeneity of distributions, which, however, is statistically significant only for PI*M1, PI*M2, PI*Z alleles. From the figures presented in Table 7 it appears that the PI*M1 allele frequency is the lowest in the Romance and Basque groups (0.68), whereas it is higher in all the other groups, especially in the Finnic group, which is without doubt due to the extremely high incidence of this allele in the Lapp speakers. The PI*M2 allele is comparatively frequent in the Romance group, which can be largely attributed to the Sardinian speakers. The lowest frequency of this allele is seen in the Lapp speakers (0.07). The incidence of PI*M3 is rather high in the Ugric (Hungarian) group (0.11), that of PI*S in the Basque group (0.11). An interesting distribution pattern appears for the PI*Z allele, as this is generally much more frequent in the Germanic group than in the others. Rare PI alleles (combined as PI*VAR) are recorded for almost all population samples from Europe. The intragroup variation of all these alleles is considerable, and it is

Table 5. Distribution of TF allele frequencies by languages

Language / Language Group	N	n	TF*C1	TF*C2	TF*C3	TF*B	TF*D	TF*VAR
Icelandic	2	287	0.8227	0.1756	0.0017	0.0000	0.0000	0.0000
Norwegian								
Swedish	6	8 757	0.7797	0.1634	0.0517	0.0011	0.0035	0.0006
Danish	1	132	0.8140	0.1860	0.0000	0.0000	0.0000	0.0000
German	25	12 022	0.7946	0.1554	0.0423	0.0053	0.0008	0.0016
English								
Dutch								
Germanic Group	**34**	**21 198**	**0.7888**	**0.1592**	**0.0454**	**0.0035**	**0.0019**	**0.0012**
Irish								
Gaelic								
Celtic Group								
Portuguese								
Spanish	10	4 285	0.7838	0.1697	0.0423	0.0030	0.0012	0.0000
French	2	503	0.7858	0.1743	0.0265	0.0134	0.0000	0.0000
Italian	36	10 951	0.7572	0.1941	0.0453	0.0014	0.0013	0.0007
Sardian	5	1 063	0.7553	0.2201	0.0236	0.0005	0.0000	0.0005
Rumanian								
Corsian								
Romance Group	**53**	**16 802**	**0.7647**	**0.1889**	**0.0426**	**0.0021**	**0.0012**	**0.0005**
Greek	**6**	**1 339**	**0.7861**	**0.1759**	**0.0343**	**0.0004**	**0.0033**	**0.0000**
Bulgarian								
Serbo-Croatian	1	94	0.7180	0.1970	0.0850	0.0000	0.0000	0.0000
Slovakian	4	682	0.7933	0.1394	0.0646	0.0000	0.0015	0.0012
Czech								
Polish	3	595	0.7709	0.1469	0.0745	0.0077	0.0000	0.0000
Russian								
Slavic Group	**8**	**1 371**	**0.7785**	**0.1466**	**0.0703**	**0.0033**	**0.0007**	**0.0006**
Latvian								
Lithuanian								
Baltic Group								
Finnish	3	1 040	0.7749	0.1012	0.0944	0.0077	0.0218	0.0000
Estonian								
Lapp	3	544	0.8221	0.1581	0.0131	0.0029	0.0038	0.0000
Finnic Group	**6**	**1 584**	**0.7913**	**0.1207**	**0.0664**	**0.0060**	**0.0156**	**0.0000**
Ugric Group	**11**	**1 280**	**0.7638**	**0.1978**	**0.0365**	**0.0000**	**0.0019**	**0.0000**
Basque								
Romany	**3**	**344**	**0.7462**	**0.2132**	**0.0393**	**0.0013**	**0.0000**	**0.0000**
Total	**121**	**43 918**	**0.7783**	**0.1708**	**0.0452**	**0.0028**	**0.0021**	**0.0008**

Table 6. Distribution of GC allele frequencies by languages

Language **Language Group**	N	n	GC*1F	GC*1S	GC*2	GC*VAR
Icelandic	2	510	0.1065	0.6207	0.2728	0.0000
Norwegian						
Swedish	6	8 406	0.1363	0.6096	0.2541	0.0000
Danish	1	1 674	0.1590	0.5720	0.2690	0.0000
German	21	13 254	0.1448	0.5745	0.2791	0.0016
English	1	100	0.1650	0.5750	0.2600	0.0000
Dutch						
Germanic Group	**31**	**23 944**	**0.1421**	**0.5876**	**0.2694**	**0.0009**
Irish						
Gaelic						
Celtic Group						
Portuguese						
Spanish	11	4 039	0.1227	0.5536	0.3236	0.0001
French	7	1 365	0.1599	0.5240	0.3117	0.0044
Italian	49	13 934	0.1492	0.5854	0.2643	0.0011
Sardian	7	1 642	0.1854	0.5514	0.2610	0.0022
Rumanian						
Corsian						
Romance Group	**74**	**20 980**	**0.1476**	**0.5727**	**0.2785**	**0.0012**
Greek	**2**	**755**	**0.1526**	**0.5921**	**0.2541**	**0.0012**
Bulgarian	1	1 134	0.1400	0.6480	0.2090	0.0030
Serbo-Croatian	4	1 067	0.1105	0.5666	0.3229	0.0000
Slovakian	4	682	0.0951	0.6057	0.2965	0.0027
Czech						
Polish	2	418	0.1303	0.5335	0.3362	0.0000
Russian	1	122	0.1920	0.6160	0.1920	0.0000
Slavic Group	**12**	**3 423**	**0.1225**	**0.5991**	**0.2769**	**0.0015**
Latvian						
Lithuanian						
Baltic Group						
Finnish	2	773	0.1297	0.6641	0.2062	0.0000
Estonian						
Lapp	1	222	0.4120	0.3670	0.2210	0.0000
Finnic Group	**3**	**995**	**0.1927**	**0.5978**	**0.2095**	**0.0000**
Ugric Group	**11**	**1 272**	**0.1239**	**0.5870**	**0.2891**	**0.0000**
Basque	**5**	**1 267**	**0.0829**	**0.5582**	**0.3589**	**0.0000**
Romany	**2**	**238**	**0.1615**	**0.5820**	**0.2565**	**0.0000**
Total	**140**	**52 874**	**0.1423**	**0.5820**	**0.2747**	**0.0010**

Table 7. Distribution of PI allele frequencies by languages

Language Language Group	N	n	PI*M1	PI*M2	PI*M3	PI*S	PI*Z	PI*VAR
Icelandic	1	42	0.8810	0.1190	0.0000	0.0000	0.0000	0.0000
Norwegian								
Swedish	2	2 931	0.7851	0.1406	0.0413	0.0157	0.0141	0.0032
Danish	1	909	0.7280	0.1360	0.0820	0.0220	0.0230	0.0090
German	16	9 572	0.7063	0.1526	0.0994	0.0233	0.0123	0.0061
English	2	1 718	0.7796	0.1185	0.0615	0.0253	0.0119	0.0032
Dutch	3	1 196	0.7116	0.1309	0.0960	0.0308	0.0101	0.0206
Germanic Group	**25**	**16 368**	**0.7303**	**0.1443**	**0.0836**	**0.0223**	**0.0130**	**0.0065**
Irish								
Gaelic								
Celtic Group								
Portuguese								
Spanish	13	4 601	0.6404	0.1624	0.0876	0.0998	0.0034	0.0064
French	5	2 543	0.6926	0.1423	0.1027	0.0468	0.0100	0.0056
Italian	40	11 053	0.6991	0.1828	0.0885	0.0209	0.0048	0.0039
Sardian	8	1 791	0.6151	0.2239	0.0873	0.0456	0.0053	0.0228
Rumanian								
Corsian								
Romance Group	**66**	**19 988**	**0.6772**	**0.1766**	**0.0900**	**0.0446**	**0.0052**	**0.0064**
Greek								
Bulgarian								
Serbo-Croatian	1	94	0.7610	0.1600	0.0640	0.0150	0.0000	0.0000
Slovakian	4	681	0.7291	0.1616	0.0993	0.0078	0.0015	0.0007
Czech								
Polish	2	423	0.6797	0.1667	0.1112	0.0165	0.0141	0.0118
Russian	4	473	0.7751	0.1362	0.0617	0.0074	0.0138	0.0058
Slavic Group	**11**	**1 671**	**0.7314**	**0.1556**	**0.0897**	**0.0103**	**0.0081**	**0.0049**
Latvian								
Lithuanian								
Baltic Group								
Finnish	4	561	0.7218	0.1414	0.1157	0.0099	0.0105	0.0007
Estonian								
Lapp	2	265	0.9261	0.0698	0.0000	0.0041	0.0000	0.0000
Finnic Group	**6**	**826**	**0.7874**	**0.1184**	**0.0786**	**0.0080**	**0.0071**	**0.0005**
Ugric Group	**10**	**1 089**	**0.7078**	**0.1560**	**0.1135**	**0.0166**	**0.0061**	**0.0000**
Basque	**7**	**2 042**	**0.6790**	**0.1469**	**0.0517**	**0.1094**	**0.0076**	**0.0054**
Romany	**3**	**424**	**0.7381**	**0.1604**	**0.0947**	**0.0020**	**0.0028**	**0.0020**
Total	**128**	**42 408**	**0.7033**	**0.1601**	**0.0861**	**0.0360**	**0.0085**	**0.0060**

Table 8. Results of the nonparametric Wilcoxon test of homogeneity of allele frequency distribution

Allele/Haplotype	χ^2	p	d.f.
A1	27.483	.0012	9
A2	77.201	.0001	9
B	62.238	.0001	9
0	87.848	.0001	9
MS	13.038	.1609	9
Ms	15.980	.0673	9
NS	45.451	.0001	9
Ns	41.956	.0001	9
CDE	21.675	.0100	9
CDe	58.118	.0001	9
Cde	23.042	.0061	9
cDE	119.830	.0001	9
cDe	68.188	.0001	9
cdE	29.480	.0005	9
cde	71.608	.0001	9
HP*1/HP*2	74.139	.0001	9
TF*C1	16.101	.0132	6
TF*C2	36.218	.0001	6
TF*C3	9.147	.1655	6
TF*B	12.419	.0532	6
TF*D	18.877	.0044	6
TF*VAR	5.116	.5290	6
(without Celtic, Baltic and Basque)			
GC*1F	23.737	.0013	7
GC*1S	5.721	.5727	7
GC*2	24.194	.0011	7
GC*VAR	7.929	.3389	7
(without Celtic and Baltic)			
PI*M1	23.718	.0006	6
PI*M2	20.598	.0022	6
PI*M3	6.634	.3560	6
PI*S	43.946	.0001	6
PI*Z	20.146	.0026	6
PI*VAR	18.163	.0058	6
(without Celtic, Greek and Baltic)			

worth mentioning that apart from the Lapp speakers it is the Basque group, which is striking by their PI allele frequency distribution pattern (Table 8).

GENETIC DISTANCE ANALYSIS

From the foregoing analyses it is seen that the allele and haplotype frequencies the seven polymorphic systems under consideration show marked distributional differences, which on the whole are clearly associated with the linguistic differentiation of the European populations. Genetic distance and principal component analyses were also performed to further analyze whether and how well the observed genetic variation corresponds with the linguistic differentiation of European populations. However, the whole set of the seven polymorphic systems could not be considered; we had to confine our genetic distance analyses to the following four: A1A2BO, MNSs, RH and HP. The result of a UPGMA clustering using Nei's (1972) standard genetic distance is shown in figure 2.

The clustering pattern, which emerges from this figure, is of considerable interest as it reveals some clear differentiation within the various European language groups. First of all it is seen that the Romany group takes an outstanding position with regard to all other European language groups. This reflects the manifold genetic peculiarities of the Romany speaking populations, which due to social and cultural factors remained largely isolated from all the other European populations.

The other European language groups are found in one big cluster, which, however, is subdivided into several subclusters. The first of them comprises the Finnic, Baltic, Ugric (Hungarian) and Slavic groups, which are all east European populations. A second subcluster, which is linked with the first one, comprises the Greek and Romance groups, all south European populations. The third subcluster finally comprises the Celtic and Germanic groups and amazingly also the Basque group. Except for the position of the Basque group, the pattern of genetic affinities based on the four polymorphic systems A1A2BO, MNSs, RH, and HP obviously corresponds well the linguistic differentiation of Europe.

This is corroborated by the results of principal component analysis as shown in figure 3. Five eigenvalues which in total explain 93% of the variance shows that the Romany group takes a distinct position, while the Romance and the Greek groups are close together. Figure 3 furthermore demonstrates that the Finnic, Ugric (Hungarian) and Slavic groups are also closely linked, and to some extent also with the Baltic group. The Germanic and the Celtic groups form another subgroup, while the Basque group shows a quite different position.

How can these associations between the linguistic differentiation of European populations and the distribution of genetic markers be explained? This is not very difficult concerning the Romany and Basque speaking populations, with whom this discussion should begin.

The Romany speaking populations of present-day Europe, especially those living in its eastern and southeastern regions, stem originally from northwest India, from where they migrated to Europe in the 14th century. During the following centuries they spread over large parts of this continent (Rishi, 1976; Vossen, 1993). However, due to many social, cultural and other factors they remained almost entirely isolated from all those European populations with whom they came into contact and hence gene flow was very restricted. Thus the genetic distinctiveness of this group was maintained, which is also valid for the genetic markers not considered in the genetic distance analysis, e.g., TF, GC and PI.

The Basque group is another one which differs obviously from all the other European population groups, linguistically and genetically. They are supposed to represent descendants of paleolithic and/or mesolithic populations of western Europe and in particular of the Iberian Peninsula and speak even today a language which is a remnant of the languages spoken in the western parts of Europe before the arrival of the Indo-European speaking populations. Although the Basques show some intragroup genetic heterogeneity, on the whole they have

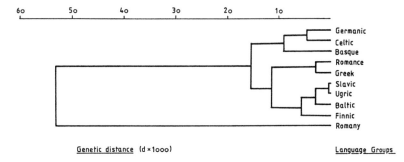

Figure 2. Genetic distances among European language groups.

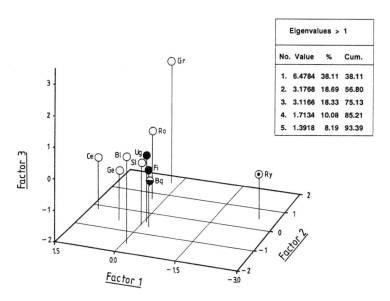

Figure 3. Principal component analysis based on four polymorphic
systems: A1A2BO, MNSs, RH and HP. First three factors.
Ce = Celtic, Ge = Germanic, Bl = Baltic, Sl = Slavic,
Ug = Ugric, Fi = Finnic, Ro = Romance, Gr = Greek,
Bq = Basque, Ry = Romany.

preserved their genetic identity. Bertranpetit and Cavalli-Sforza (1991), who presented a genetic reconstruction of the history of the population of the Iberian Peninsula came to the conclusion that "prolonged cultural preservation of successful food economy, of ancient language and customs have helped preserve the Basque identity by a relative closure to later comers, which was both cultural and genetic" (p. 65). That the Basque speakers are genetically closer to the Celtic and Germanic groups as is seen from figure 2 can be attributed to the frequencies of the cde haplotype as well as that of the 0 allele, which are generally more frequent in these two groups.

The other languages spoken in present-day Europe can be grouped into two great language families: Indo-European and Uralic-Altaic. Germanic, Celtic, Romance, Greek, Slavic and Baltic languages belong to the first language family; Finnic and Ugric (Hungarian) to the second one. The polutions belonging to these various language groups show some clear genetic differentiations.

The Indo-European languages can be traced to a so-called Proto-Indo-European language, which evolved more than 6000 years ago in the southwestern parts of Asia, from where it spread westwards (and eastwards), splitting gradually into many language groups and – within them – languages, some of which are extinct already. This expansion seems to be closely related to the expansion of agriculture in Europe (Ammermann and Cavalli-Sforza, 1984; Renfrew 1989), and began most likely in the fourth millenium B.C. (Gamkrelidse and Iwanow, 1990). It is assumed that since then differentiation of the various Indo-European language groups took place. As these groups were more or less geographically isolated from each other, peoples belonging to different language groups could develop specific genetic profiles, which could be maintained until today since, in general, even now no marked genetic contacts and thus gene-flow exist among them. The pattern of genetic differentiation depicted in figures 2 and 3 can be explained on the basis of the history of Germanic, Celtic, Romance, Greek and Slavic speaking populations.

The Finnic and Ugric (Hungarian) languages – part of the Uralic-Altaic language family – are spoken by populations which because of their geographic origin belong to quite another genetic stock than the Indo-Europeans. According to Meinander (1986) and Fodor and Czeizel (1991) the Finnic and Ugric (Hungarian) speaking populations originate from the Oka and Volga region (Finnic) and Volga and Kama region (Ugric), respectively, in western Siberia, from where they migrated at different times and routes into their present territories (for details see Meinander, 1986; Fodor and Czeizel, 1991). These groups cluster with the Slavic and Baltic language groups (Figures 2 and 3) by gene flow among them, which is evident the geographic distribution and from the history of the Finnic and Ugric (Hungarian) speaking populations. Also considering the history and ethnogenesis of the Slavic, Ugric (Hungarian), Baltic and Finnic language speaking population groups it is not surprising that they are genetically quite different from all other European populations as emerges from figures 2 and 3.

CONCLUDING REMARKS

In summary it can be pointed out that the haplotype and allele frequencies, respectively, of the polymorhic systems A1A2B0, MNSs, RH, HP, TF, GC and PI are not randomly distributed among European populations but show more or less clear patterns of distribution, which are associated with the linguistic differentiation of Europe. A quite distinct distribution pattern was found for the Romany and Basque speaking populations, which can be explained by their specific histories and by the marked and long-standing genetic isolation of these two groups from all the other European populations. The Slavic, Ugric (Hungarian), Baltic and Finnic language speaking populations are genetically closer to each other as compared to Germanic, Celtic, Romance and Greek speaking populations. It is worth mentioning that Germanic and Celtic speakers on the one hand and Romance and Greek speakers on the other are genetically more homogeneous. It is reasonable to assume that this language-associated genetic differentiation of European populations reflects the ethnohistorical processes, which took place in Europe during the last 5000 years.

REFERENCES

Ammermann, A.J., Cavalli- Sforza, L.L.,1984, "The Neolithic Transition and the genetics of Populations in Europe," Princeton University Press, Princeton, N.J.

Bertranpetit, J., Cavalli-Sforza, L.L., 1991, A genetic reconstruction of the history of the population of the Iberian Peninsula, *Ann. Hum. Genet.* 55: 51.

Cavalli-Sforza, L.L., 1992, Stammbäume von Völkern and Sprachen, *Spektrum der Wissenschaft* 1 January: 90.

Cavalli-Sforza, L.L., Bodmer, W.F., 1971, "The Genetics of Human Populations," W.H. Freeman and Company, San Francisco.

Fodor, J., Czeizel, A., 1991, The origins of the Hungarian population, *in* "Genetics of the Hungarian Population. Ethnic Aspects, Genetic Markers, Ecogenetics and Disease Spectrum," Czeizel, A., Benkmann, H.-G., Goedde, H.W., eds., Springer-Verlag, Berlin. p. 7.

Gamkrelidse, Th. W., Iwanow, W.W., 1990, Die Frühgeschichte der indoeuropäischen Sprachen, *Spektrum der Wissenschaft* 5 May: 130.

Haldane, J.B.S., 1940, The Blood-group frequencies of European Peoples, and racial origins. *Hum. Biol.* 12: 457.

Harrison, G.A., 1988, Human genetics and variation, *in* "Human Biology: An Introduction to Human Evolution, Variation, Growth, and Adaptability," Harrison, G.A., Tanner, J.M., Pilbeam, D.R., Baker, P.T., eds., Oxford University Press, Oxford, p. 145.

Mastana, S.S., and Papiha, S.S., 1992, Origin of the Romany Gypsies — genetic evidence, *Z. Morph. Anthrop.* 79: 43.

Meinander, C.F., 1986, Die ethnogenese der Finno-Ugrier aus der Sicht der Vor – und Frühgeschichte, *in* "Ethnogenese europäischer Völker," Bernhard, W., Kandler-Pálsson, A., Hg., Gustav Fischer Verlag, Stuttgart. p. 363.

Mourant, A.E., 1983, "Blood Relations, Blood Groups and Anthropology," Oxford University Press, Oxford.

Mourant, A.E., Kopec, A.C., Domaniewska-Sobczak, K., 1976, "The Distribution of the Human Blood Groups and other Polymorphisms," Second edition, Oxford University Press, Oxford.

Nei, M., 1972, Genetic distance between populations. *Amer. Nat.* 106: 283.

Renfrew, C., 1989, Der ursprung der indoeuropäischen Sprachfamilie. *Spektrum der Wissenschaft* 12 December: 114.

Rishi, W.R., 1976, "Roma: The Panjabi Emigrants in Europe, Central and Middle Asia, the USSR and the Americas," Punjabi University, Patiala.

Roychoudhury, A.K., Nei, M., 1988, "Human Polymorphic Genes, World Distribution," Oxford University Press, Oxford.

Ruhlen, M., 1987, "A Guide to the World's Languages – Vol. 1: Classification," Standford University Press, Standord.

Sokal, R.R., 1991, Ancient movement patterns determine modern genetic variances in Europe, *Hum. Biol.* 63: 589.

Sokal, R.R., Oden, N.L., Thomson, B.A., 1988, Genetic changes across language boundaries in Europe, *Amer. J. Phys. Anthrop.* 76: 337.

Sokal, R.R., Oden, N.L., Legendre, P., Fortin, M.-J., Kim, J., Vandor, A., 1989a. Genetic differences among language families in Europe, *Amer. J. Phys. Anthrop.* 79: 489.

Sokal, R.R., Harding, R.M., Oden, N.L., 1989b, Spatial patterns of human gene frequencies in Europe, *Amer. J. Phys. Anthrop.* 80: 267.

Stevenson, J.C., Buell, P.D., Schanfield, M.S., 1988, European population movements and genetic marker distributions, *Int. J. Anthrop.* 3: 319.

Tauszik, T., Hummel, K., 1991, Blood groups and the results of certain blood group tests in Hungary, *in* "Genetics of the Hungarian Population. Ethnic Aspects, Genetic markers, Ecogenetics and Disease Spectrum," Czeizel, A., Benkmann, H.-G., Goedde, H.W., eds., Springer-Verlag, Berlin. p. 116.

The Cambridge Encyclopedia, 1991, Crystal, D., ed., Cambridge University Press, Cambridge.

Tills, D., Kopec, A.C., Tills, R.E., 1983, "The Distribution of the Human Blood Groups and other Polymorphisms, Supplement 1," Oxford University Press, Oxford.

Vogel, F., Motulsky, A.G., 1986, "Human Genetics. Problems and Approaches," Second edition. Springer-Verlag, Berlin.

Vossen, R., Zigeuner, 1983, "Roma, Sinti, Gitanos, Gypsies zwischen Verfolgung und Romantisierung," Ullstein, Frankfurt.

Walter, H., 1962, "Die Bedeutung der Serologischen Merkmale für die neue Rassenkunde," Gustav Fischer Verlag, Stuttgart.

Walter, H., Danker-Hopfe, H., Bhasin, M.K., 1991, "Anthropologie Indiens. Untersuchungen zur genetischen Variabilität der Bevölkerung Indiens mit besonderer Berücksichtigung ihrer regionalen, ethno-sozialen und sprachlichen Gliederung," Gustav Fischer Verlag, Stuttgart.

GENETIC DIVERSITY AND AFFINITIES AMONG INDIAN POPULATIONS: AN OVERVIEW

Partha P. Majumder and B.N. Mukherjee

Anthropometry and Human Genetics Unit
Indian Statistical Institute
Calcutta 700035, India

INTRODUCTION

One of the primary objectives of the study of human biology is to understand the nature and extent of biological variation among human populations. Genetic characters are commonly used for this purpose. Genetic diversity among Indian populations is large (Table 1). From Table 1 it is seen that the variation in gene frequencies among populations of South Asia (comprising primarily the Indian subcontinent) is the highest compared to the populations resident in the other Asian regions. Using gene frequency data on ten polymorphic loci, Roychoudhury (1983) obtained relationships among some castes and tribes of India and some neighbouring populations. He found that the south Indian castes are genetically very close to the Sinhalese of Sri Lanka. This is because of past large-scale migrations of Tamil-speaking south Indian people to Sri Lanka; the linguistic and cultural similarities of Sinhalese with the south Indian people (especially those of Tamilnadu) has also resulted in a good deal of genetic admixture. On the whole, the caste groups of India were found to be closer to the Iranians and Afghans than to Mongoloid populations of Malaya and China. The south Indian tribal groups were found to be genetically close to the tribal Veddahs of Sri Lanka. He also found the tribal groups of Malaya and the aborigines of New Guinea and Australia to be genetically unrelated to the south Indian tribes. Using data on four polymorphic loci (Tf, Gc, Gm and Km), Walter (1986) arrived at similar conclusions, and additionally found that the east Indian populations are more similar to the Mongoloid populations of Nepal, Japan and China.

The foregoing discussion clearly points out that, by and large, the patterns of genetic similarities of the Indian populations with the populations of the neighbouring countries are in concordance with the hypothesis of immigration and subsequent admixture of Caucasoids and Mongoloids from other regions of Asia. The objective of this paper is to examine closely the extent and pattern of genetic variation and to identify factors (ethnological, cultural, geographical, etc.) that contribute to this variation.

Human Population Genetics, Edited by P.P. Majumder
Plenum Press, New York, 1993

Table 1. Variances of allele frequencies at four genetic loci in different regions of Asia

| Region | ABO | | | MN | HP | PGM1 |
	A	B	O	M	HP*1	PGM1*1
West Asia	.0025	.0042	.0022	.0046	.0023	.0022
South Asia (comprising primarily the Indian subcontinent)	.0050	.0080	.0156	.0108	.0041	.0054
East Asia	.0017	.0004	.0010	.0102	.0028	.0037
Inner & North Asia	.0016	.0022	.0023	.0041	.0962	.0002

Computations are based on data given in Mourant et al. (1976).

CLASSIFICATION OF INDIA: ETHNO-CULTURAL AND GEOGRAPHICAL

The populations of India can be broadly classified as tribes, castes and religious groups. The tribal populations of India are supposedly the first settlers, and number about 400. These tribes are scattered all over India, and vary greatly in their numerical strength — the Gonds and Bhils are about 4,000,000 each, while the Andamanese number only about 25. Linguistically they belong to one of the four broad linguistic families — Austro-Asiatic (Khasi, Santhali, Ho, Mundari, Gondi, etc.), Tibeto-Burman (Bhotia, Lepcha, Miri, Kachari, Angami, Lushai, etc.), Dravidian (Korwa, Yerukula, etc.) and Indo-European (Hajong, Bhili, etc.). Culturally they can broadly be classified as hunter-gatherers, hill cultivators, agriculturists, artisans and industrial labourers (in some urban areas). The caste system in India was formed primarily on the philosophy of division of labour — with cooperation rather than competition. This consequently has led to a hierarchical subdivision — the Brahmins who have the traditional occupation of preaching belong to the top rung, and the so-called "scheduled castes" who are supposed to perform menial tasks such as scavenging, occupy the bottom rung. The religion of all the caste groups is Hinduism; the tribal groups do not have a uniform religion and many are ancestor-worshippers. The major languages spoken by the caste groups of the northern half of India belong to the Indo-European family, while those spoken by caste groups of the southern half belong to the Dravidian family. Apart from the castes and tribes, there are several other religious groups — Muslims, Christians, Parsees, Iranis, Sikhs, Buddhists, etc. A majority of these religious groups are local converts, while some (e.g., Parsees, Iranis, etc.) are recent migrants. All of these factors — language, culture, religion, etc. — have affected gene flow and genetic diversity within and between populations.

Morphologically the people of India has been broadly classified into four types (Malhotra, 1978). Details are provided in Table 2.

Table 2 . Morphological types, features and present geographical distribution of the people of India

Type	Morphological Characterisation	Present Distribution
Negrito	Short stature, frizzly hair, fine hair texture,brachycephalic head, dark complexion,short and protruding face, broad nose, thick and everted lips.	Andaman Islands, Nilgiri Hills of Tamilnadu. (May have had a wide distribution at one time.)
Australoid	Short stature (taller than Negrito), wavy to curly hair, hyperdolichocephalic to dolichocephalic head, dark complexion, stout brow ridges, sunken nasal root, abundance of body hair.	Chiefly in central and southern regions
Mongoloid	Short stature, broad shoulder, scanty facial and body hair, brachycephalic head, flat face, prominent cheek bones, flat nose, epicanthic fold.	Sub-himalayan and north-east regions
Caucasoid	Generally tall stature, dolichocephalic to brachycephalic head, light complexion, straight to wavy hair, sometimes light eyes, arched forehead, long face with well-developed chin, narrow and prominent nose.	Found in most regions

The Negritos are claimed to be the real autochthones, and the Australoids stand next in order of antiquity. The earliest ancestors of the Australoids can be traced back to Palestine (Vidyarthi, 1983), and prehistoric material culture evidence points to their presence for at least 50,000 years (Malhotra, 1978). The Caucasoids are the result of several migrations through the north-west at different times and perhaps from different regions of western and central Asia. Consequently, they have a variable antiquity, but from material culture evidence it seems that they have been occupying the different regions of India for the past 10,000 - 20,000 years. The Mongoloids are also immigrants and entered India from inner and eastern Asia through the north-eastern frontier. The major non-Hindu, non-tribal religious groups of India — the Muslims and the Christians — are mostly local converts, and reveal, depending upon the area of habitation, all morphological types, except the Negrito. Since most caste groups are morphologically Caucasoid, in the present discussion we have used a hierarchical classification to group the castes, and have treated religious groups separately. We have, thus, used the ethno-cultural classification given in Table 3.

The geographical classification of India poses considerable problems. From an ecological perspective, India needs to be divided into several zones to achieve ecological homogeneity within each zone. For purposes of understanding geographical patterns of genetic variation it would be ideal to study patterns within and between homogeneous ecological zones, since variable pressures of natural selection because of differences in ecology may be expected to be a cause of genetic variation. Unfortunately, the number of zones that need to be defined to achieve ecological homogeneity is too large to perform any meaningful statistical analysis of genetic data; there would be many zones from which no, or very few, data are available. Another, perhaps more serious, problem is that homogeneous ecological zones would cut across State/Union-Territory (political/administrative unit) boundaries, and since exact locations of sampling are often not available, such a zonal classification would force one to discard a large number of data sets. Further, defining many zones would also lead to other problems because the habitats of several populations are large, and, therefore, would not necessarily be confined to one geographical/ecological zone. At the risk of introducing some ecological heterogeneity, but to avoid some of the problems listed above, we have divided India into 5 geographical zones as defined below:

Zone	Constituent State/Union Territory
Northern	Jammu and Kashmir, Himachal Pradesh, Punjab, Delhi, Haryana, Uttar Pradesh
Western	Gujarat, Rajasthan, Maharashtra, Goa, Daman, Diu
Southern	Karnataka, Kerala, Tamilnadu, Andhra Pradesh
Eastern	Bihar, Orissa, West Bengal, Sikkim, Assam, Meghalaya, Tripura, Mizoram, Manipur, Nagaland, Arunachal Pradesh
Central	Madhya Pradesh

It may be noted that the Lakshadweep archipelago in the Arabian Sea and the Andaman and Nicobar archipelago in the Bay of Bengal have not been classified in any zone; they have been treated separately.

EXTENT OF GENETIC VARIATION AND PATTERN OF AFFINITIES AT THE MACRO-LEVEL

A measure of within-population genetic diversity is average heterozygosity, H. In Table 4 are presented the average heterozygosity values in various populations. It is seen that the level of heterozygosity among Indians is similar to that in any of the three major races of man. Further, in any geographical zone of India the average heterozygosity is also high. This indicates that there is a great amount of genetic diversity among the population groups of India, and the level of diversity is comparable to that existing within the major races of man (Nei and Roychoudhury, 1972).

As mentioned earlier, there has been large-scale admixture of immigrant Caucasoids and Mongoloids with the autochthones of India. There is also speculation that the Negrito element was introduced in India from Africa, and that the Negritos were widespread in distribution at one time and were pushed to the restricted locations where they are presently found. It is, therefore, pertinent to ask how the Indians are genetically positioned relative to Caucasoids, Mongoloids and Negroids. Such an exercise was performed by Roychoudhury (1977b) using gene frequency data of 29 loci. This study showed that the Indians (represented by north Indians, Bengalis from eastern India and a tribal population — Kadar — from southern India) are the closest to the Mongoloids and farthest from the Negroids; the Caucasoids being at the intermediate level. [A similar conclusion was also reached by Walter (1971) using a more restricted data set.] This is in conformity with the historical evidence of large Caucasoid and Mongoloid admixture with the autochthones of India, but is in discordance with the hypothesis of widespread distribution of the Negroids. It is of course possible that despite a widespread distribution there was little genetic admixture with Negroids or that because they were pushed to restricted geographical regions, the genes that were introduced through admixture have been lost from most hybrid populations; both these theories are, however, unlikely. The most parsimonious explanation of the fact that the Indians show the smallest affinity with the Negroids is that there were only sporadic, numerically small, waves of African immigration to restricted geographical regions of India. Such historical evidence has been found by M. Vijayakumar (personal communication) in course of his investigations among Siddis of northern Karnataka, who are alleged to have some African admixture.

Table 3.   Ethno-cultural classification of the people of India

1.	Tribe	-	1.1	Australoid
			1.2	Mongoloid
			1.3	Caucasoid
			1.4	Negrito
2.	Caste	-	2.1	Upper
			2.2	Middle
			2.3	Lower
3.	Religious Group	-	3.1	Christian
			3.2	Buddhist
			3.3	Muslim
			3.4	Parsi
			3.5	Sikh

Table 4. Average heterozygosities (H) in some Indian and neighbouring populations

Population	No. of loci	H
Indians (All India)	29	0.104
Caucasoids (West Asia)	29	0.125
Mongoloids (East and Central Asia)	29	0.089
Negroid (American Blacks)	29	0.098
Indian Geographical Zones		
Northern Zone:		
North Indians	10	0.141
Punjabis	10	0.141
Brahmin	18	0.081
Arora	18	0.099
Khatri	18	0.093
Rajput	18	0.097
Lower Caste	18	0.085
Western Zone:		
Marathi	21	0.145
Gujarati	21	0.153
Vania Soni	21	0.146
Parsi	21	0.164
Irani	21	0.143
Southern Zone:		
South Indians	10	0.127
Siddi	16	0.299
Havig Brahmin	15	0.306
Muslim	11	0.319
Eastern Zone:		
Bengali	10	0.135
Brahmin	10	0.197
Bagdi	10	0.212
Munda	10	0.221
Central Zone:		
Hindu	10	0.152
Bhil	15	0.301

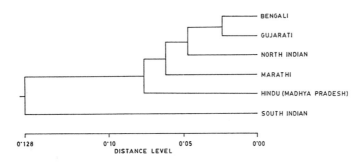

Figure 1. Genetic affinities among Indian populations

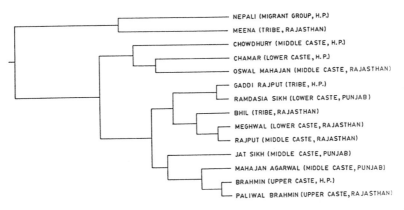

Figure 2. Genetic affinities among some population groups of north India

Within India, the broad pattern of genetic affinities is depicted in Figure 1. This dendrogram was obtained by the single-linkage method using the matrix of genetic distances presented by Roychoudhury (1977b) based on the gene frequency data for 10 loci. From Figure 1 it is seen that the populations of southern India stand apart from the populations of the rest of India. The populations of north, west, central and east India are genetically fairly close. This is in concordance with the sharp linguistic separation of southern India from the rest of India. The languages spoken in the southern region of India belong to the Dravidian linguistic family, while most of the languages spoken in the northern region belong to the Indo-Aryan family. The pattern also supports the theory of admixture of Aryan-speaking immigrants from west Asia with indigenous populations of India and that the immigrants did not penetrate the southern regions of India. The Mongoloids generally show a closer genetic affinity to the people of the northern region than to the people of the southern region.

VARIATION AND AFFINITIES WITHIN SMALLER GEOGRAPHICAL REGIONS

In the previous section we have examined variation of and affinities at the all-India level. In this section, we shall report findings of studies conducted within smaller geographical regions.

Northern Zone: The pattern of genetic differentiation among the north Indian populations indicates that although there is a broad Caucasoid-Mongoloid separation, no clear correlation between genetic distances and ethno-historical relationships seems to exist. As a part of a large collaborative project between the Indian Statistical Institute, Calcutta, and the Institute of Ethnography, Moscow, 14 endogamous groups were screened in respect of 5 blood group and 6 red-cell enzyme loci (Papiha et al., 1982). These endogamous groups belonged to the various strata of the Hindu caste system and some were tribal groups. The samples were drawn from the States of Punjab (3 groups), Himachal Pradesh (5 groups) and Rajasthan (6 groups). Intra- State comparison of population groups showed that heterogeneity for the 4 polymorphic enzyme systems (ADA, AK, PGM1, and ESD) was not significant at the 5% level, but was significant for each of the 5 blood group loci. Significant genetic differences were observed among populations between States. The genetic relationships among the 14 population groups are depicted in Figure 2, which shows that the two Brahmin groups are genetically close, although they are from two separate States. The Mongoloid Nepalis from the cis-Himalayan region of Himachal Pradesh show close genetic affinity with the Meena tribals of Rajasthan. There is no ethno-historical evidence to support this affinity, although it is noteworthy that both these groups are characterised by the complete absence of the RH-negative phenotype. The remaining clusters do not exhibit any remarkable correlation between genetic and ethno-social affinities. Thus, in the northern (north-western) India, although there seems to be a clear separation of the Caucasoid high caste and Mongoloid tribal groups, there is no definite pattern of genetic clustering for the intermediate middle and low caste groups and non- Mongoloid tribals.

Western Zone: Sanghvi and Khanolkar (1949) and Sanghvi (1954) reported results of genetic studies on four endogamous populations of Maharashtra. The populations studied belong to the Hindu caste system and were all classified under the same "physical type" by Risley (1915) and the same "race" (Marathi race) by Guha (1935). Two of the populations – Desasth Brahmin and Koknasth Brahmin – belong to the upper stratum of the Hindu caste system, while the other two – Chandrasenya Kayasth Prabhu (CKP) and Maratha – belong to the middle rank. The Desasth Brahmins are further socially divided into two subgroups –

Desasth Rgvedi Brahmin (DRB) and Desasth Yajurvedi Brahmin (DYB). Ethno-historical evidence suggests that the Desasth Brahmins and Marathas have been residing in this region for a long period of time; Marathas probably dating back to the pre-Christian era. CKP and the Koknasth Brahmins are more recent occupants; CKP can be traced in this region from the 10th century onwards, while the Koknasth Brahmins did not arrive until about the 17th century. The Marathas and Koknasth Brahmins were the ruling dynasties successively in the 17th and 18th centuries. Until their rise to power, the other Brahmin groups of the region held the Koknasths in low social esteem. Based on 6 genetic traits, the following patterns of affinity have been observed: (i) DRB and DYB are homogeneous; (ii) the Desasth Brahmins (DRB + DYB) are very close to the Marathas, and show no significant difference for any of the six traits; (iii) both Desasth Brahmins and Marathas differ significantly from the Koknasth Brahmins; (iv) the CKPs show dissimilarities with all the three other groups; and, (v) the Koknasth Brahmins and the CKPs are most dissimilar. The genetic evidences, therefore, support the ethno-historical facts to some extent, but also exhibit discordance in some respects. In spite of the fact that socially the two Brahmin groups are closer, genetic evidence points to a closer affinity between the Desasth Brahmins and the Marathas. Although the position of the Koknasth Brahmins is unclear, the genetic evidence that CKPs are dissimilar with all the other groups is in concordance with the historical fact that it is a migrant group to this region. In fact, in another genetic study (Mukherjee et al., 1979) conducted in the same State (Maharashtra), the CKPs are found to cluster (based on 16 genetic loci) with another migrant group – Parsees. It is, however, unclear why the Parsees show similarities with the CKP, but it is significant to note that both these migrant groups show dissimilarities with the populations that have been living in this region for a long time.

In a study conducted by the Indian Statistical Institute (Mukherjee et al., 1979), comprising nine endogamous population groups – Bhil, Pawara, Katkari (all tribal), Chitpavan and Desasth Rgvedi Brahmins (both upper caste), Maratha and C.K.P. (both middle caste), Nava-Budha (low caste) and Parsee (migrant) – revealed some interesting patterns of affinity as depicted in Figure 3. It is seen that the tribal groups are well-differentiated from the caste groups. The most striking difference between the tribal and caste groups is that the percentage of G-6-PD deficiency (3.5% - 8%) and the sickle cell haemoglobin allele frequency (3% - 11%) are high among the tribals but are absent or near-absent among the caste groups. The close relationship of the CKP and Parsee has already been mentioned. Among the remaining four caste groups belonging to one cluster, there is no clear concordance between genetic distance and social rank. Within the cluster of the three tribal groups, the Bhils and Pawaras are close, while the Katkaris are fairly distant. The reason for this may be that while the Bhils and Pawaras were sampled from the same district (Dhulia), the Katkaris live in a region that is over 300 km. away from this district. Thus, we again see that increase in geographical distance between habitats leads to increase in genetic distance.

Southern Zone: The Indian Statistical Institute conducted genetic surveys among several endogamous groups of south India, including the States of Andhra Pradesh, Karnataka, Tamilnadu and Kerala (Banerjee et al., 1988; Vijayakumar et al., 1987). In this study we have analyzed the data pertaining to A1A2BO, RH, ACP, ESD and HB loci of 15 populations. The populations included from the various States are — Andhra Pradesh: Vaidik Brahmin (upper caste), Vysya (middle caste), Kamma (middle caste); Karnataka: Lingayat (upper caste), Vokkaliga (middle caste), Jenu Kuruba (tribe), Siddi (subdivided into three religious isolates – Hindu, Muslim and Christian; brought into India as slaves primarily from east Africa; Negroid); Tamilnadu: Chettier (middle caste), Pariah (low caste), Kallan (low caste); Kerala: Namboodiri Brahmin (upper caste), Ezhava (middle caste), Pulayan (low caste), Urali (tribe). Except for the Siddis, all the other groups speak Dravidian languages. The Siddis are multilingual and speak both Indo-European and Dravidian languages.

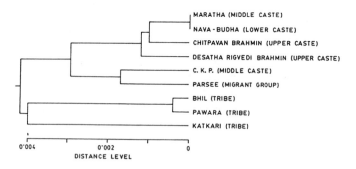

Figure 3. Genetic affinities among some population groups of western India

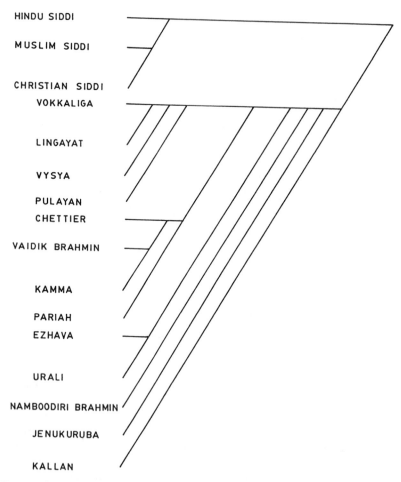

Figure 4. Genetic affinities among some population groups of southern India

Figure 4 depicts the genetic relationships among these 15 populations, from which it is clear that the Siddis are genetically distinct from the other south Indian populations. This is expected because the Siddis are immigrants into India from Africa; they have retained their genetic identity because of restricted admixture with indigenous populations of southern India. Figure 4 also shows that there is no correspondence between genetic and socio-cultural affinities. In fact, there are indications, albeit weak, that genetic similarity in this zone decreases with geographical distance.

In a study conducted among several populations of Tamilnadu by Sanghvi et al. (1981), genetic data were collected from 10 populations: Upper Caste – Iyer, Iyengar; Middle caste – Vellalan, Naidu, Mudaliar; Lower caste – Vanniyan, Kallan, Paraiyan, Chakkiliyan; and, Tribe – Irula. The genetic data did not show concordance with social rank. One of the upper castes – Iyer – appears to be distinct from the other populations, but the other upper caste group – Iyengar – is close to a cluster of population groups comprising middle caste Vellalans and Naidus, and the tribal Irulas. The low caste Paraiyans and Chakkiliyans are genetically the closest among all the populations. The Vellalans and Kallans are also close to the Paraiyans. Scanty ethno-historical data on the origins of these populations do not provide an adequate explanation of this pattern of affinities. It is possible that some of the groups are immigrants and that there has been considerable admixture among these populations.

Eastern Zone: Based on genetic surveys carried out primarily by the Indian Statistical Institute in the eastern Indian States of Bihar, Orissa, West Bengal, Assam and Manipur, we have attempted to investigate the genetic relationships of populations in this zone. Twentyfour population groups have been included in this analysis. These are — Bihar (Saha et al., 1992; and other unpublished data): Brahmin (upper caste), Chamar (low caste), Santal (tribe), Bhuiya (tribe), Muslim; Orissa (Reddy et al., 1989): Vadabalija (low caste; subdivided into two isolates at Penticotta and Vadapeta), Jalari (low caste), Kevto (low caste); West Bengal (Banerjee et al., 1992): Poliya, Deshi, Tiyor (all low caste and descendants of the Koch); Assam (Walter et al., 1986; Das et al., 1987; Mukherjee et al., 1989): Brahmin (upper caste), Kalita (middle caste), Kaibarta (low caste), Muslim, Ahom (ancestral population of Assam; migrants from Myanmar), Karbi (tribe), Kachari (tribe), Sonowal (a division of Kachari), Rajbanshi (low caste), Chutiya (ancestral population of Assam; tribe); Manipur (Singh et al., 1986): Brahmin (upper caste), Meitei (middle caste). Based on gene frequency data of 5 loci — ABO, HP, ACP, ESD, AK — a dendrogram of genetic relationships was constructed (Figure 5). Figure 5 indicates that, by and large, the genetic clustering of these populations is based on geographical contiguity of habitats. The two population samples from Manipur cluster together. The four populations samples from Orissa also cluster together. The low caste (Chamar) and tribal (Bhuiya and Santal) of Bihar cluster together. The high caste Brahmins and Muslims of Bihar show close genetic affinity. In this context, it may be mentioned that Majumder and Rao (1958) had also noted anthropometric similarities of the Muslims of some regions of Bengal with the higher castes, and concluded that the conversion of Hindus to Muslims (which is a historical fact) had taken place not only from the lower castes but also from the higher castes. It is unclear why the Brahmins and Muslims of Bihar show close genetic affinities with the Chutias of Assam. It is, however, noteworthy that the Chutiyas who are an ancient population of Assam, were subjugated by the migrant Ahoms and were pushed to their present localized territory in upper Assam. This may have resulted in certain peculiarities in their genetic profile which distinguishes them from the other populations of Assam. The three descendant populations of the Koch (who ruled vast areas of Assam and northern West Bengal until about 300 years ago) — Poliya, Deshi and Tiyor — show close genetic affinities among themselves and also with the Karbi of Assam. The genetic affinities among the population samples of Assam do not show any large degree of congruence with socio-cultural hierarchy.

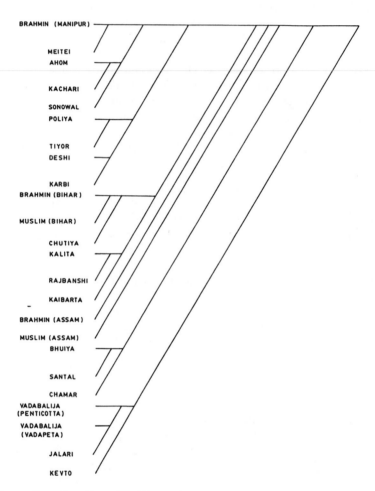

Figure 5. Genetic affinities among some population groups of eastern India

GENETIC DIFFERENTIATION AT THE MICRO-LEVEL

The conclusion thus far is that the effects of geographical distance and ethnosocial backgrounds on the genetic relationships among populations are at best equivocal. While in certain geographical regions, ethnohistory seems to be an important determinant of genetic affinity, in other regions no such pattern is observed. Similarly, geographical proximity seems to be an important determinant of genetic affinity within some ethnosocial categories, while in some other categories no geographical effect is discernible. It is thus not possible to identify either geographical distance or past ethnic/social connections are the more important determinant of genetic similarity at an all-India level or even at the level of smaller geographical zones. It is, of course, possible that the geographical zones are too broad to be ecologically homogeneous, resulting in inconsistent patterns of relative effects of distance and ethnohistory on genetic affinities. To investigate this, it is useful to consider genetic differentiation at a micro-level. In other words, we shall consider the nature and extent of genetic variation among populations that inhabit a geographically restricted area (say, a State or a few districts within a State). This, we believe, will throw light on the patterns of micro-differentiation which, in turn, may point to factors other than geographical distance and ethnic/social status that control the process of genetic differentiation.

Kanets and Kolis of Kinnaur District, Himachal Pradesh: The Kanets and Kolis are the two main endogamous populations of the Kinnaur district, which shares its bordrs with Tibet (China), of Himachal Pradesh in northern India. The Kanets are agriculturists, while the Kolis are a low caste engaged in menial work and agricultural labour. Blood samples were collected and analysed (Papiha, 1985) from four settlements of Kinnaur – Puh, Kalpa, Sangla and Nachar. In terms of geographical distance from Tibet, Puh is the closest and Nachar is the farthest; Kalpa and Sangla are more-or-less equidistant. There are clear evidences of genetic admixture with the Tibetans, and as is expected, the proportion of Tibetan genes in the gene pools of these four settlements decrease with increase in geographical distance from Tibet. Thus, it is estimated that the Puh Kanets have 76% of Tibetan genes, while Nachar Kanets have 51%. The Kolis (data of four locations pooled for lack of adequate number of samples from individual locations) possess only 15% Tibetan genes in their gene pool. Analysis of data on 24 single gene traits (Papiha, 1985) showed that there is little genetic differentiation among the Kanets of Nachar, Sangla and Kalpa. The Kolis are fairly distinct from this cluster of Kanets, and the Puh Kanets are the farthest. Thus we see that the pattern of differentiation in this small geographical region is such that the two endogamous groups of Kolis and Kanets are distinguishable. Within the Kanet settlements the pattern of affinities is highly correlated with the amount of admixture with the neighbouring Tibetans.

Dhangars of Maharashtra: The Dhangars are traditionally shepherds, although many are associated with allied professions, e.g., woollen-blanket weaving. They number about 3 million and are distributed over all the 26 districts of Maharashtra. The Dhangars are a caste-cluster and comprise 23 endogamous castes. The castes show considerable variation in population size. Some of the castes have been living in their present habitat for the last several centuries, while a few are recent immigrants to Maharashtra. Ethnographic data show that at least some of the castes have resulted from fission, that is, from breakdown of larger castes. The estimated rate of inter-caste marriage among the Dhangars is about 1 in 10,000 marriages (Malhotra, 1984). Genetic data (of 7 loci) were collected from 14 of these 23 castes. The geographical locations of these 14 castes are shown in Figure 6. In comparison with castes of different ranks or regions, the extent of genetic differentiation among the Dhangar castes is small, although significantly different (Malhotra et al., 1978),

267

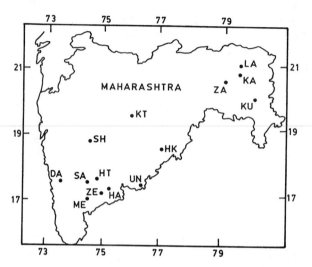

Figure 6. Geographical locations of 14 Dhangar castes of Maharashtra

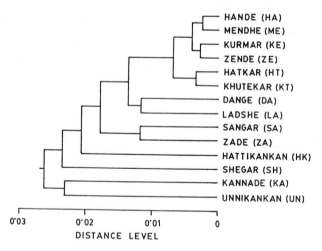

Figure 7. Genetic affinities among Dhangar castes of Maharashtra

which, of course, is expected because of both geographical and social contiguity. Further, archaeologic and ethnographic evidence suggests that the Dhangars are immigrants from north-west India (corroborated by the presence of haemoglobin D and J among them and also by the absence of Hb S) and split into the various castes about 250-500 generations ago. Despite the limited extent of differentiation, is the pattern of affinities among them in concordance with geographical distance or ethno-history ? The genetic affinities are given in Figure 7, which, if considered in conjunction with Figure 6, shows that geographical contiguity does not imply genetic affinity; the Handes and the Unnikankans live in geographical proximity but are genetically distant, while the Kurmars and Zendes although separated by a distance of about 700 km. are genetically close. Some of the patterns of genetic affinity are, however, corroborated by ethno-historical data. The Kannades and the Unnikankans are both relatively recent migrants from Karnataka. This fact, it may be noted, only lends support to the contention that geographical contiguity leads to biological affinity; these two castes perhaps lived in close proximity in Karnataka before they moved to their geographically distant habitats in Maharashtra. Ethno-historical data do not provide any additional interpretive insight, and, thus, geographical contiguity of habitat does not seem to determine genetic similarity among Dhangar castes. It is seen that the probability of identity of two genes drawn from two different geographical locations (of Dhangars) decreases as the distance increases, but the rate of this decrease is small, indicating that the extent of gene differentiation among the Dhangars is small, and there are considerable deviations from this pattern. Further, only about 7% of the total genetic variation is attributable to variation between castes; most variation (about 93%) is within castes.

Koya Doras of Three Districts of Andhra Pradesh: The Koya Doras are a major tribal community of Andhra Pradesh inhabiting the districts of East and West Godavari, Adilabad, Warangal and Khammam. They are claimed to have migrated from the Bastar area of Madhya Pradesh because of famines and disputes. They have a dark skin, coarse black hair, flat and broad nose and high cheek bones, and are classified as Australoid. Blood samples from the Koya Doras inhabiting three districts: West Godavari and Warangal districts on the western side of the Godavari river, and the East Godavari district located on the eastern bank, were obtained. Analysis of the genetic data of 5 protein and enzyme loci showed that there is a significant genetic variation among Koya Doras of the three districts (Walter et al., 1981). It is also seen that the Koya Doras of West Godavari and Warangal are genetically closer than those of East Godavari district. This pattern of affinities can be easily explained by hypothesizing that the Godavari river acts as a barrier to gene flow, so that the two subgroups living on the west bank of the river are genetically closer. Unfortunately, data relevant to testing this hypothesis regarding gene flow are not available.

Endogamous Groups of West Bengal: Variation and affinities among ten endogamous population groups of West Bengal have been examined in respect of 12 genetic loci by Chakraborty et al. (1986). The ten endogamous groups surveyed were: (i) Rarhi Brahmin – upper caste, may have resulted from intermixing of orthodox Brahmin and non-Brahmin families; (ii) Vaidya – middle caste, show similarities in culture with Brahmins, some claim that they have resulted from intermixture of Brahmins with low caste groups, most engaged in white-collar jobs; (iii) Rajbanshi – largest low caste community in West Bengal, may have a Mongoloid affinity, primarily agriculturists; (iv) Bagdi – low caste, may be remnant of an aboriginal tribe who, by intermarrying with low caste Hindus, left their tribal way of life and engaged themselves in fishing, cultivating and menial work, avoid inbreeding; (v) Jalia Kaibarta – low caste, considered to be one of the earliest inhabitants of West Bengal, fishermen, some evidence of intermixture with Brahmins; (vi) Rabha – Mongoloid tribe, settled agriculturists; (vii) Garo – Mongoloid tribe, perhaps resulted from intermixture of Tibetans and Hindus, show affiliation with the Rabhas, mostly agriculturists; (viii) Mech – Mongoloid tribe, numerically small, most practise shifting cultivation; (ix) Munda – Australoid tribe, numerically large, engaged in agriculture and plantation work, avoid inbreeding; (x) Lodha – numerically small Australoid tribe, hunter-gatherers, some engaged as agricultural labourers, avoid inbreeding. Thus, the endogamous groups can be clustered ethno-socially as: Upper caste – Rarhi Brahmin; Middle caste – Vaidya; Low caste – Rajbanshi,Bagdi, Jalia Kaibarta; Mongoloid tribe – Rabha, Garo, Mech; and, Australoid tribe – Munda, Lodha. In terms of geographical proximity there are three main clusters: Rarhi Brahmin, Vaidya, Jalia Kaibarta, Bagdi, Munda, Lodha, and Rajbanshi, Rabha, Garo, Mech. The maximum within-cluster geographical distance is 40 kms. The populations belonging to the first and second geographical clusters are from two southern districts of West Bengal (Midnapore and 24- Parganas), while those of the third cluster are from two northern districts (Cooch Behar and Jalpaiguri). Thus, populations of the first and second clusters are geographically closer than to populations of the third cluster. The ten endogamous groups differ markedly in respect of allele frequencies at 9 of the 12 loci screened. The amount of genetic diversity within each population is fairly high (average heterozygosity = 0.197 - 0.230). The extent of genetic differentiation between populations is very low; in fact, it is about one-half of the amount of diversity found within the subcastes of the Dhangar caste-cluster. However, as in the Dhangars, the extent of between-population diversity is much smaller than the extent of within-population diversity. The genetic relationships among the populations is depicted in Figure 8, which shows three main clusters. The two Australoid tribes, Munda and Lodha, forms a cluster. The four populations, one low caste (Rajbanshi) and three tribes (Mech, Rabha, Garo), which are claimed to have Mongoloid affinities and which are also geographically clustered, show a high degree of genetic resemblance. The remaining four populations – Rarhi Brahmin, Jalia Kaibarta, Vaidya and Bagdi – forms a separate cluster. The fact that genetically the low caste groups (Jalia Kaibarta and Bagdi) are close to the higher castes (Rarhi Brahmin and Vaidya) lends support to the claim that the Rarhi Brahmins and Vaidyas are a result of intermarriage between people of high and low castes. The clustering pattern obtained on the basis of genetic data agree with both ethno-history and geographical clustering. Unfortunately, the geographical and ethno-historical clusters, in the present case, overlap so greatly that it is not possible to say which of the two factors (ethnicity and geographical proximity) plays a more substantial role in bringing about genetic similarity. However, it is seen that although the probability of gene identity decreases significantly with increase in geographic distance, the proportion of variation in gene identities among these populations explained by differences in geographical distance is only 14.4%. This indicates that geographical distance perhaps plays a small role in causing genetic differentiation among the 10 populations inhabiting the restricted region.

Gonds of Central India: The Gondi-speaking Gonds are one of the largest tribal populations of Central India, and number about 4 million which is about 13% of the total tribal population of India. The Gondi language belongs to the Dravidian linguistic family. The Gonds occupy the region that forms the junction of the three States – Andhra Pradesh, Madhya Pradesh and Maharashtra. They have been classified as proto-Australoid both by Guha (1935) and Majumdar (1958). The Gonds are subdivided into several endogamous subpopulations. Pingle (1984) reported results of her anthropometric and genetic investigations among five Gond subpopulations living in Adilabad district of Andhra Pradesh and Chanda district of Maharashtra. The Raj Gonds and Kolams of Adilabad district are cultivators, but while the Kolams are engaged in shifting cultivation on hilltops, the Raj Gonds cultivate the valley bottoms. Two other Gond subpopulations inhabiting relatively plain ar-

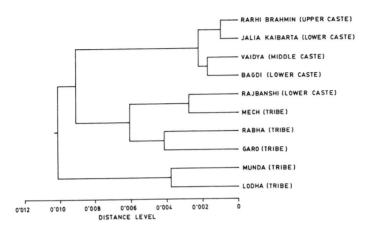

Figure 8. Genetic affinities among some populations of West Bengal

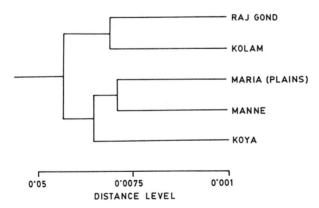

Figure 9. Genetic affinities among Gond subpopulations of central India

eas in the same district are Koyas and Mannes. The Koyas practise settled cultivation while the Mannes are shifting cultivators. The Maria Gond is the fifth subpopulation studied and they occupy a wider region, but have been sampled from Chanda district (Maharashtra). They practise both settled and shifting cultivation. The origin of these subpopulations is uncertain and, in fact, it is also uncertain whether the Gonds are the original settlers to this area. Perhaps the subpopulations have arisen because groups of individuals may have moved away from their original region of settlement (which may have been small) in times of economic adversity, and/or because of the practice of excommunication. The subgroups thus formed came in contact with Hindu populations and adopted many Hindu customs including endogamy (Fuchs 1973). Based on data of 7 genetic loci, Pingle (1984) found considerable heterogeneity in gene frequencies among the 5 subpopulations of Gonds. The genetic relationships among the subpopulations are presented in Figure 9. The genetic similarities are in broad agreement with geographical proximity. In terms of geographical distance of habitat, there are two clusters: Raj Gond, Kolam and (Manne, Koya), Maria Gond; the Mannes and Koyas are closer within the second cluster.

CONCLUDING REMARKS

The people of India exhibit a great deal of genetic diversity. Major reasons for this are: (i) large-scale admixture with Caucasoids of west Asia, Mongoloids of east and central Asia, and, to a lesser extent, with Negroids of Africa; (ii) social customs restricting free intermarriage resulting in a large number of endogamous populations; (iii) social and physical barriers restricting mobility and often leading to isolated populations of small size thereby resulting in greater diversity by genetic drift; and, (v) great ecological diversity leading to differential pressures of natural selection. The fact that many caste and tribal populations of northern, western and central India show close genetic affinities with the people of west Asia, and that, in general, the people of southern India are biologically fairly distant from people of other regions, bear testimony to immigration (and subsequent admixture with autochthones) of the so-called Aryans of west Asia to India and to their being restricted to the northern half of the country. The genetic affinity that many populations of eartern India exhibit with east Asian populations bears testimony to immigration and admixture of Mongoloids of east Asia. Some populations (e.g., Siddis of Karnataka) show biological evidence of Negroid admixture (Vijayakumar et al., 1987). Indeed Indians seem to be genetically about equidistant from Caucasoids and Mongoloids, and are farthest from the Negroids. Although some anthropologists have contended that the Negritos were, at one time, widespread in India and have subsequently been pushed to their restricted areas, genetic evidence does not support this contention.

Geographical distance has been found to correlate negatively with genetic similarity in many regions of the world. The unique social organization of Indian popopulations also provides opportunity of increasing genetic diversity by restricting gene flow. However, these factors, whether considered in isolation or in conjunction, is not found to completely explain the pattern of genetic differentiation or affinities. On a broad level, past ethno-historic connections seem to be discernible from gene frequencies; the tribes and castes seem to form genetically distinct clusters. Genetic variation within ethno-social categories, e.g., among castes, does not correlate well with social rank. Geographically, no clear clines of gene frequencies are observed at any genetic locus, although some clinal trends are present at some loci. Geographical proximity does not also always correlate well with genetic affinity. However, in some cases, physical barriers to gene flow are seen to determine genetic similarity, e.g., among the Koya Doras of Andhra Pradesh, or among the Mongoloid tribes of eastern India. It thus seems that there is no clear pattern of genetic variation and affinities in Indian populations that is attributable to any particular factor. The fact that genetic variation within most Indian populations is greater than that between populations argues against any genetic classification of the people of India. Genetic affinities among populations within smaller geographical regions seem to be primarily determined by the amount of local gene flow. The endogamous groups seem to be subdivided into local subpopulations (facilitating the operation of genetic drift), as a result of which the pattern of affinities at the micro-geographical level even within an endogamous group seems to be determined largely by local geographical barriers.

ACKNOWLEDGMENTS

We thank Mr. R.N. Das for some help with data analysis, and Professor A. Basu and Dr. R. Gupta for comments on portions of this manuscript.

REFERENCES

Banerjee, S., Roy, M., Dey, B., Mukherjee, B.N., and Bhattacharjee, S.K.,1988, Genetic polymorphism of red cell antigen, enzyme, haemoglobin and serum protein in fifteen endogamous groups of south India, *Jour. Indian Anthrop. Soc.* 23: 250.

Banerjee, S., Das, M.K., Das, K., Dey, B., Roy, M., and Mukherjee, B.N.,1992, Genetic differentiation among the three Hindu low caste groups of Malda district, West Bengal, India, *Internat. Jour. Anthrop.* 7: 7.

Chakraborty, R., Walter, H., Mukherjee, B.N., Malhotra, K.C., Sauber,P., Banerjee, S., and Roy, M.,1986, Gene differentiation among ten endogamous groups of West Bengal, India, *Am. J. Phys. Anthrop.* 71: 295.

Das, B.M., Walter, H., Gilbert, K., Lindenberg, P., Malhotra, K.C., Mukherjee, B.N., Deka, R., and Chakraborty, R.,1987, Genetic variation of five blood group polymorphisms in ten populations of Assam, India, *Internat. Jour. Anthrop.* 2: 325.

Fuchs, S.,1973, "The Aboriginal Tribes of India," Macmillan India, New Delhi.

Guha, B.S.,1935, "The Racial Affinities of the People of India," Census of India, 1931, Vol. I, Part IIIA.

Majumdar,D.N.,1958, "Races and Cultures of India," Asia Publishing House, Bombay.

Majumder, D.N., and Rao, C.R.,1958, Bengal anthropometric survey, 1945: A statistical study, *Sankhya* 19: 201.

Malhotra, K.C.,1978, Morphological composition of the people of India, *Jour. Hum. Evol.* 7: 45.

Malhotra, K.C.,1984, Population structure among the Dhangar caste-cluster of Maharashtra, India, *in:* "The People of South Asia," Lukacs, J.R., ed., Plenum Press, New York, p. 295.

Malhotra, K.C., Chakraborty, R., and Chakravarti, A.,1978, Gene differentiation among the Dhangar caste-cluster of Maharashtra, India, *Hum. Hered.* 28: 26.

Mourant, A.E., Kopec, A.C., and Domaniewska-Sobczak, K.,1976, "The Distribution of Human Blood Groups and other Polymorphisms," Oxford University Press, London.

Mukherjee, B.N., Majumder, P.P., Malhotra, K.C., Das, S.K., Kate, S.L., and Chakraborty, R.,1979, Genetic distance analysis among nine endogamous population groups of Maharashtra, India, *Jour. Hum. Evol.* 8: 567.

Mukherjee, B.N., Malhotra, K.C., Roy, M., Banerjee, S., Walter, H., and Chakraborty, R.,1989, Genetic heterogeneity and population structure in eastern India: Red cell enzyme variability in ten Assamese populations, *Z. Morph. Anthrop.* 77: 287.

Nei, M., and Roychoudhury, A.K.,1972, Gene differences between Caucasians, Negro and Japanese populations, *Science* 177: 434.

Papiha, S.S.,1985, Genetic structure and microdifferentiation among populations of Kinnaur district, Himachal Pradesh, India, in: "Genetic Microdifferentiation in Human and other Animal Populations," Ahuja, Y.R. and Neel, J.V., eds., Indian Anthropological Association, Delhi, p. 80.

Papiha, S.S., Mukherjee, B.N., Chahal, S.M.S., Malhotra, K.C., and Roberts, D.F.,1982, Genetic heterogeneity and population structure in north-west India, *Ann. Hum. Biol.* 9: 235.

Pingle, U.,1984, Morphological and genetic composition of the Gonds of central India: A statistical study, in: "Human Genetics and Adaptation, Vol. 1," Malhotra, K.C., and Basu, A.,eds., Statistical Publishing Society, Calcutta, p. 391.

Reddy, B.M., Chopra, V.P., Rodewald, A., Mukherjee, B.N., and Malhotra, K.C.,1989, Genetic differentiation among four groups of fishermen of the eastern coast, India, *Ann. Hum. Biol.* 16: 321.

Risley, H.H.,1915, "The People of India," Thacker, Spink and Company, Calcutta.

Roychoudhury, A.K.,1977b, Genetic diversity in Indian populations, *Hum. Genet.* 40: 99.

Roychoudhury, A.K.,1983, Genetic polymorphisms in human populations in India, in: "Peoples of India: Some Genetical Aspects," Satyavati, G.V.,ed., Indian Council of Medical Research, New Delhi, p. 1.

Saha, N., Tay, J.S.H., Roy, A.C., Das, M.K., Das, K., Roy, M., Dey, B., Banerjee, S., and Mukherjee, B.N.,1992, Genetic study of five populations of Bihar, India, *Hum. Biol.* 64: 175.

Sanghvi, L.D.,1954, Genetic diversity in the people of Western India, *Eugenics Quarterly* 1: 235.

Sanghvi, L.D., Balakrishnan, V., and Karve, I.,1981, "Biology of the People of Tamil Nadu," Indian Society of Human Genetics, Pune, and Indian Anthropological Society, Calcutta.

Sanghvi, L.D., and Khanolkar, V.R.,1949, Data relating to seven genetical characters in six endogamous groups in Bombay, *Ann. Eugen.* 15: 52

Singh, K.S., Mukherjee, B.N., Walter, H., Lindenberg, P., Gilbert, K., Dannewitz, A., Malhotra, K.C., Banerjee, S., Roy, M., and Dey, B.,1986, Genetic markers among Meiteis and Brahmins of Manipur, India, *Hum. Hered.* 36: 177.

Vidyarthi, L.P.,1983, Tribes of India, in: "Peoples of India: Some Genetical Aspects," Satyavati, G.V.,ed., Indian Council of Medical Research, New Delhi, p. 85.

Vijayakumar, M., Malhotra, K.C., Walter, H., Gilbert, K., Lindenberg, P., Dannewitz, A., Sorensen, A., Chakraborty, R., Reddy, A.P., and Mukherjee, B.N., 1987, Genetic studies among Siddis of Karnataka, India: A migrant population from Africa, *Anthropologischer Anzeiger* 77: 92.

Walter, H.,1971, Notes on the distribution of serum protein polymorphism in India, *in*: "Proceedings of the International Symposium on Human Genetics," Chakravartti, M.R.,ed., Andhra University Press, Waltair, p. 205.

Walter, H.,1986, Genetic differentiation processes among the populations of India, *Internat. Jour. Anthrop.* 1: 297.

Walter, H., Pahl, K.-P., Hilling, M., Veerraju, P., Goud, J.D., Naidu, J.M., Babu, M.S., and Kishan, G.J.,1981, Genetic markers in eight endogamous population groups from Andhra Pradesh (South India), *Z. Morph. Anthrop.* 72: 325.

RECENT DEVELOPMENTS IN GENETIC EPIDEMIOLOGY

Newton E. Morton

CRC Genetic Epidemiology Research Group
The Princess Anne Hospital
SOUTHAMPTON, SO9 4XY
U.K.

INTRODUCTION

The foundations of genetic epidemiology were discussed several years ago (Morton, 1986), at a time when the field could easily be summarized (Morton, 1982; King et al., 1984). Recent publications have dealt with hearing impairment (Morton, 1991), cancer (Lynch and Tautu, 1991), coronary heart disease (Bearn, 1992), and metabolic disease (Tikkanen and Tuomilehto, 1992). Many developments are reflected in the journal Genetic Epidemiology inaugurated in 1984, but most of the literature is in less specialized journals. A review of the past decade is overdue and will be attempted here.

The primary division of population genetics is between evolutionary inference and observations on contemporary, often health-related problems. This distinction is recognized in the conventional definition of genetic epidemiology as "a science that deals with the etiology, distribution, and control of disease in groups of relatives and with inherited causes of disease in populations". However, the methods of formal genetics do not distinguish between diseases and other traits. Markers unrelated to disease are required for positional cloning, and it would be unreasonable to treat locations of diseases and markers separately. Population structure plays the same role for diseases, polymorphisms, and even surnames, and forensic use of DNA markers involves the same formal genetics and analysis of population structure as disease genes. Therefore genetic epidemiology should be broadly defined to embrace all aspects of population genetics except evolution, including phenotypes, gene-environment interactions, and modes of transmission related to health or to location of disease genes, or requiring methods of analysis developed for genetic determinants of disease. No topic or approach with biological significance should be excluded by a narrow definition of the field of study.

Several other disciplines are subsumed by this definition, including ecogenetics, behaviour genetics, and demographic genetics. Molecular epidemiology is an unstable amalgam of DNA technology and genetic epidemiology (Dorman, 1992). There is no immunological, histological, or haematological epidemiology, and we do not recognise molecular paleontology, histology or dermatology. What distinguishes these nonexistent fields from molecular epidemiology? Genetic epidemiology is clearly molecular, but differs from the rest of epidemiology by its risk factors, sampling designs, hypotheses that are entertained, and ways they are tested. It takes more than an appreciation of DNA technology to create a science, and so molecular epidemiology as it relates to inherited risk factors is contained within genetic epidemiology.

FORMAL GENETICS

The outstanding development of the past decade has been mapping of genes for disease as a prelude to cloning and sequencing. Success has been phenomenal with genes of large effect, even when indistinguishable phenotypes are produced by different genes in different families. However, this method has been much less successful in two limiting cases: when the number of genes with large effects is great although individually rare (as for nonspecific mental retardation and deafness) and when several common genes interact to produce affection (as for schizophrenia and type II diabetes). Neither the search for homologous loci in the mouse nor identity by descent in affected relatives (Lander and Botstein, 1987) has so far been of much help. Better candidate loci and more efficient genome searches may be useful, although the already acute problem of excessive type 1 errors will persist. We are so familiar with the reliability of lods for monogenic phenotypes that unreliability for complex diseases is shocking.

Apart from failure to score phenotypes and markers blindly, the central problem seems to be that validity of a lod depends on the truth of the null hypothesis. Let $P(S; \Omega, \frac{1}{2})$ denote the probability of a sample under the null hypothesis H_0 that specifies free recombination ($\theta = \frac{1}{2}$) but also a number of other conditions Ω that include mode of inheritance, gene frequencies, absence of typing errors, and correct specification of parents. Let $P(S; \Omega, \theta_1)$ denote the probability of the sample under an alternative value of recombination θ_1 and the specified conditions. Then the lod is

$$z (S; \Omega, \theta_1) = \log \lambda, \text{ where } \lambda = [P(S; \Omega, \theta_1)/P(S; \Omega, \frac{1}{2})]$$

and so

$$P(\lambda > A \mid H_0) < 1/A \qquad (1)$$

a principle discovered independently by Haldane and Smith (1947) and Wald (1947). Since it depends only on the truth of H_0, θ_1 may be replaced by its maximum likelihood

value $\hat{\theta}$ (Collins and Morton, 1991). The effect of replacing $\hat{\theta}$ by sex-specific values $\hat{\theta}_m$,

$\hat{\theta}_f$ is not great (Lander and Lincoln, 1988), and nuisance parameters in Ω may be efficiently estimated under the two hypotheses without ill effect. However, misspecification of the null hypothesis invalidates Eq 1 and all other statistics that test for linkage.

The contribution of Haldane to gene mapping in man has not been properly recognized. In 1919 he derived the relationship between map distance and recombination under the assumption of no interference, showed that his formula did not fit Drosophila data, and gave a much more accurate result using an empirically determined mapping parameter. This seminal work was neglected for nearly 60 years until Rao et al. (1977) applied it to man, and it is still ignored in multipoint mapping. Haldane (1922) recognized that genetic map distances are sex-specific and that the homogametic sex usually has more recombination per unit of physical distance. The classic paper of Haldane and Smith (1947) came close to developing lod scores but was diverted by Bayesian arguments. It must have seemed to Haldane that his likelihoods had destroyed sib pair methods (Penrose, 1935), which are inexact and inefficient. Unfortunately a new generation of human geneticists rediscovered them. What could be simpler than to classify pairs of affected sibs into four classes as in table 1, with equal probabilities under

Table 1. Probabilities of the four outcomes for pairs of sibs when $\theta = \frac{1}{2}$

		sharing maternal allele	
		+	-
sharing paternal allele	+	¼	¼
	-	¼	¼

the null hypothesis of no linkage? However this encounters serious difficulties:

1. Under many circumstances (including an untested parent, a diallelic intercross, and dominance), identity by descent cannot be inferred with certainty.

2. Typing a homozygous parent as a heterozygote gives spurious evidence of linkage.

3. Affection status of parents is ignored.

4. There is no distinction between genetic heterogenity and recombination, and so no useful estimate of recombination rate.

5. The method does not extend to more than a single pair of affected relatives.

To appreciate the last point, consider a mating in which all of R affected sibs receive the same parental allele. The probability of this event if there is no linkage is $(\frac{1}{2})^{R-1}$.

However, if all possible pairs are made, the probability $(\frac{1}{2})^{R(R-1)/2}$ will be assigned, which is clearly incorrect for $R > 2$. Affected pairs of relatives are not only inefficient but they give an excess of type 1 errors unless each individual enters into only one pair or the dependency of multiple pairs is allowed for in a more complicated way. The apparent simplicity of affected pairs is deceptive and does not provide an acceptable alternative to likelihood analysis.

Penetrance and other parameters for monogenic disease can be estimated by segregation analysis and assumed in linkage. However, this has not been successful for complex disorders where linkage can help to establish the mode of inheritance. Combined linkage and segregation analysis has been developed for two disease loci and a single marker locus (Morton et al., 1991) with allowance for ascertainment and graded susceptibility in affected (severity) and unaffected (diathesis). In this way the normality assumption for quantitative trait is avoided. If the marker locus is a strong candidate for disease liability the hypothesis of no recombination tests deviation from the model. Regressive models have also been developed for combined segregation and linkage analysis (Bonney, 1986). They are evolving, but at the present time have a number of limitations:

1. Only the class D model corresponds to any genetic mechanism, and then only for quantitative data in nuclear families. Depending on how the model is implemented, known parental phenotypes and constant number of sibs may be required for exact correspondence.

2. Normality of deviations is assumed and false evidence for a major locus may be obtained when that assumption is violated.

3. There is no adequate correction for ascertainment.

4. A strictly ordered polychotomy for diathesis or severity is not implemented.

So far none of the applications for regressive models to real data has led to firm conclusions; attempts with qualitative traits have been particularly inclusive. Regressive models may be predictive, but they violate the principle of efficient and final causes expressed with typical cogency by Haldane (1965): "It is not true that I have large hairs on my nose because my father had them". The "law of ancestral heredity" was rejected in Haldane's youth. He would be surprised that the tribe of Weldon survived to war against the tribe of Bateson.

Association studies require meticulous controls, which may be taken from haplotypes not transmitted to affected relatives (Falk and Rubinstein, 1987), but this is biased against common alleles. Association is not informative unless there is tight linkage to the candidate locus, and then is not sensitive unless there is strong linkage disequilibrium. Therefore estimates of variance components due to association are uninformative except as lower limits to the effect of the candidate region. For example, liability to ankylosing spondylitis is determined by B27, and association with other HLA factors is deceptively weak.

I have said nothing about path analysis, which is treated elsewhere (Vogler and Rao, this volume). Its function is to test simple hypotheses about genetic and cultural inheritance, excluding those that are clearly wrong but not analyzing in detail the acceptable models, which increase with the number of variables. For that reason I am sceptical of extensions to multivariate analysis and longitudinal studies: this is holistic at a time when the great successes of genetics are reductionist.

MAPPING STRATEGIES

Localization of disease genes requires a good map of markers. If typing errors are negligible, there is no interference, and all markers are typed on the same families and analyzed without partition, the best map would be obtained by multilocus analysis. When any of these conditions fail it is better to use multiple pairwise analysis of lods, which does not in general require access to the raw data (Morton, 1988).

Given a disease that has not been localized, markers should initially be chosen to be highly polymorphic and located in regions that are of interest either because of a candidate locus or because other markers in that region have not been tested. Markers giving negative lods will be set aside pending a search for more promising markers. Investment to exclude a region (subject to sometimes dubious assumptions) is not worthwhile unless it contains a strong candidate locus, since the object is not exclusion but detection of linkage. Once a suggestion of linkage has been obtained, other markers in the region should be tested even if they are not highly polymorphic (Terwilliger et al., 1992). The selection of "index" markers on the basis of location and polymorphism is of limited utility and changes with techniques and more complete sampling of the genome. Meetings in which dozens of people spend much of their time choosing index or reference markers are not cost-effective. Mapping panels analogous to recombinant inbred lines have been proposed, but their efficiency by comparison to other strategies has not been evaluated: I do not think it can be high, unless established from sperm.

Deletions, chromosome rearrangements, and sequencing of candidate loci are the most powerful methods for physical mapping of diseases. The success of sequencing depends on two conditions: that the candidate locus is correct and that a change of sequence within its exons is causal of disease, for which other mechanisms in the region (exon duplication and mutations in introns or controlling elements or other loci of a cluster) are not responsible. Alzheimer's disease (Lawrence et al., 1992) and persistent fetal haemoglobin provide examples of disease loci outside but closely linked to selected exons. Although heuristic, sequencing must be supplemented by linkage studies to confirm that a detected mutation is causal of disease and to detect closely linked disease loci.

INTEGRATION OF MAPS

Any serious attempt to construct maps must integrate different sources within and between genetic and physical maps, projecting all information into a composite physical map by priority and weighting algorithms (Morton et al., 1992). The Human Genome Project pays lip service to this effort but contributes nothing, plodding along in an outdated five-year plan reminiscent of Stalin (Jordan, 1992). History may record that its main result was to direct royalties for sequence patents to France and Japan.

Integration of maps requires a location database in which the position of a locus is given as distance from an origin, conventionally taken as pter. The problem of long-range connectivity is central to a location database, and several approaches have been tried. The best is radiation hybrids, which give physical location on the assumption that breakage is uniform along the chromosome. It is unnecessary to assume that retention is uniform, and in fact the highest retention is usually near the centromere and the lowest retention near the telomere (Lawrence et al., 1991). It is not clear whether this is due to the requirement for a centromere or to selection in rodent cells against expressed human sequences which are richest near the telomeres. In any case, methods that assume a uniform retention rate are unrealistic. Over smaller distances, pushmi-pullyu hybrids exhibit a gradient of retention frequencies from a flanking marker on one side, which is selected for, to the flanking marker on the other side, which is selected against. Except for instances of point mutation or deletion, the hybrids have undergone breakage in the intervening region regardless of the X radiation dosage applied. A large part of the information resides in the retention frequencies and is recovered by partitioning the data into marginal retention and conditional retention or loss. Attempts have been made to specify an optimal dosage (Lange and Boehnke, 1992) but they depend on unsupported assumptions and are unnecessary both for long-range mapping where the dosage of about 8000 R works well, and for pushmi-pullyu hybrids over smaller distances where recovery of recombinants depends more on selection than dosage.

Regions defined by deletions and somatic cells, by in situ hybridization in balanced translocations, or in flow-sorted chromosomes can give good estimates of physical location if the panel is large. Unfortunately there is no consensus on the composition of such panels, the methods to be used to assay presence of a locus, means to provide public access to the panel, or designation of the defined regions. Only for parts of some chromosomes is regional assignment competitive with other methods for long-range physical connectivity.

A few years ago it seemed possible that rare-cutter enzymes like Not1 might be used to regionalize whole chromosomes, but this effort failed for technical reasons. Emphasis of molecular biologists is now on contigs of YACS and large cosmids, which give long-range connectivity only in the limit. Sequence Tagged Sites (STSs) unify different methods of physical mapping and provide loci small enough so that overlap is not a problem, but they are resisted by some investigators despite brilliant successes with a proportion of loci (Chumakov et al., 1992; Foote et al., 1992). In this vacuum, statistical methods to project data onto the physical map offer some promise. Whether the genetic data come from linkage or chiasma maps, they have good connectivity but low precision. The same is true for estimates of physical location from assignment to chromosome bands or regions. However, the approximate location determined by regressing physical on genetic location is automatically refined as partial maps are accumulated.

MUTATION

A number of studies have failed to show any difference in mutation rate on the X chromosome between eggs and sperm (Morton and Lalouel, 1979; Williams et al., 1983). This is also true in the mouse if each mutant in a oogonial cluster is counted separately as is appropriate since generations are delineated by the haplophase (BEIR, 1990). However, Barbujani et al. (1990) has reported several studies of Duchenne muscular dystrophy, none published in detail, that appear to contradict this. The conclusions depend critically on two points: (1) overestimating the ascertainment probability π by counting examined secondary cases as probands and (2) attributing a proband to some sibships with only secondary cases that by death, residence, or year of birth are outside the sampling frame and would not have been ascertained without a proband elsewhere in the pedigree. This tends to inflate the proportion of sibships without mutants and to underestimate the frequency of sporadic cases, on which the mutation rate in eggs is based. The ascertainment problem has so far dominated advances in carrier detection, and an excess of X-chromosome mutation in sperm is not proven. There is an excess for some dominant lethals and chromosome rearrangments, while the full mutation in the fragile X syndrome occurs only in eggs (Morton and Macpherson, 1992).

Various estimates of spontaneous detrimental mutation rates are in surprisingly good agreement. However, induced mutation rates are controversial because the gap between single locus tests in the mouse and infant morbidity in man has not been closed. Estimates in the mouse are based largely on heroic experiments at Oak Ridge with visibles at 7 loci. Neel et al. (1990) have questioned whether these loci are representative of the mouse genome. In man there is uncertainty about the fraction of mutations that are expressed in heterozygotes. It is generally believed that there are 100,000 structural loci in man and mouse. With much less assurance, perhaps 10,000 of these would cause lethality if homozygous for certain mutations and perhaps 1000 are haplo-insufficient and could be detected by early death or severe morbidity in F1 progeny of exposed individuals. However these numbers are no more than educated guesses and they do not specify the proportion of mutations that are dominant. Dominant lethals in Habrobracon increase as the 1.7th power of the dose, indicating a significant proportion of two-hit deletions at high acute dose. Another difficulty in evaluating the human evidence is that dosage is always uncertain, and the uncertainty cannot be evaluated. The Atomic Energy Commission revised its estimates of gamma and neutron dosage for the Hiroshima and Nagasaki bombs at least once, and shielding factors based on recollection were being revised many years after the event.

The first studies of sex ratio indicated a decrease of sons with exposure of their mothers, as expected for sex-linked mutation (Neel et al., 1953). However, with addition of subsequent data and new estimates of dosage, the direction of the deviation apparently changed (Schull et al., 1966). There has been no test for the temporal trend which cell selection or mutation repair would produce. The Oak Ridge experiments did not attach enough importance to sex ratio as an indicator of mutational damage to report sex in their experiments, nor did they give any weight to the swept radius approach advocated by Haldane (1964) which uses the inbred load in the neighbourhood of a marker locus. As noted in the early days of the Atomic Bomb Casualty Commission, it was not likely that the quantity and quality of information in Hiroshima and Nagasaki would allow detection of induced mutations, unless man were much more sensitive than the mouse. Fortunately that is not the case. However, uncertainties about dosage, dominance, and temporal trend in sex ratio make the human data less critical than the mouse experiments despite their flaws. Therefore I see no practical alternative to acceptance of 50 rad as the acute doubling dose for both organisms, pending more conclusive information about man, which will not come from accidents like Chernobyl nor regions of high background radiation. As Haldane (1965) presciently remarked, "I doubt whether the scheme proposed for the radioactive areas in Kerala (South India) will yield any more information on the mutagenic activity of radioactive elements than did the Atomic Bomb Casualty Commission in Japan. It is perhaps unlikely to yield as much".

INBREEDING EFFECTS

The factors that determine mating frequencies also determine population structure. This subject was actively pursued in the seventies and early eighties, but became quiescent when all the problems of general interest appeared to have been solved (Morton, 1982). In the interim evolutionary studies were devoted to inferring phylogeny and admixture by methods that do not permit comparison of different population structures. This situation has changed recently for two reasons. On the one hand, migrants from populations with preferential consanguineous marriage into countries where this practice was rare have raised questions about biological effects of inbreeding (Bittles and Roberts, 1992). On the other hand, population structure has become central to use of DNA markers for identification (Morton, 1991).

There are three ways by which inbreeding has biological effects:

1. Confounding with socioeconomic factors that must be controlled if genetic factors are to be identified;

2. Homozygosity for rare recessive genes;

3. More complicated and obscure genetic mechanisms, including interaction of two or more loci (epistasis) and outliers in a continuous distribution of cumulative genetic effects (polygenes).

The theory of genetic loads is based on the second mechanism, for which the probability of a particular type of morbidity is

$$1\text{-}e^{-(A+BF)} \approx A+BF \tag{2}$$

where F is an inbreeding coefficient, A is the panmictic load and B is the inbred load. The expressed load is $A+BF$ and the relative risk associated with F is $1+(B/A)F$ (Morton et al., 1956). Evidence for rare recessive genes comes from clinical genetics, segregation analysis, and high relative risk. In Drosophila more complicated genetic mechanisms become important only for high levels of inbreeding ($F > \frac{1}{4}$), where they produce phenodeviants expressed by quadratic and higher order terms in F (Lerner, 1954).

There have been many studies of inbreeding effects on specific morbidity and mortality, using neighbours, outmarrying sibs, or covariance analysis to control confounding of socioeconomic factors, with variable success. The panmictic load includes an environmental component that is large for mortality and some types of morbidity, among which both B and A vary. B/A ratios are unstable among populations, but B is less variable, especially for disorders with a large component due to rare recessive genes and a correspondingly high B/A ratio (Morton, 1991).

Some geneticists and more laymen have suggested that inbreeding effects are greatly reduced by many generations of preferential consanguineous marriage (Sanghvi, 1966). This is not supported by careful studies (Bundey et al., 1991), but some reduction is expected for the inbred load on theoretical grounds. Estimates of long-term inbreeding under preferential consanguineous marriage suggest that B might be reduced by 50 per cent compared with populations that tend to avoid consanguineous marriage (Morton, 1982). However, this estimate is based on current frequencies of inbreeding in an expanding population, and very different frequencies must be observed in stable or declining populations. Only meticulous studies can determine accurate and specific risks for consanguineous marriage in different populations. As Haldane (1965) succinctly wrote, "High inbreeding has been going on for many centuries, perhaps many millennia, but has not yet become harmless". This has a practical application in clinical genetics and public health and a theoretical importance in determining gene frequencies, mutation rates, and numbers of loci for specific disorders (Morton, 1961).

IDENTIFICATION WITH DNA MARKERS

An expert witness who testifies that two DNA samples do not match must defend his molecular technology, but population genetics is not at issue. On the contrary, evidence that two samples match requires an assessment of the probability that two random individuals from an appropriate population have indistinguishable genotypes at the loci tested. If d denotes the defendant, c the culprit, R is their relationship and the culprit belongs to population k, then the probability that the culprit has the same genotype $G_r G_s$ as the defendant is

$$C_{G_r G_s} \equiv P(c = G_r G_s \mid d = G_r G_s, R, k). \tag{3}$$

There is a generally accepted theory that defines such a matching probability and derives it from genotype frequencies within a population (Morton, 1991). There is also a generally accepted theory to calculate genotype probabilities from estimates of gene frequency and kinship, and a large body of evidence about kinship in various populations. Valid use of this theory and evidence depends on several self-evident principles:

1. There is no connection between a matching probability in one population and gene frequencies in an unrelated population.

2. If C_i is the matching probability for genotype i and there is an upper bound B_i such that $C_i < B_i$ for all i, then B_i is not a probability.

3. Of the indefinitely large number of ways in which such a bound B_i may be estimated, few conform to generally accepted theory.

4. For every specific matching probability C_i there is a confidence interval, a mean matching probability over all genotypes, a set of estimates based on independent samples, and a deviance based on fragment sizes without defining alleles. These values protect adequately against excessive weight on a small value of C_i.

5. An acceptable bound must not violate statistical or genetic principles or known values of gene frequency or kinship.

6. Gene frequencies should be estimated in large samples to minimise sampling error.

7. In the absence of evidence to the contrary, the suspect and culprit should be assumed to be randomly drawn from a forensic population. Contrary evidence may be accommodated by the affinal model in which the culprit is related to the suspect as closely as a spouse would be and by an appropriate estimate of kinship, without altering gene frequencies in the reference population.

These principles have been violated in several ways. Nichols and Balding (1991) took as a bound for Eq. 3

$$B_{G_r G_s} \equiv 2\, P(G_r G_r \mid G_r)\, P(G_s G_s \mid G_s) = 2\, [Q_r + (1-Q_r)\varphi]\, [Q_s + (1-Q_s)\varphi]$$

where φ is deliberately overestimated as .05, a value that has been observed within race only in a few isolates. This is extremely conservative, although there is no formal proof that it holds under all population structures and errors in gene frequency estimation. However, it violates principles 2, 3, 4, 5, and 7.

Following Lenin, Lewontin and Hartl (1991) developed their argument in a passage inauspiciously headed "What is to be done?". They proposed that for every locus a sample be sought in which the frequency of alleles manifest in the suspect is maximal, perhaps through sampling error or mistyping. Thus the culprit might be assumed to be a Basque for one locus and a Maltese for another. This provides no solution, since however many populations are sampled a more extreme one might be found or a finer partition might be demanded. The ethnic populations are constrained to be "relevant", but this is undefined.

In general there is no reason to suppose that the population of the suspect places a narrow constraint on the population of the culprit unless the suspect is guilty.

They support their position, which violates all 7 principles, by citing a number of gene frequencies, allegedly "typical" values from a 1954 book. Their choices are extremely biased, and less than 2 per cent of the diversity selected by Lewontin and Hartl is due to the national kinship to which they attribute it, little of which persists in regional forensic samples. Hypervariable loci used for forensic identification have less diversity than blood groups, isozymes, and oligoallelic DNA markers (Morton et al., 1992).

Lander (1991) proposed that "regardless of the defendant's ethnic background, each allele frequency used should be the maximum observed in various ethnic samples". The latter are not specified, except that they be "a dozen or so well-separated ethnic population samples", not necessarily including the population of the suspect. It is not clear how to sample an ethnic group rather than a local population. At a single locus the culprit might be assumed to be a Lapp for one allele and a Hottentot for another. However extensively our species might be sampled, an ethnic group could always be found with an apparently higher gene frequency, either through drift, mistyping, or sampling error. This violates all 7 principles and is the most extreme of all attempts to preempt the court's responsibility for assessing weight of evidence. It was adopted without discussion by a recent National Research Council committee, which added the stipulation than an arbitrary value (.05 or .1) be substituted for any gene frequency too damaging to the defendant (NRC, 1992). If the suspect has a wooden leg and evidence were presented that the culprit had a wooden leg, it would be immaterial if there is a Cambodian village where wooden legs are common. Any juror can understand that the same principle applies to DNA, and so a court in Kansas or Cornwall will be reluctant to assign guilt on the basis of gene frequencies in Cambodia or simply invented by pseudo-scientific mumbo-jumbo that conceals evidence and discredits expert testimony. There is no excuse or legality for calculations without theoretical basis or empirical support.

SUCCESSFUL PARADIGMS

Although much remains to be done in genetic epidemiology, progress in certain areas has been remarkable. With complete sequencing of the mitochondrion a number of diseases have been identified. The molecular basis of the fragile-X syndrome is understood well enough to explain the Sherman paradox, and other diseases caused by excessive amplification of a repeat sequence that leads to anticipation have been identified. Evidently these regions are under selective constraints, and high copy number can interfere with transcription of flanking genes. The causes of meiotic nondisjunction, a surprisingly frequent event in man, are beginning to be understood. They include failure of preferential recombination in the pseudoautosomal region, aberrant recombination near the centromere, nonconjunction, and a process (perhaps failure of the spindlefibre or kinetochore) that increases with maternal age independently of recombination. With coronary heart disease the successful paradigm uses apolipoproteins as candidate loci, but controlling elements and factors other than hyperlipidaemia have not been studied. With hearing impairment the paradigm takes syndromes that are much more homogenous than nonspecific deafness and so are mappable. With cancer the major paradigm invokes tumour suppressors, the amorphic allele being inherited as a dominant but requiring mutation or deletion of the normal homolog for expression. This mechanism accounts for a proportion of cancer of the breast, ovary and colon as well as many rarer tumours and family cancer syndromes, especially with precocious onset. In the minor paradigm protooncogenes initiate tumorigenesis by dysregulating through translocation a locus whose unchecked transcription can immortalise a cell, and mutations of protooncogenes control tumour progression and metastasis although seldom in an orderly sequence. Gonadal mosaicism, imprinting, uniparental disomy and other mechanisms previously undemonstrable are beginning to be studied. As genetic epidemiology assimilates molecular biology it adds its own mathematical and epidemiological dimension.

ENVOI

I have quoted several times the address Haldane gave to the International Congress of Genetics in 1963. I can still hear his distinctive voice and have always admired the verve with which he attacked in a few sentences the pope and hereditary monarchs, including the patron of the Congress. His paper was entitled "The implications of genetics for human society", which he summarized as follows:

"Geneticists must consider their social function, which differs in some respects from that of other scientists. The practical applications of genetics, such as the production of hybrid maize, have perhaps been uniformly desirable. But the influence of genetical ideas, such as that of racial purity, has often been disastrous. We have the duty to spread what we believe to be the truth, but it is extremely hard to avoid exaggerating the genetical evidence which we may think favours what, on non-genetical grounds, we regard as ethically or politically desirable aims .... Human geneticists have a double duty, to discover new truths, and to spread old ones. It does not much matter if many people know nothing about hybrid maize or progeny tests for milk yield. It matters a great deal if they know nothing about human genetics, because it is a topic of interest to all human beings, and the gaps in their knowledge will be filled by superstition or intellectually dishonest propaganda. So long as this is the case, we must be teachers and popularizers as well as researchers."

Haldane lived by that philosophy, which led after some indecision to his break with Lysenko and the Communist Party (Paul, 1983) . Not all geneticists have been as honest, and the literature on the XYY phenotype and heritability of IQ has some disgraceful infusions of bias from the political Left and Right. Committees and government projects are especially sensitive to self-aggrandisement and sloppy thinking, which reach their apogee in action groups like Science for the People and its successor the Council for Responsibility in Genetics. The hubris that transforms expert testimony into perjury is only an exaggeration of the self-conceit that sets vague and trivial goals for the Human Genome Project. Now more than ever science needs curmudgeons like Haldane.

REFERENCES

Barbujani, G., Russo, A., Danieli, G.A., Spiegler, A.W.J., Borkowska, J., and Petrusewicz, I.H., 1990, Segregation analysis of 1885 DMD families: significant departure from the expected proportion of sporadic cases, *Hum. Genet.* 84:522.

Bearn, A.G., 1992, "Genetics of Coronary Heart Disease", Institute of Clinical Genetics, University of Oslo, Oslo.

BEIR V, 1990, "Health Effects of Exposure to Low Levels of Ionizing Radiation", National Academy Press, Washington D.C.

Bittles, A.H., and Roberts, D.F., 1992, "Minority Populations. Genetics, Demography, and Health", The Galton Institute, McMillan Press, London.

Bonney, G.E., 1986, Regressive logistic models for familial disease and other binary traits, *Biometrics* 42:611.

Bundey, S., Alson, H., Kaur, A., Mi, S., and Lancashire, R., 1991, Why do UK-born Pakistani babies have high perinatal and neonatal mortality rates?, *Paed. Biomat. Epid.* 5:101.

Chumakov, I., Rigault, P., Guillou, S., Ougen, P., Billaut, A., Guasconi, G., Gervy, P., Legall, I., Soularue, P., Grinas, L., Bougueleret, L., Bellannechantelot, C., Lacroix, B., Barillot, E., Gesnouin, P., Pook, S., Vaysseix, G., Frelat, G., Schmitz, A., Sambucy, J.L., Bosch, A., Estivill, X., Weissenback, J., Vignal, A., Riethman, H., Cox, D., Patterson, D., Gardiner, K., Hattori, M., Sakaki, Y., Ichikawa, H., Ohki, M., Lepaslier, D., Heilig, R., Antonarakis, S., and Cohen, D., 1992, Continuum of overlapping clones spanning the entire human chromosome 21q, *Nature* 359:380.

Collins, and Morton, N.E., 1991, Significance of maximal lods, *Ann. Hum. Genet.*, 55:39.

Dorman, J.S., 1992, Molecular epidemiology: applications of molecular biology to epidemiologic research (submitted).

Falk, C.T., and Rubinstein, P., 1987, Haplotype relative risks: an easy way to construct a proper control sample for risk calculations, *Ann. Hum. Genet.* 51:227.

Foote, S., Vollrath, D., Hilton, A., and Page, D.C., 1992, The human Y chromosome: overlapping DNA clones spanning the euchromatic region, *Science* 258:60.

Haldane, J.B.S., 1919, The combination of linkage values, and the calculation of distance between loci of linked factors, *J. Genet.* 8:299.

Haldane, J.B.S., 1922, Sex ratio and unisexual sterility in hybrid animals, *J. Genet.* 12: 101.

Haldane, J.B.S., 1964, A defense of beanbag genetics, *Perspect. Biol. Med.* 7:343.

Haldane, J.B.S., 1965, The implications of genetics for human society, *in*: "Genetics Today Proc. XI Int. Cong. Genet." Geerts, S.J., ed., Pergamon Press, Oxford.

Haldane, J.B.S., and Smith, C.A.B., 1947, A new estimate of the linkage between the genes for colour blindness and haemophilia in man, *Ann. Eugen.* 14:10.

Jordan, E., 1992, Invited Editorial: the Human Genome Project: where did it come from, where is it going?, *Am. J. Hum. Genet.* 51:1.

King, M.C., Lee, G.M., Spinner, N.B., Thomson, G., and Wrensch, M.R., 1984, Genetic epidemiology. *Ann. Rev. Public Health* 5:1.

Lander, E.S., 1991, Invited editorial: Research on DNA catching up with courtroom application, *Am. J. Hum. Genet.* 48:819.

Lander, E.S., and Botstein, D., 1987, Homozygosity mapping: a way to map human recessive traits with the DNA of inbred children, *Science* 236:1567.

Lander, E.S., Lincoln, S.E., 1988, The appropriate threshold for declaring linkage when allowing sex-specific recombination rates, *Am. J. Hum. Genet.*, 43:396.

Lange, K., and Boehnke, M., 1992, Bayesian methods and optimal experimental design for gene mapping by radiation hybrids, *Ann. Hum. Genet.* 56:119.

Lawrence, S., Keats, B.J., and Morton, N.E., 1992, The AD1 locus in familial Alzheimer disease, *Ann. Hum. Genet.* 56:295.

Lawrence, S., Morton, N.E., and Cox, D.R., 1991, Radiation hybrid mapping. *Proc. Nat. Acad. Sci. USA* 88:7477.

Lerner, I.M., 1954, "Genetic Homeostasis", John Wiley & Sons, New York.

Lewontin, R.C., and Hartl, D.L., 1991, Population genetics in forensic DNA typing, *Science* 254:1745.

Lynch, H.T., and Tautu, P., 1991, "Recent Progress in the Genetic Epidemiology of Cancer", Spring-Verlag, Berlin.

Morton, N.E., 1961, Morbidity of children from consanguineous marriages, *Progress in Medical Genetics* 1:261.

Morton, N.E., 1982, Kinship and inbreeding in populations of Middle Eastern origin and controls, *in* "Current Developments in Anthropological Genetics", Crawford, H., Mielke, J.H., eds., Plenum Press, New York, vol 2, pp. 449.

Morton, N.E., 1986, Foundations of genetic epidemiology, *J. Genet.* 65:202.

Morton, N.E. 1988, Multipoint mapping and the emperor's clothes, *Ann. Hum. Genet.* 52:309.

Morton, N.E., 1991, Genetic structure of forensic populations, *Proc. Natl. Acad. Sci. USA* 89:2556.

Morton, N.E., 1991, Genetic epidemiology of hearing impairment, *in:* "Genetics of Hearing Impairment", Ruben, R.J., Van De Water, T.R., Steel, K.P. eds., *Ann. N. Y. Acad. Sci.* 630:16.

Morton, N.E., 1982, "Outline of Genetic Epidemiology" S. Karger, Basel.

Morton, N.E., Collins, A., and Balazs, I., 1992, Bioassay of kinship for hypervariable loci in Blacks and Caucasians, *Proc. Natl. Acad. Sci. USA* (in press).

Morton, N.E., Collins, A., Lawrence, S., and Shields, D.C., 1992, Algorithms for a location database, *Ann. Hum. Genet.* 56:223.

Morton, N.E., Crow, J.F., and Muller, H.J., 1956, An estimate of the mutational damage in man from data on consanguineous marriages, *Proc. Natl. Acad. Sci. USA.* 41:855.

Morton, N.E., and Lalouel, J.-M., 1979, Genetic counseling in sex linkage, *Birth Defects: Original Artic Series* 15 (5C):9.

Morton, N.E., and Macpherson, J.N., 1992, Population genetics of the fragile-X syndrome: multiallelic model for the FMR1 locus, *Proc. Nat. Acad. Sci. USA* 89:4215.

Morton, N.E., Shields, D.C., and Collins, A., 1991, Genetic epidemiology of complex phenotypes, *Ann. Hum. Genet.* 55:301.

Neel, J.V., Schull, W.J., McDonald, D.J., Morton, N.E., Kodani, M., Takeshima, K., Anderson, R.C., Wood, J., Brewer, R., Wright, S., Yamazaki, J., Suzuki, M., and Kitamura, S., 1953, The effect of exposure of parents to the atomic bombs on the first generation offspring in Hiroshima and Nagasaki: (preliminary report), *Jap. J. Genet.* 28:211.

Neel, J.V., Schull, W.J., Awa, A.A., Satoh, C., Kato, H., Otake, M., and Yoshimoto, Y., 1990, The children of parents exposed to atomic bombs: estimates of the genetic doubling dose of radiation for humans, *Am. J. Hum. Genet.* 46:1053.

Nichols, R.A., and Balding, D.J., 1991, Effects of population structure on DNA fingerprint analysis in forensic science, *Heredity* 66:297.

NRC Committee on DNA Technology in Forensic Science, 1992, "DNA Technology in Forensic Science", National Academy Press, Washington DC.

Paul, D.B., 1983, A war on two fronts: J.B.S. Haldane and the response to Lysenkoism in Britain, *J. Hist. Biol.* 16:1.

Penrose, L.S., 1935, The detection of autosomal linkage in data which consist of pairs of brothers and sisters of unspecified parentage, *Ann. Eugen.* 6:133.

Rao, D.C., Morton, N.E., Lindsten, J., Hulten, M., and Yee, S., 1977, A mapping function for man, *Hum. Hered.* 27:99.

Sanghvi, L.D., 1966, Inbreeding in India, *Eugen. Quart.* 13:291.

Schull, W.J., Neel, J.V., and Hashizume, A., 1966, Some further observations on the sex ratio among infants born to survivors of the atomic bombings of Hiroshima and Nagasaki, *Am. J. Hum. Genet.* 18:328.

Terwilliger, J.D., Dang, Y., and Ott, J., 1992, On the relative importance of marker heterozygosity and intermarker distance in gene mapping, *Genomics* 13:951.

Tikkanen, M.J., and Perheentupa, J., (Eds) 1992, Special section in: "Molecular Genetics and Genetic Epidemiology of Cardiovascular Disease and Diabetes", *Annals of Medicine* 24:313.

Wald, A., 1947, "Sequential Analysis", Wiley, New York.

Williams, W.R., Thompson, M.W., and Morton, N.E., 1983, Complex segregation analysis and computer-assisted risk assessment for Duchenne muscular dystrophy, *Am. J. Med. Genet.* 14:315.

PATH ANALYSIS IN GENETIC EPIDEMIOLOGY: THEORY AND APPLICATIONS

George P. Vogler and D. C. Rao

Division of Biostatistics
Washington University School of Medicine
Box 8067, 660 S. Euclid Avenue
St. Louis, MO 63110

INTRODUCTION

Causal influences on complex phenotypes cannot always be directly observed, but familial influences can be inferred from patterns of interrelationships among a set of variables (i.e., phenotypes observed on different classes of relatives). Path analysis was developed as a method of using structural linear regression analysis to explain the observed interrelationships among variables in the context of a specified model and to determine the impact of hypothesized causal influences on the observed variable (Wright, 1921, 1978; Li, 1975).

Whereas regression analysis is the appropriate statistical method to use when the goal is simply prediction, path analysis serves a different function by providing a plausible model for explaining the structural relationships among variables. It is the nature of this methodology that different models can be developed which explain the interrelationships among a set of variables equally well. This flexibility is viewed by some data analysts as a fundamental limitation of the method, particularly in the absence of supporting evidence in favor of one model over others. However, when there is a theoretical basis for selecting a particular model, path analysis provides a formal method for testing whether the theortical model is consistent with the observed interrelationships among variables. In genetic epidemiology, the principles of Mendelian genetics, extended to multifactorial inheritance, provide a reasonably firm theoretical basis for a causal model of familial resemblance.

PATH MODELS IN GENETIC EPIDEMIOLOGY

Basic path models of familial resemblance have been extended to various complex data structures, including nuclear families and other family configurations, multiple phenotypes, with or without a measure of the familial environment, and multiple time points. Appropriate statistical methods of analysis have been developed for both random and nonrandomly ascertained samples.

Human Population Genetics, Edited by P.P. Majumder
Plenum Press, New York, 1993

Basic Models of Familial Resemblance

In the context of a model of familial resemblance, an individual's phenotype (P) can be considered to be a function of familial influences (T) that are transmitted from parents to children and environmental influences (E) that are unique to each individual. In terms of a linear additive model,

$$P = T + E,$$

where all variables have zero means, and the transmissible effects and the environment are assumed to be uncorrelated. This model of familial resemblance has been termed the TAU model by Rice et al. (1978). In the TAU model, additive polygenic effects (i.e., effects of numerous additive genes each of small effect) and familial environmental effects that are shared by family members are not distinguished. Such a model is appropriate for nuclear families because there is insufficient information in a simple nuclear family design to resolve these effects (in the absence of a direct measure of the familial environment or the genotype).

The phenotypic variance (σ^2_P) is given by:

$$\sigma^2_P = \sigma^2_T + \sigma^2_E$$

where σ^2_T is the variance of T and σ^2_E is the variance of E. If the variables are all standardized, the structural equation for the TAU model becomes

$$P = tT + eE$$

where t and e are standardized partial regression coefficients called path coefficients whose squares denote the proportions of the phenotypic variance due to transmissible or unique environmental effects, respectively. Formally, $t^2 = \sigma^2_T / \sigma^2_P$ and $e^2 = \sigma^2_E / \sigma^2_P$, and $t^2 + e^2 = 1$, which is called the equation for complete determination.

Path coefficients may be estimated from nuclear family data using one of the statistical methods discussed under "Statistical Methods of Analysis" and the nuclear family TAU model shown in Figure 1. Four members of a nuclear family are shown: father (F), mother (M), and two children $(C1$ and $C2)$. In addition to the path coefficients defined above, additional path coefficients are needed for describing resemblance among relatives: u is the correlation between spousal transmissible influences; b is the correlation between sibling transmissible influences, and τ_M and τ_F denote the effects of the transmissible factors of mother and father, respectively, on that of a child. Since there is insufficient information in nuclear family data to estimate all of these path coefficients, τ_F may be fixed at the value of one-half (expected under a polygenic model of transmissibility), whereas τ_M can differ from one-half to allow for maternal influences. Alternatively, if the sibling phenotypic correlation is less than both parent-child correlations, one may set $b = 0$ and estimate τ_M and τ_F in addition to the other parameters (t and u). As yet another alternative, investigators use a pseudo-polygenic model by fixing $\tau_M = \tau_F = 1/2$ and estimate only t, u, and b.

The basic TAU model is readily extended to the analysis of multiple phenotypes measured on each individual. In this model, the covariance among phenotypes is

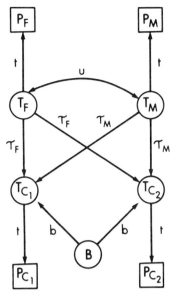

Figure 1. TAU path model of resemblance among relatives in nuclear families. P denotes a phenotype, T indicates transmissible influences, and the subscripts M, F, C1, and C2 denote the mother, father, and two children, respectively. B denotes nontransmitted common environment for a sibship. Path coefficients are defined in the text.

expressed in terms of the covariance due to familial influences and that due to nonfamilial influences:

$$COV\ (P_i,\ P_j) = COV\ (T_i,\ T_j) + COV\ (E_i,\ E_j)$$

or, when the two phenotypes P_i and P_j are standardized,

$$r_{P_iP_j} = t_it_jr_{Tij} + e_ie_jr_{Eij}$$

where r_{Tij} and r_{Eij} are the correlations, within an individual, between the transmissible influences and the nontransmitted environmental influences, respectively, of the two phenotypes i and j.

A more informative model which requires a more complex data structure for resolution has been termed the BETA model (Rice et al., 1978) and includes an index of the familial environment which permits differentiation between additive genetic effects and familial environmental effects. In the BETA model,

$$P = G + C + E$$

where G is the additive polygenic genotype, C denotes the familial environment (or culturally transmitted effects), and E is the nontransmitted environment. In standardized notation,

$$P = hG + cC + eE$$

where h^2 is termed genetic heritability, defined as the proportion of phenotypic variance explained by the genotype; cultural heritability (c^2) is the proportion of the phenotypic variance explained by the familial environment; and e^2 is the proportion of the phenotypic variance explained by residual environmental effects (including random error). Thus, the standardized variance of P is

$$1 = h^2 + c^2 + e^2.$$

As with the TAU model, genetic and familial environmental effects cannot be differentiated using phenotypic data on nuclear family members alone. One approach around this limitation is to derive an additional variable on each family member called the *environmental index* (Morton, 1974; Rao et al., 1982). This is a second observable measure obtained on each individual in a nuclear family but which is selected to be as close to a measure of the family environment, free of genetic influences, as is possible. This index I can be a direct measure or can be constructed for quantitative phenotypes by regressing the phenotype on a battery of potentially relevant environmental variables $(X_1, X_2,...,X_n)$:

$$\hat{P} = \hat{\alpha} + \Sigma \hat{\beta}_i X_i + \varepsilon_i$$

where the index is defined as $I = \hat{P} = \hat{\alpha} + \Sigma \hat{\beta}_i X_i$, with only significant terms from the regression analysis retained. Providing that the variables in the set of potential environmental variables are in fact strictly environmental, this approach resolves the BETA model with valid inferences. However, if in addition to the familial environment, the "environmental variables" consist of a genetic component that is correlated with the genetic component of the phenotype, inferences regarding genetic and cultural heritability can be incorrect. This is because the estimate of genetic heritability from such a situation will reflect only the residual genetic effects on the phenotype after adjusting for the correlated genetic effects on the index and thus is underestimated. Similarly, cultural heritability will include genetic influences and thus will be overestimated. However, estimates from such an analysis can be taken as a lower bound estimate of genetic heritability and an upper bound estimate of cultural heritability (Rao et al., 1984).

Parameters of the BETA model are estimated from nuclear family data if an index of the family environment is available. The path diagram of the nuclear family BETA model is presented in Figure 2. Additional parameters that can be estimated are summarized below. More detailed exposition of the BETA model and its theoretical development have been presented by Rao et al. (1974), Wright (1978), Cloninger et al. (1979), Rao et al. (1979), Cloninger (1980), and Rao et al. (1984).

The correlation between the familial environments of the spouses is denoted by u, which is one of several possible models of spouse resemblance. Intergenerational differences are modelled using z, a multiplier of h in parents permitting differences in genetic heritability; and y, a multiplier of c in parents allowing for differences in cultural heritability. The sibling familial environments are correlated partly due to influences transmitted by parents, and partly due to nontransmitted factors, the latter denoted by b. Cultural transmission is modelled using f_f and f_m, the paternal and maternal cultural transmission parameters. Parameters relating the environmental index to the latent familial environment are i and v.

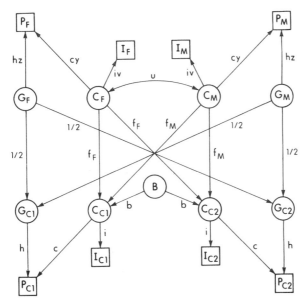

Figure 2. BETA path model of familial resemblance in nuclear families. P, G, C, and I denote phenotype, genotype, indexed familial environment, and index, respectively. Subscripts F and M denote father and mother, and C1 and C2 denote two children. B is the nontransmitted common environment for a sibship. Parameters of the model are defined in the text.

As with the TAU model, the BETA model can be readily extended to include multiple phenotypes measured on each individual (McGue et al., 1983; Vogler et al., 1987). This approach is particularly useful for analyzing correlated risk factors for a disease in order to understand the interrelationships among the multiple risk factors. In addition to the inferences that can be made from univariate analyses, multivariate analyses provide information on the coaggregation of risk factors. In theory, one can obtain an environmental index for each phenotype and analyze a problem with n phenotypes and n indices. Practical limitations related to the computational burden of such a problem severely restrict the number of variables that can be included in a multivariate analysis. One way to circumvent this problem if a number of related phenotypes are analyzed is to estimate a single environmental index that is appropriate for all of the phenotypes.

STATISTICAL METHODS OF ANALYSIS FOR NUCLEAR FAMILIES

When a study is undertaken to understand a process at the population level, random sampling is the most straightforward and representative strategy. Many studies of familial resemblance are carefully designed to avoid complications introduced by nonrandom ascertainment. In contrast, large-scale studies are often designed to focus on individuals who are at the extremes of the population distribution and consequently at elevated risk for a disease, or families ascertained through probands with elevated levels of risk factors or who manifest a disease state. Such nonrandomly-ascertained samples can also be used for making inferences regarding the structure of genetic and environmental influences on a phenotype in the population if the nonrandom ascertainment is systematically taken into account in the statistical methodology.

Statistical methods for analyzing random samples are presented in the next section. This is followed by a discussion of a general method for analyzing nonrandom samples and a brief discussion of methods for analyzing data obtained through more specialized sampling schemes. The methods are general and applicable to the analysis of data regardless of the path model used.

Analysis of Random Sample Data

A random sample of nuclear families with variable sibship sizes will consist of vectors of data on a father, mother, and s children. Let $\mathbf{X}' = (X_{11}, X_{21}, X_{31}, X_{32}, ..., X_{3s})$ denote the row vector of phenotypes for the father (X_{11}), mother (X_{21}), and children (X_{3k}, $k = 1,...,s$). In a model in which there is more than one measure for each individual, such as the BETA model with a phenotype and environmental index or a multivariate model with multiple phenotypes, X_{jk} for an individual is replaced by a vector of measurements on that individual. Assuming that the means and variances are equal for all children, the vector of means μ' of $\mathbf{X}'$ is denoted by $(\mu_1, \mu_2, \mu_3, ..., \mu_3)$, where μ_1, μ_2, and μ_3 are means for the father, mother, and any child, and the corresponding variances are σ_1^2, σ_2^2, and σ_3^2. The vector σ is defined as $(\sigma_1, \sigma_2, \sigma_3, ..., \sigma_3)$. A correlation matrix $\mathbf{R}$ is defined which consists of the intercorrelations among all variables in $\mathbf{X}'$. Depending on the model, many of the elements of the correlation matrix can be further equated. For example, in the nuclear family TAU model, there are four independent correlations:

spouse: $\rho(X_{11}, X_{21})$,
father-child: $\rho(X_{11}, X_{3k})$, $k = 1,...,s$
mother-child: $\rho(X_{21}, X_{3k})$, $k = 1,...,s$
sibling: $\rho(X_{3k}, X_{3l})$, $k \neq l = 1,...,s$.

In path analysis, these correlations are expressed as functions of the path coefficients of the model (e.g., see Rao et al., 1984). The covariance matrix Σ is defined in terms of the correlations and variances as $\Sigma = \sigma' \mathbf{R} \sigma$.

Assuming that the data vector $\mathbf{X}$ for a given family follows a multivariate normal distribution, the log-likelihood function for the family is

$$\ln L = -1/2 \left[\ln|\Sigma| + (\mathbf{X} - \mu)' \Sigma^{-1} (\mathbf{X} - \mu) \right] + \text{constant}$$

where $|\Sigma|$ is the determinant and Σ^{-1} is the inverse of the covariance matrix Σ. The total log-likelihood function for n families is given by:

$$\ln L = \sum_{i=1}^{n} \ln L_i.$$

Parameters of the model, including the path coefficients or correlations, means, and variances, are estimated simultaneously by maximizing $\ln L$. Tests of hypotheses are

296

conducted using a likelihood ratio test in which the difference between $-2 \ln L$ with $k + w$ parameters estimated and $-2 \ln L$ when only k of the parameters are estimated is distributed as Chi-square with w degrees of freedom.

Analysis of Nonrandom Sample Data

The maximum likelihood technique described for random samples can be used to analyze data that have been collected under an ascertainment scheme that provides a selected sample if adjustment for ascertainment is incorporated into the likelihood function. Boehnke and Lange (1984), building on the work of Cannings and Thompson (1977) and Hopper and Mathews (1982), proposed a general method for adjusting for ascertainment through a proband by conditioning the likelihood function of a particular family on the phenotypic value of the proband. The conditional likelihood of a family given the proband's phenotypic value is written as:

$$L\ (\mathbf{X}|X_p) = f_k(\mathbf{X})/f_1(X_p),$$

where $\mathbf{X}$ is the m-dimensional vector of data for a family, X_p is the proband's phenotypic value, and f_v is the v-variate normal density function. This conditional likelihood approach is generally used for any sample ascertained through a single proband, and for that reason, is termed the *generic* method (Rao et al., 1988). If a particular variable (such as the father's phenotype) is the variable on which selection occurs in every family, then there is no information for estimating the mean and variance of that variable and these parameters must be estimated from other sources.

Under certain ascertainment schemes, there is additional information regarding ascertainment which can be used to increase the efficiency of estimation and power of hypothesis tests over that provided by the conditional likelihood approach (Rao & Wette, 1987; Rao et al., 1988). Such additional information is available from nonrandom samples in which selection of probands is through truncation on the distribution of a phenotype or correlated trait such that probands whose selection variable value lies above a defined threshold on the distribution of the trait are included in the sample, otherwise they are excluded. Truncation can be *direct* on the phenotype itself, *indirect* on a correlated quantitative trait, or *latent* on a binary variable such as disease state for which a latent liability distribution is assumed. Methods to correct for ascertainment through truncation (Rao & Wette, 1987) provide more efficient parameter estimates and more powerful hypothesis tests provided the region of truncation is known and adherence to the sampling scheme is maintained, providing a *specific* method (Rao et al., 1988). If the region of truncation is not known and must be estimated, then the generic method and the specific method are approximately equal in efficiency and power. The generic method is computationally simpler and easier to implement and is more robust to departures from adherence to the truncation sampling scheme.

Another method for adjusting for nonrandom ascertainment was developed by Hanis and Chakraborty (1984). This method yields accurate estimates of parameters for most types of nonrandom sampling, including complex ascertainment, and is robust against departures from multivariate normality. However, it is computationally intensive and provides only approximate tests of hypotheses, and while it can be theoretically applied to multiple ascertainment schemes, the sample sizes required for implementation become impractically large (Rao & Wette, 1989).

STUDY DESIGNS FOR RESOLVING GENETIC AND CULTURAL INHERITANCE

Nuclear Families

Nuclear family studies use the most representative sample of all genetic epidemiology study designs. As indicated in the discussion of the TAU model, phenotypic data alone in nuclear families do not provide sufficient information to resolve genetic and cultural inheritance. The inclusion of an index of the familial environment, discussed above, in nuclear family data provides the information for resolution of genetic and cultural inheritance since it is assumed that the similarity among relatives on the environmental index is due solely to cultural influences, whereas similarity among relatives on the phenotype is attributable to both genetic and cultural influences. The use of the regression procedure (Rao et al., 1974) guarantees that the environmental index includes only influences that are related to influences on the phenotype, excluding family environmental influences that occur for the index variables but not for the phenotype. There is no procedure, however, that can guarantee that the familial influences on the environmental index exclude all genetic influences correlated with the phenotype. Thus, an investigator must be careful in selecting variables to include in the environmental index in order to minimize the potential biases that could result if correlated genetic influences on the index are present.

Arbitrary Pedigrees Containing Multiple Relationships

An alternative approach which eliminates the uncertainties regarding the use of an environmental index is to analyze in addition to nuclear family data the more atypical relationships such as twins and adopted relatives (McGue et al., 1985). These relationships provide information to resolve genetic and cultural inheritance under certain assumptions, but without resorting to environmental indices. When twins are used, it is assumed that shared environmental influences are of the same magnitude for monozygotic (MZ) twins as for dizygotic (DZ) twins (and absence of assortative mating if twins alone are used). Twins have the advantage of being fairly easy to obtain and to analyze. For adopted relatives, it is usually necessary to assume that there is no selective placement and that the biological and adoptive families of adoptees are representative of the population. Provided that the assumptions hold, resolution of genetic and cultural inheritance in an adoption study is extremely powerful, but adoption data has become very difficult to obtain. Despite the strong assumptions that must be made in using these relationships in genetic analyses, they remain appealing (especially in cases where it is difficult to obtain an index of the environment for a nuclear family study).

It is possible to analyze any combination of relationships to enhance the power to resolve genetic and cultural inheritance and to provide cross-checks on the assumptions underlying the individual paradigms. A general method can be used to perform likelihood analysis of quantitative data on pedigrees of arbitrary structure, either by estimating correlations directly or by estimating parameters of a path model (Rao et al., 1987). Quite simply, the log-likelihood function defined above for the analysis of random samples is applied, where the phenotypic vector X, and consequently the correlation matrix R and covariance matrix Σ can differ for each pedigree. Methods for adjustment for ascertainment can be applied to this approach with complete generality.

Nuclear Families with Twin Offspring

One study design which represents a compromise between the representativeness of nuclear families and the resolving power of twins is a design which includes twins, their siblings, and parents (Fulker, 1982). This method combines a model of assortative mating based on spouse correlations with the resolving power of the twin design while simultaneously providing parameter estimates that are maximally consistent with both twin data and parent-offspring data.

Analysis of Correlations Obtained from Independent Samples

If familial correlations $r_1, r_2,...,r_m$ are obtained from independent samples of size n_1, $n_2,...,n_m$, the data can be compiled into a single analysis using a different procedure. First, each sample correlation coefficient is converted into Fisher's z-transformation (Fisher, 1921), z_i. Assuming asymptotic normality of the z's, the log-likelihood function is

$$\ln L = -1/2 \; \chi^2 + \text{constant, where, approximately, } \chi^2 = \Sigma \; n_i(z_i - \bar{z}_i)^2.$$

As with the pedigree likelihood function, correlations can be defined in terms of parameters of a path model and tests of hypotheses can be conducted using the likelihood ratio test (see Rao et al., 1974). This method provides unbiased parameter estimates and valid tests of hypotheses when the sample correlations are independent. If this method is applied to pedigree data, however, the assumption of independence is violated. Simulation studies have shown that application of this method to nonindependent data results in unbiased parameter estimates but invalid tests of hypotheses, generally in the direction of reduced power (McGue et al., 1987).

MODELLING COMPLEXITIES

A number of assumptions underlie models of familial resemblance and charac-teristics of the data. These include linearity, additivity, normality, and absence of gene-gene interactions (dominance and epistasis) and gene-environment interactions.

Linearity, Additivity, and Normality

In the linear structural models, each effect variable is a linear function of its causal variables. Multiple causal variables are assumed to act additively on the effect. Wright (1983) and Cloninger et al. (1983) consider the implications of the assumptions of linearity and additivity and do not find evidence that these assumptions are excessive in practical applications, especially when inference is sought to a degree of approximation. Multivariate normality of the data is critical in the maximum likelihood procedures advocated for analysis of family data. McGue et al. (1987) evaluated the effects of departures from multivariate normality in a simulation study. The pedigree method of maximum likelihood analysis was found to be robust against minor departures from normality, whereas the independent correlation method was found to be highly conservative under all conditions of departures from multivariate normality.

One approach to handling observed deviations from normality is to use a transformation, $t_i(X_i)$, proposed by MacLean et al. (1976) as a variation of the Box and Cox (1964) transformation:

$$t_i(X_i) = Y_i = \begin{cases} r_i/p_i\,[(X_i/r_i + 1)^{p_i} - 1],\ p_i = 0, \\[2mm] r_i \ln (X_i/r_i + 1),\ p_i = 0, \end{cases}$$

where r_i is a scaling constant to ensure that $(X_i/r_i + 1)$ is positive, and p_i is an unknown parameter that tends to eliminate skewness. Although this procedure does not guarantee multivariate normality, it tends to minimize the impact of deviations from normality (Rao et al., 1984).

Genotype-by-Genotype and Genotype-by-Environment Interactions

Gene-gene interactions can result in dominance (interactions among alleles at individual loci) or epistasis (interaction among multiple loci). These effects are largely assumed to be absent in multifactorial models of familial resemblance because of the difficulty in detecting or resolving these effects in nonexperimental genetic studies. Certain designs, such as twin studies, can provide secondary evidence of nonadditive genetic effects. While dominance is confounded with cultural transmission in twin data, parameter estimates that are outside of the range of acceptable values for a model of cultural transmission but within acceptable range for a model which includes dominance can be taken as evidence of the presence of dominance. However, dominance and cultural transmission are not resolvable separately in the twin design. Gene-gene interactions represent a largely unexplored area in genetic epidemiology. It may become more practical to consider these issues as the ability to examine multiple measured genotypes develops.

Genotype-environment interaction, if present in data but ignored in data analysis, can lead to overestimation of heritabilities (Lathrop et al., 1984). Under certain circumstances, genotype-environment interaction can be detected using multivariate data on adult twins and their spouses and offspring (Heath et al., 1992). However, the sample sizes required are beyond the reach of most studies and detection is possible only when differential heritability occurs for multiple phenotypes. Studies designed to use direct models of genotype-environment interaction, for example, models which exploit differential exposure to environmental risk factors, have been largely underutilized, especially in the context of multifactorial models.

Models of Temporal Trends and Development

Models have been developed which reflect temporal effects on familial aggregation. Province and Rao (1985, 1988) develop a method to model changes in the population over time by defining critical model parameters as a mathematical function of time. Such temporal trends occur in variables such as blood pressure and CHD morbidity and mortality which have been strongly influenced by environmental intervention and lifestyle changes over time.

Developmental models of changes within an individual over time (Eaves et al., 1986) are used to characterize a phenotype which is not static over an individual's lifetime but rather which is a function of a developmental process. This kind of approach has seen limited application because it requires a large amount of longitudinal data, which is expensive and time-consuming to collect.

APPLICATIONS

The effects of analyzing nuclear family data using a TAU model without environmental indices and a BETA model with environmental indices are illustrated by applying both models to the analysis of cholesterol and triglyceride levels of subjects in a single data set. Nuclear family data were obtained from five Lipid Research Clinics (Cincinnati, Iowa, Minnesota, Oklahoma, and Stanford) as part of the National Heart, Lung, and Blood Institute's LRC Family Study (Heiss et al., 1980; Lipid Research Clinics Program Epidemiology Committee, 1979).

The nuclear families included 2,818 observations from 747 families who were selected through random probands. The cholesterol and triglyceride values were adjusted for the effects of age, sex, age group (adult greater than 20 years of age versus pediatric 20 years of age or younger), clinic, and oral contraceptive usage. Adjusted z-scores were then normalized using Blom's (1958) inverse normal transformation of the ranks of the adjusted phenotype.

The environmental index was constructed separately for the adult and pediatric samples and separately for each phenotype. Each phenotype was regressed on a number of variables, including body mass index, educational level, special diet usage, smoking and drinking behaviors, skinfold, measures of dietary intake, physical activity, and graded exercise tests for adults only. Age- and sex-adjusted, normalized regression scores constitute the index.

Phenotype and index data were analyzed using the BETA model with seven path coefficients (h, c, u, f_F, f_M, b, and i) and the offspring's variances estimated. Phenotype data alone were analyzed using the TAU model with three path coefficients (t, u, and b) and the offspring's variance estimated. A pseudopolygenic TAU model was used in which the parameters τ_F and τ_M were fixed at values of 1/2. Parameter estimates and their standard errors are summarized in Table 1. Estimates of the contribution of familial influences to the phenotype are consistent between models, especially for cholesterol which did not show evidence of important effects of the familial environment under the BETA model. The largest fluctuations in parameter estimates are for b and u. This may be attributed to two factors. First, the BETA model, under which the effects of familial

Table 1. Parameter estimates from analysis of cholesterol and triglyceride from LRC family data using the BETA and TAU models

	BETA Model			TAU Model	
Parameter	Cholesterol	Triglyceride	Parameter	Cholesterol	Triglyceride
h	0.75 + 0.02	0.60 + 0.03	t	0.73 + 0.03	0.55 + 0.06
c	0.13 + 0.05	0.28 + 0.02			
u	0.29 + 0.16	0.22 + 0.05	u	0.07 + 0.09	0.26 + 0.09
i	0.70 + 0.16	1.00 (bound)			
b	0.55 + 0.12	0.45 + 0.03	b	0.17 + 0.16	0.22 + 0.24
f_F	0.09 + 0.10	0.16 + 0.04			
f_M	0.41 + 0.18	0.23 + 0.03			

influences on the phenotype (c^2) are estimated to be small, indicates that the familial environments between siblings and between spouses are moderately correlated but their effect on the phenotype is small. Under the TAU model, the influence of total familial effects on the phenotypes (t^2) are fairly large, so consequently estimates of b and u are estimated to be smaller. Second, the BETA model differentiates between familial environment and genetic effects, and b and u denote the correlation between only familial envrionments of two individuals. The TAU model cannot differentiate between these two effects, and therefore b and u denote the correlation between both familial environment and genetic effects. Although the parameters are very similar in the two models, their interpretation is slightly different.

DISCUSSION

Investigations of complex quantitative phenotypes cannot yet take full advantage of the rapidly advancing techniques of molecular genetics and linkage. The complex interplay of multifactorial genetic and environmental influences requires appropriate theoretical models for disentangling the causal effects on such phenotypes. Although current models are capable of providing important insights, further work is needed to fully encapture the complex effects due to gene-gene and gene-environment interactions, as well as developmental effects.

Path analysis and similar methods serve some important purposes. First, resolution of genetic and cultural heritabilities is useful as a first step. Evidence of strong biological influences might be suggestive of the value of designing pharmacological intervention programs for disease amelioration. In contrast, evidence of strong environmental influences on disease risk factors would suggest that active environmental intervention strategies would be more effective. However, as awareness of the role of risk factors in disease increases and efforts are made to modify disease risk factors, genetic and cultural heritabilities will change. Thus, it may be necessary to evaluate the appropriateness of various intervention strategies from time to time.

Another important use of path analysis lies in motivating searches for individual genes underlying the phenotype. Evidence of genetic effects from path analysis and similar studies provides motivation for understanding much more complex and computer-intensive segregation analyses (Borecki et al., 1990) that can suggest major gene effects. Additionally, one may argue that such evidence provides ample justification for conducting candidate gene studies or genomic searches for individual genes affecting the phenotype.

The concept of heritability is dependent on the assumption that effects on a phenotype are multifactorial, or a function of many small effects acting additively. Molecular genetics holds the promise for identification of the underlying genes. These measured genotypes can then be included as covariates of the phenotype, and any additional (residual) genetic effects can be investigated. This approach, termed the measured genotype approach (Sing & Moll, 1990) is likely to be most effective in oligogenic systems (phenotypes that are affected by a few genes of moderate effect). Direct assessment of allelic effects and the interactions among alleles may provide a promising means for detecting nonadditive genetic effects. To date, statistical techniques of adequate power to perform such analyses have not been developed.

One occasionally hears the argument that molecular-based studies will soon make population-based family studies which rely on statistical techniques obsolete. However, the most complex, and arguably the most interesting, phenotypes are functions of far more complex systems than molecular approaches are capable of characterizing. Complex influences of multiple genes, multiple environmental effects, and the interplay among these effects must be addressed directly. Although molecular genetics might contribute to direct genotype assessment, this will not be the end of the story for complex phenotypes. An optimistic projection is that the coupling of molecular studies with population-based epidemiological approaches will permit the incorporation of some complexities discussed in this paper into more realistic models of familial resemblance.

ACKNOWLEDGMENTS

This work was partly supported by NIH grants GM-28719, HD-18281, HL-47317, and NIMH Grant MH-31302, and MRC of Canada grant PG-11811.

REFERENCES

Blom, G., 1958, "Statistical Estimation and Transformed Beta Variables," John Wiley & Sons, New York.

Boehnke, M., and Lange, K., 1984, Ascertainment and goodness of fit of variance component models for pedigree data, *in*: "Genetic Epidemiology of Coronary Heart Disease: Past, Present, and Future," D.C. Rao, R.C. Elston, L.H. Kuller, M. Feinleib, C. Carter, and R. Havlik, eds., Alan R. Liss, New York.

Box, A.E.P., and Cox, D.R., 1964, An analysis of transformations, *J. Roy. Statist. Soc. Ser. B* 26:211-252.

Cannings, C., and Thompson, E.A., 1977, Ascertainment in the sequential sampling of pedigrees. *Clin. Genet.* 12:208-212.

Cloninger, C.R., 1980, Interpretation of intrinsic and extrinsic structural relations by path analysis: Theory and applications to assortative mating, *Genet. Res.* 36:133-145.

Cloninger, C.R., Rice, J., and Reich, T., 1979, Multifactorial inheritance with cultural transmission and assortative mating. II. A general model of combined polygenic and cultural inheritance. *Am. J. Hum. Genet.*31:176-198.

Cloninger, C.R., Rao, D.C., Rice, J., Reich, T., and Morton, N.E., 1983, A defense of path analysis in genetic epidemiology, *Am. J. Hum. Genet.* 35:733-756.

Eaves, L.J., Long, J., and Heath, A.C., 1986, A theory of developmental change in quantitative phenotypes applied to cognitive development. *Behav. Genet.* 16:143-162.

Fisher, R.A., 1921, On the probable error of a coefficient of correlation deduced from a small sample, *Metron*, 1:1-32.

Fulker, D.W., 1982, Extensions of the classical twin method, *in*: "Human Genetics, Part A: The Unfolding Genome," B. Bonne-Tamir, T. Cohen, and R.M. Goodman, eds., New York, Alan R. Liss.

Hanis, C.L., and Chakraborty, R., 1984, Nonrandom sampling in human genetics: Familial correlations, *IMA J. Mathematics Appl. Med. Biol.* 1:193-213.

Heath, A.C., Kessler, R.C., Neale, M.C., Hewitt, J.K., Eaves, L.J., and Kendler, K.S., 1992, Testing hypotheses about direction-of-causation using cross-sectional family data. (Manuscript in review).

Heiss, G., Tamir, I., Davis, C.E., Tyroler, H.A., Rifkind, B.M., Schonfeld, G., Jacobs, D., and Frantz, I.D., 1980, Lipoprotein-cholesterol distributions in selected North American populations: The Lipid Research Clinics Program Prevalence Study, *Circulation* 61:302-315.

Hopper, J.L., and Mathews, J.D., 1982, Extensions to multivariate normal models for pedigree analysis, *Ann. Hum. Genet.* 46:373-383.

Lathrop, G.M., Lalouel, J.M., and Jacquard, A., 1984, Path analysis of family resemblance and gene-environment interactions, *Biometrics* 40:611-625.

Li, C.C., 1975, "Path Analysis: A Primer," Boxwood Press, Pacific Grove, CA.

Lipid Research Clinics Program Epidemiology Committee, 1979, Plasma lipid distributions in selected North American populations: The Lipid Research Clinics program prevalence study, *Circulation* 60:426-439.

MacLean, C.J., Morton, N.E., Elston, R.C., & Yee, S., 1976, Skewness in commingled distributions, *Biometrics* 32:695-699.

McGue, M., Rao, D.C., Reich, T., Laskarzewski, P., Glueck, C.J, and Russell, J.M., The Cincinnati Lipid Research Clinic family study: Bivariate path analysis of lipoprotein concentrations, *Genet. Res.* 41:117-135.

McGue, M., Rao, D.C., Iselius, L., and Russell, J.M., 1985, Resolution of genetic and cultural inheritance in twin families by path analysis, *Am. J. Hum. Genet.* 37:998-1014.

McGue, M., Wette, R., and Rao, D.C., 1987, A Monte Carlo evaluation of three statistical methods used in path analysis, *Genet. Epidemiol.* 4:129-155.

Morton, N.E., 1974, Analysis of familial resemblance. I. Introduction, *Am. J. Hum. Genet.* 26:318-330.

Province, M.A., and Rao, D.C., 1985, A new model for the resolution of cultural and biological inheritance in the presence of temporal trends: Application to systolic blood pressure, *Genet. Epidemiol.* 2:363-374.

Province, M.A., and Rao, D.C., 1988, Familial aggregation in the presence of temporal trends, *Statistics in Medicine* 7:185-198.

Rao, D.C., Morton, N.E., and Yee, S, 1974, Analysis of family resemblance. II. A linear model for familial correlation, *Am. J. Hum. Genet.* 26:767-772.

Rao, D. C., Morton, N.E., and Cloninger, C.R., 1979, Path analysis under generalized assortative mating. I. Theory, *Genet. Res.* 33:175-188.

Rao, D.C., McGue, M., Wette, R., and Glueck, C.J., 1984, Path analysis in genetic epidemiology, *in*: "Human Population Genetics: The Pittsburgh Symposium," A. Chakravarti, ed., Van Nostrand Reinhold, New York.

Rao, D.C., Province, M.A., Wette, R., and Glueck, C. J.,1984, The role of path analysis in coronary heart disease research, *in*: "Genetic Epidemiology of Coronary Heart Disease: Past, Present, and Future," D.C. Rao, R.C. Elston, L.H. Kuller, M. Feinleib, C. Carter, and R. Havlik, eds., Alan R. Liss, New York.

Rao, D.C., Vogler, G.P., McGue, M., & Russell, J.M., 1987, Maximum likelihood estimation of familial correlations from quantitative data on pedigrees: A general method and examples, *Am. J. Hum. Genet.* 41:1104-1116.

Rao, D.C., and Wette, R., 1987, Nonrandom sampling in genetic epidemiology: Maximum likelihood methods for multifactorial analysis of quantitative data, *Genet. Epidemiol.* 4:357-376.

Rao, D.C., Wette, R., and Ewens, W.J., 1988, Multifactorial analysis of family data ascertained through truncation: A comparative evaluation of two methods of statistical inference, *Am. J. Hum. Genet.* 42:505-515.

Rice, J., Cloninger, C.R., and Reich, T., 1978, Multifactorial inheritance with cultural transmission and assortative mating. I. Description and basic properties of the unitary models, *Am. J. Hum. Genet.* 30:618-643.

Sing, C.F., and Moll, P.P., 1990, Genetics of atherosclerosis, *Ann. Rev. Genet.* 24:171-187.

Vogler, G.P., Rao, D.C., Laskarzewski, P.M., Glueck, C.J., and Russell, J.M., 1987, Multivariate analysis of lipoprotein cholesterol fractions, *Am. J. Epidemiol.* 125:706-719.

Wright, S., 1921, Correlation and causation, *J. Agr. Res.* 20:557-585.

Wright, S., 1978, "Evolution and the Genetics of Populations: Vol. 4. Variability Within and Among Natural Populations," University of Chicago Press, Chicago.

NEURODEGENERATIVE DISORDERS AMONG THE CHAMORRO PEOPLE OF GUAM: STUDIES OF AN ISOLATE IN TRANSITION

Ralph M. Garruto[1] and Chris C. Plato[2]

[1]Laboratory of Central Nervous System Studies, NINDS
[2]Applied Physiology Section, NIA
National Institutes of Health
Bethesda, MD 20892 USA

INTRODUCTION

During the past three decades we have recognized a unique opportunity to study high-incidence foci of amyotrophic lateral sclerosis (ALS) and parkinsonism-dementia (PD) in non-western anthropological populations in the Pacific Basin aimed at identifying their etiology and mechanism of pathogenesis and their relationship to other neuro-degenerative disorders, such as Alzheimer's disease, parkinsonism, and early neuronal aging. Pacific ALS and PD are two progressive and fatal neurological disorders that occur hyperendemically in different cultures, in different ecological zones and among genetically divergent populations in the Mariana Islands, in the Kii Peninsula of Japan and in southern West New Guinea. Our cross-disciplinary approach to these intriguing neurological disorders and the accumulated epidemiological, cellular and molecular evidence strongly implicate environmental factors in their causation, specifically the role of aluminum and its interaction with calcium. The remaining question is whether the clinical expressions of these disorders are solely environmentally induced or whether they represent a genetic susceptibility combined with an "environmental trigger". The dramatic decline of these disorders in all three high-incidence foci over the past 40 years can be supported by both models. A clearer understanding of these natural paradigms will surely have far-reaching implications for Alzheimer's disease and other disorders with long latency and slow progression. In this paper we concentrate our efforts on the Guamanian focus of ALS and PD which is far and away the best studied of the three high incidence Pacific foci and summarize the results of familial and genetic studies conducted during the past 30 years on Guam.

ETHNOHISTORIC AND DEMOGRAPHIC NOTES

Guam is the largest of the Mariana Islands in the western Pacific with an area of 215 square miles. In 1960, when genetic studies of ALS and PD were initiated, the population of Guam was 67,044[1]. Of these, 34,762 were native Guamanians (Chamorros), 8,580 Filipinos, 20,724 Caucasians (mostly Americans), and about 3,000 Carolinians from the Yap and Truk Islands of Micronesia. The original Chamorros are presumed to have immigrated to Guam from the Malay Archipelago, as early as 3517 B.P. $\pm$ 200 years [2]. Their first encounter with Europeans was in 1521 when Magellan landed near the village of Umatac. In 1565, Guam was officially proclaimed a Spanish territory. During the following one hundred years the island remained in relative obscurity until 1668 when Father Sanvitores and his assistants arrived from Mexico and established a Jesuit mission. At this time, the native Chamorro population of Guam was estimated to have been about 50,000[3], a figure which may represent an overestimation.

The discovery and subsequent colonization of Guam by the Spanish brought about radical genetic changes to the indigenous Chamorro population. This sparked a series of Chamorro uprisings in which thousands of natives lost their lives. The frequent Chamorro–Spanish wars, two very destructive typhoons, a smallpox epidemic in 1688, and other infectious diseases introduced by the Spanish and other groups that visited the island from time to time contributed to the decimation of the native population, so that by 1783 there were only 1,500 Chamorros living on the island[3]. The reduction in the native population resulted in a shortage of laborers which necessitated the introduction of large groups of Filipino males, most of whom remained on Guam, many marrying Chamorro women. The intermarriage between Chamorro women and the new male immigrants, as well as the genetic contributions by whalers, privateers and pirates resulted in a genetic hybridization of the Guamanian Chamorro population. The intermarriage was so extensive that "had it not been for the intermarriage between Chamorros and non–natives, particularly Filipinos, the Mariana Islands might have eventually been totally depopulated" [3].

With the beginning of the nineteenth century, the population of Guam increased steadily, although the number of pure Chamorros continued to diminish. According to Thompson[4], by the middle of the nineteenth century no "full blooded" Chamorro existed. Thus by the end of the nineteenth century a new hybrid Chamorro population had been established, comprised of an admixture of the original Chamorro gene pool in combination with Filipino, Mexican and Spanish genes. This hybrid gene pool remained stable and for the most part free of the introduction of significant new genes until the mid–twentieth century. Detailed census reports since 1910 indicate that the abrupt increase in population was primarily due to a high birth/death ratio rather than due to additional immigration waves. According to the 1980 US Census Report, the population of Guam was 106,000. Of these about 60,000 were Chamorros, 20,000 Filipinos and 26,000 other ethnic groups (mostly American military personnel and their dependents). Genetic studies of the Chamorros and neighboring populations support the ethnohistorical record and the conclusion that the present day Guamanian population is genetically admixed.

CLINICAL AND NEUROPATHOLOGICAL CHANGES

In 1953 Arnold and his associates reported an unusually high incidence of amyotrophic lateral sclerosis (ALS) among the Chamorros of Guam[5]. Subsequent reports[6-9] verified and expanded Arnold's findings. Mulder and colleagues[9] also reported on 22 patients with parkinsonism, some of whom demonstrated dementia. This combination of parkinsonism with dementia was subsequently studied in detail, clinically and neuropathologically, by Hirano and colleagues[10,11], who concluded that it represented a unique disease, parkinsonism–dementia (PD). Furthermore, it was shown that a number of Guamanian patients had clinical and neuropathological symptoms of both ALS and PD,[12,13] and a number of families had patients with ALS, PD or both diseases[8,14,15,16]. These findings provided the impetus for the National Institute of Neurological Diseases and Blindness in 1956 to establish the Guam Research Center. Scientists at the Center and their colleagues on the mainland embarked on an intensive multidisciplinary research effort towards understanding the etiology of ALS and PD, with an ultimate goal of treatment and prevention of these disorders which have plagued the Guamanian people for at least a century.

ALS on Guam, known locally as *lytico* or *paralytico* is a neuromuscular disorder which affects the motor neurons of the brain and spinal cord. Clinically, the disease is indistinguishable from classical sporadic ALS encountered worldwide. It is characterized by progressive muscle weakness, atrophy, fasciculations (muscle jumping) and spasticity, with the latter symptom being indicative of the degree of upper and lower motor neuron involvement. A year or so after the appearance of the initial symptoms, the disease results in extensive muscle wasting, paralysis and eventually death. Neuropathologically, Guamanian ALS differs from classical sporatic ALS found worldwide. In addition to the loss of neurons, Guamanian ALS is characterized by the extensive development of neurofibrillary tangles in brain and spinal cord neurons[17]. The neurofibrillary tangles seen in ALS (as well as in PD) appear at the cellular and molecular level to be the same or very similar to those found in the brain of Alzheimer's disease patients[18,19]. They are composed of β-amyloid and a number of cytoskeletal proteins which are thought to be directly involved in the disease process.

PD, known locally as *bodig* or *rayput*, occurs in high incidence in the same villages and same sibships as ALS, and occasionally a patient may have both disorders[20]. PD is an extrapyramidal disease characterized clinically by parkinsonian features, including mask–like facies, muscular rigidity, tremor, bradykinesia (slow voluntary movement) and a typical shuffling gait. It is also coupled with onset of a progressive dementia as an early clinical symptom. Neuropathologically, PD, like ALS on Guam, demonstrates the accumulation of abundant Alzheimer neurofibrillary tangles in the brain[11,21]. Both ALS and PD are late onset, uniformly fatal disorders with a long latency and slow progression.

EPIDEMIOLOGICAL FINDINGS

Since the initiation of systematic studies of ALS and PD on Guam over 35 years ago, a number of epidemiological changes have occurred. When ALS on Guam was first described in the mid–1950's, the incidence rate of the disease was reported to be more than 50-100 times higher than the rate of 1-2 /100,000 population for the continental United States[7]. Today the risk for developing either disease on Guam has dramatically declined to levels only several fold higher than that for non-Chamorro residents living in the continental United States[20,22,23]. Besides the decrease in the overall incidence of these disorders, there has also been a specific geographic decline in the incidence of ALS and PD on the island[20]. The sharpest decline in incidence occurred in the southern villages where the disorders were most prevalent. The moderate incidence regions of central and northern Guam has been slower to decline. The western part of the island has maintained a low incidence for ALS and PD throughout the 35 year surveillance period[17]. The male:female ratio of ALS and PD has also changed. In the early stages of our studies (in the 1950s) the male:female ratio was 2:1 for ALS and almost 3:1 for PD. Presently the sex ratio for both ALS and PD approaches unity, with the most significant changes occurring in males[17]. The age of onset for both ALS and PD has progressively increased during the past three decades[20].

Equally important has been the declining incidence rates of ALS and PD in two other Pacific foci. In the Kii peninsula focus of Japan, incidence rates in the Hobara subfocus have decreased from 55 to 14 per 100,000 population, and no new cases have been seen in the Kozagawa subfocus during the past ten years[24]. In the remote West New Guinea focus, with village prevalence rates as high as 1,300 per 100,000 population, ALS and PD have also disappeared during the past two decades in one village and declined in several others that have had increasing western contact and introduction of new foodstuffs[25,26]. These dramatic epidemiological changes strongly support the notion of a cohort effect, and the involvement of environmental factors in the etiology and pathogenesis of these disorders. It also suggests that if genetics are involved, they are likely to be secondary to some environmental insult, although a recent study by Bailey-Wilson and colleagues[15,16] could not exclude a major gene effect in combination with environmental factors.

FAMILIAL AND GENETIC STUDIES

One of the earliest objectives of the Guam Research Center was to investigate the epidemiological and familial aspects of these diseases through a long term prospective case–control study (ALS/PD Registry) which was initiated in 1958[14]. As a corollary to the epidemiological efforts, a number of genetic studies were undertaken which included detailed pedigree evaluations of ALS and PD patients[14] as well as an exhaustive genealogical study of the southern village of Umatac[27], which consistently demonstrated the highest incidence of ALS and PD on the island. In addition to studies of patients and their families, extensive efforts were made to determine the genetic structure of the Guamanian population as well as

Table. List of genetic studies of Guamanian Chamorro with amyotrophic lateral sclerosis (ALS) and parkinsonism–dementia (PD)

I. Prospective Case–Control Study: Registry
    a. Prospective follow–up of all index cases and controls registered between 1957 and 1962
    b. Prospective follow–up of all first degree relatives (parents, sibs and offspring) and spouses of patients and controls

II. Pedigree Analyses
    a. Individual family pedigrees of all index cases and controls
    b. Pedigree analysis of the highest incidence village on Guam (Umatac)
    c. Calculation of coefficients of inbreeding
    d. Segregation analysis

III. Offspring Study of Conjugal Pairs
    a. Both spouses affected (ALSxALS, PDxPD, ALSxPD matings)
    b. One spouse affected
    c. Neither spouse affected (controls)

IV. Genetic Markers
    a. Major histocompatibility system (HLA–A and B loci)
    b. Blood group systems
    c. Hemoglobin variants
    d. Red cell enzymes
    e. Serum proteins and immunoglobulin allotypes
    f. Glucose–6–Phosphate Dehydrogenase deficiency
    g. Red–green color blindness
    h. Dermatoglyphics

V. Molecular Genetics
    a. Isolation and purification of genomic DNA for restriction fragment length polymorphism (RFLP) and DNA Fingerprinting
    b. Beta amyloid gene duplication studies
    c. Messenger RNA studies for amyloid β–protein
    d. Metal–gene and metal–enzyme interaction studies

that of neighboring Carolinian populations in Micronesia and to search for molecular genetic mechanisms that may be involved in the disease process (Table).

The Registry

In 1958, a prospective patient–control Registry was established in an effort to ascertain if first degree (0.5) relatives and spouses of patients with ALS and /or PD had a higher risk of developing disease than corresponding relatives of non–affected controls who were individually matched to patients for age, sex and village of residence. The Registry was comprised of four independent panels: ALS patients, ALS controls, PD patients and PD controls. Each panel included clinical and epidemiological data on the index ALS and PD cases, matched index controls and their respective first degree relatives (parents, sibs, and offspring) and spouses. When the Registry was established in 1958, ALS and PD were treated as two separate clinical entities. However, the Registry data were also analyzed under the assumption that ALS and PD were two forms of the same disease process. The Registry was closed in February, 1963 (five years after its inception), although it has been maintained and updated periodically during the last 30 years. The Registry data were first evaluated soon after its closing in 1963, and again in 1983, some 25 years since its inception. A detailed description of the Registry and methods of ascertainment are given in our original report[14].

At the time of its closing, the Registry included 126 patients (77 ALS, 42 PD and 7 with both ALS and PD) and an equal number of controls; 994 living first degree relatives of patients and 1218 of controls; and 88 living spouses of patients and 101 of controls. A recent analysis of this prospective registry by Plato and colleagues[22] a quarter of a century after its inception demonstrated a significantly increased risk of developing ALS or PD among parents, siblings and spouses of patients, but not among relatives of the controls. However, offspring of both patients and controls showed no significantly increased risk of developing disease. The increased risk among spouses of patients and the lack of an increased risk among their offspring, lend additional support to the contention that a cohort effect exists and confirms that exogenous factors are strong contributors to the etiology of ALS and PD. A recent retrospective study by Reed and colleagues[23] also confirmed the decline in ALS, but their data for the decline in PD was less convincing. An increase in longevity in the Guam population and the more frequent diagnosis of PD in those over the age of 70 has been observed, factors which may have contributed to the less dramatic decline in PD, particularly among women, during the past decade. It is also suspected that some of these "so called" PD cases are actually Alzheimer's disease patients. Forty years ago, the dramatic number of deaths due to ALS and PD over age 25, coupled with a relatively shorter longevity in the general population compared to recent years, did not allow the later onset clinical expression of Alzheimer's disease in susceptible individuals who died of other causes.

Individual Patient Pedigrees

Concurrent with the establishment and maintenance of the Registry, more elaborate pedigrees of individual patients were constructed and updated through the years. A recent analysis of these data using mathematic models of segregation analysis, indicated that under both the normal and logistic models, purely environmental and purely Mendelian dominant and recessive gene hypotheses for ALS and PD could be rejected, although a two allele additive major locus hypothesis could not be rejected[16]. If suported by further data, the segregation analysis models would suggest that a major gene was "necessary" but "not sufficient" to produce clinical disease and that an environmental factor(s) were necessary to trigger disease expression.

Complete Genealogical Study of Umatac Village

At about the same time that the Registry and individual patient pedigree collections were initiated in the late 1950s and early 1960s, the village of Umatac, located in the southern part of the island, was selected for detailed genetic analysis. The village of Umatac was selected because incidence rates for of ALS and PD were ten times higher than for any other village on Guam.

During the 1950s and 1960s Umatac was a small isolated community of less than 750 people. With the exception of a narrow coastal road, it could be reached only by boat. As a result of its geographical isolation, Umatac became a relatively closed breeding population with very little intermarriage with individuals from other Guamanian villages. Through an exhaustive house to house survey (which included neurological examination of every adult over the age of 20), church baptismal records, birth, marriage and death certificates and through numerous personal interviews, almost 95% of all individuals born in the village of Umatac since 1830 could be incorporated into a single pedigree comprised of 1,450 individuals and 262 sibships through eight generations[27]. The results of the initial Umatac pedigree segregation analysis argued against the involvement of a recessive gene in the etiology of ALS and PD, although the action of a dominant gene concomitant with environmental factors could not be excluded[27]. The Umatac pedigree, like the Registry and patient pedigrees, has been maintained and updated periodically.

Other Gene Marker Systems

In addition to the genetic studies described above, a number of genetic markers were used to assess the association between various markers and ALS and PD as well as to describe their distribution in the general Guamanian and neighboring populations free of disease. Among those studied were numerous blood group systems[28-30], serum proteins[28,31], red cell enzymes[31], abnormal hemoglobins[32], carbonic anhydrase[32], histocompatibility (HLA) antigens[33], Glucose–6 Phosphate Dehydrogenase deficiency[34], red–green color blindness[34], Xg^a blood group[34] and dermatoglyphics[29,35] (Table). Although both ALS and PD are known to occur together in the same Chamorro family, the same sibship, and even occassionally in the same individual, intensive genetic studies including pedigree analyses of high-incidence villages, calculation of inbreeding coefficients for such villages, and the identification of selected gene markers, have not yielded a satisfactory genetic explanation.

Furthermore, changes in the population gene pool can not account for the pattern of decline of ALS and PD on Guam. Neither out-migration of high-risk individuals nor changes in mating patterns (which remained unusually stable until 1965 or 1970, when the decline was already apparent) or exhaustion of genetic susceptibles accounts for the decline of ALS and PD. Additionally, onset of both ALS and PD usually occurs during the postreproductive years, thereby not affecting completed fertility. Although the pattern of inheritance does not appear to be Mendelian, and individuals born after 1920 are at much lower risk of developing disease, a genetic susceptibility in combination with an "environmental trigger" remains a possibility, but if true, its value in a practical sense remains minimal since removal of the "environmental trigger" will (and has) result in a decline of these disorders to levels found sporadically worldwide.

ENVIRONMENTAL, ECONOMIC and CULTURAL FACTORS

Ongoing epidemiological studies together with intensive genetic studies of individual patient and village pedigrees, as well as islandwide population surveys, while demonstrating strong familial aggregations for the disease, have failed to identify a single genetic factor or mechanism solely responsible for the etiology of ALS or PD. The evidence thus far excludes primary genetic factors and many environmental possibilities, but strongly suggest a cohort effect and a geographic residential distribution of disease (i.e. household effect) involving exposure to some environmental factor(s). The involvement of environmental factors contributing to the expression of the disease has become a virtual certainty in view of the striking decline in incidence rates.

Accumulating epidemiological, genetic and environmental data indicate that toxic metals and essential minerals interact in such a way as to provoke a chronic degeneration of central nervous system neurons in ALS and PD patients[36]. These data support the hypothesis that a basic defect in mineral metabolism induces a form of secondary hyperparathyroidism, provoked by chronic nutritional deficiencies of calcium and magnesium leading to enhanced gastrointestinal absorption of bioactive aluminum and the deposition of calcium and silicon as hydroxyapatites or aluminosilicates in neurons. Through alterations in normal biosynthesis or catabolism of cytoskeletal proteins and brain amyloid, this metal deposition may lead to a fundamental disruption in the neuronal cytoskeleton. Furthermore, it is postulated that the metabolic defect, with resultant deposition of these elements in the central nervous system, may occur *in utero* , during infancy, childhood, adolescence or even later, but long before the onset of clinical disease. Later intake of calcium and subsequent correction of any secondary hyperparathyroidism is unlikely to prevent or reverse the neuronal damage and cell death that these deposits are likely to have produced.

The changing disease patterns are consistent with a population undergoing a significant cultural, nutritional and economic transition. The disappearance of the high incidence of ALS and PD on Guam has occurred during the aggressive acculturation from a previously agricultural and fishfolk subsistence society, immediately following World War II, to an almost completely westernized culture, having a cash economy today. Economic developments occurred faster in the northern and central regions of the island, where large military istallations, major harbor facilities and the capital city of Agana are located. The high–incidence southern region was more impervious to change, but increasingly, men living in the south went to work in the more economically developed northern and central regions. Women in this oceanic Latin culture were much slower to follow this trend and continued to remain at home. This may account for the earlier and sharper decline in disease incidence among men, particularly those living in traditional southern villages.

The Western Pacific foci of ALS and PD have provided, and continue to provide, a unique opportunity to study the natural history and basic mechanisms of pathogenesis of these disorders to an extent that would not be possible in large cosmopolitan western communities. These natural paradigms also provide insights for understanding other neurological disorders, including Alzheimer's disease, Parkinson's disease and motor neuron disease[36]. It is reasonable to suggest, therefore, that appropriate neuroepidemiological studies of Alzheimer's disease, parkinsonism and related disorders be conducted in such a way as to identify high risk individuals worldwide based on the epidemiological and ecological studies of events and exposures early in life, even *in utero* or prezygotically. It is also concluded that to the extent possible gene-environment interactions be explored in much greater depth than it has been in the past, for it represents an emerging and critically important area of research on late onset, uniformly fatal disorders with long latency, slow progression and of unknown etiology.

REFERENCES

1.   U.S. Bureau of the Census, U.S. Census of Population: general population characteristics, Guam. U.S. Government Printing Office (1960).
2.   A. Spoehr, Marianas prehistory: Archeological Survey and Excavations on Saipan, Tinian and Rota. In "Fieldiana: Anthropology vol 48", Chicago Natural History Museum, Chicago (1957).

3. P. Carano and P. Sanchez, A complete History of Guam. Charles Tuttle Company, Rutland, Vermont (1964).

4. L. Thompson, "Guam and its people", Princeton University Press, Princeton, N.J. (1947).

5. A. Arnold, D.C Edgren and V.S. Palladino, Amyotrophic lateral sclerosis: 50 cases observed on Guam, *J Nerv Ment Dis* . 117:135 (1953).

6. S. Tillema and C.J. Wynberg, "Endemic" amyotrophic lateral sclerosis on Guam; Epidemiological data: A preliminary report. *Doc Med Geogr Trop* . 5:366 (1953).

7. L.T. Kurland and D.W. Mulder, Epidemiologic investigations of amyotrophic lateral sclerosis. 1. Preliminary report on geographic distribution, with special reference to the Mariana islands, including clinical and pathological observations, *Neurology*. 4:355–78, 438 (1954).

8. L.T. Kurland and D.W. Mulder, Epidemiologic investigations of amyotrophic lateral sclerosis. 2. Familial aggregations indicative of dominant inheritance. *Neurology*. 5:182–96, 249 (1955).

9. D.W. Mulder, L.T. Kurland, and L.L.G.Iriarte, Neurologic diseases on the island of Guam. *U.S. Armed Forces Med J* . 5:1724 (1954).

10. A. Hirano, L.T. Kurland, R.S. Krooth, and S. Lessel, Parkinsonism–dementia complex, an endemic disease on the island of Guam. 1. Clinical Features. *Brain* . 84:642 (1961).

11. A. Hirano, N. Malamud, and L.T. Kurland, Amyotrophic lateral sclerosis and parkinsonism dementia complex of Guam. 2. Pathological Features. *Brain*. 84:662 (1961).

12. A. Hirano, N. Malamud, T.S. Elizan, and L.T. Kurland, Amyotrophic lateral sclerosis and parkinsonism–dementia complex on Guam. *Arch Neurol* . 15:35 (1966).

13. T.S. Elizan, A. Hirano, B.M. Abrams, R.L. Need, C. Van Nuis, and L.T. Kurland, Amyotrophic lateral sclerosis/parkinsonism dementia complex of Guam. Neurological re–evaluation. *Arch Neurol* . 14:356 (1966).

14. C.C. Plato, D.M. Reed, T.S. Elizan, and L.T. Kurland, Amyotrophic lateral sclerosis/parkinsonism –dementia complex of Guam. IV. Familial and genetic studies. *Am J Hum Genet*. 19:617 (1967).

15. J.E. Bailey–Wilson, R.C. Elston , C.C. Plato, and R.M. Garruto, Segregation analysis of amyotrophic lateral sclerosis (ALS) and parkinsonism dementia (PD) in the Western Pacific. *Am J Hum Genet*. 54: supplement, A23 (1989).

16. J.E. Bailey-Wilson, C.C. Plato, R.C. Elston, and R.M. Garruto, Potential role of an additive genetic component in the etiology of amyotrophic lateral sclerosis and parkinsonism-dementia in the western Pacific. *Am J Med Genet*. (in press).

17. R.M. Garruto, Amyotrophic lateral sclerosis and parkinsonism–dementia of Guam: Clinical, epidemiological and genetic patterns. *Am J Hum Biol* . 1:367 (1989).

18. D.C. Guiroy, M. Miyazaki, G. Multhaup, P. Fisher, R.M. Garruto, K. Beyreuther, C. Masters, G. Simms, C.J. Gibbs Jr, and D.C. Gajdusek, Amyloid of neurofibrillary tangles of Guamanian parkinsonism–dementia and Alzheimer disease share an identical amino acid sequence. *Proc Natl Acad Sci USA*. 84:2073 (1987).

19. S. Shankar, R. Yanagihara, R.M. Garruto, K. Iqbal, K.S. Kosik, and D.C. Gajdusek, Immunocytochemical characterization of neurofibrillary tangles in amyotrophic lateral sclerosis and parkinsonism–dementia on Guam. *Ann Neurol*. 25:146 (1989).

20. R.M. Garruto, R. Yanagihara, and D.C. Gajdusek, Disappearance of high incidence amyotrophic lateral sclerosis and Parkinsonism–dementia on Guam. *Neurology*. 35:193 (1985).

21. P. Rodgers–Johnson, R.M. Garruto, R. Yanagihara, K-M Chen, D.C. Gajdusek , and C.J. Gibbs Jr., Amyotrophic lateral sclerosis and parkinsonism dementia on Guam: A 30 year evaluation and clinical and neuropathological trends. *Neurology*. 36:7 (1986).

22. C.C. Plato, R.M. Garruto, K.M. Fox , and D.C. Gajdusek, Amyotrophic lateral sclerosis and parkinsonism–dementia on Guam: A 25–year prospective case–control study. *Am J Epidemiol* . 124:643 (1986).

23. D.M. Reed, D. Labarthe, K-M Chen, and R. Stallones, A cohort study of amyotrophic lateral sclerosis and parkinsonism–dementia on Guam and Rota. *Am J Epidemiol*. 125:92 (1987).

24. R.M. Garruto and Y. Yase, Neurodegenerative disorders of the Western Pacific: the search for mechanisms of pathogenesis. *Trends Neurosci*. 9:368 (1986).

25. D.C. Gajdusek, Environmental factors provoking physiological changes which induce motor neuron disease and early neuronal aging in high incidence foci in the Western Pacific. *In:* "Research Progress in Motor Neuron Diseases", F.C. Rose, ed. Pitman Books, Kent (1984).

26. D.C. Gajdusek and A.M. Salazar, Amyotrophic lateral sclerosis and parkinsonism syndromes in high incidence among the Auyu and Jakai people in West New Guinea. *Neurology*. 32:107 (1982).

27. C.C. Plato, M.T. Cruz, and L.T. Kurland, Amyotrophic lateral sclerosis/parkinsonism dementia complex of Guam: Further genetic investigations. *Am J Hum Genet* . 21:133 (1969).

28. C.C. Plato and M.T. Cruz, Blood group and haptoglobin frequencies of the Chamorros of Guam. *Am J Hum Genet*. 19:722 (1967).

29. R.M. Garruto, C.C. Plato, N.C. Myrianthopoulos, M.S. Schanfield, and D.C. Gajdusek, Blood groups, immunoglobulin allotypes and dermatoglyphic features of patients with amyotrophic lateral sclerosis and parkinsonism–dementia of Guam. *Am J Med Genet*.14:289 (1983).

30. C.C. Plato, D.L. Rucknagel, and L.T. Kurland, Blood group investigations on the Carolinians and Chamorros of Saipan. *Am J Phys Anthropol*. 24:147 (1966).

31. N.M. Blake, R.L. Kirk, S.R. Wilson, R.M. Garruto, D.C. Gajdusek, C.J. Gibbs Jr., and P. Hoffman, Search for a red cell enzyme or serum protein marker in amyotrophic lateral sclerosis and parkinsonism–dementia of Guam. *Am J Med Genet* . 14:299 (1988).

32. R.E. Tashian, C.C. Plato, and T.B. Shows Jr, Inherited variant of erythrocyte carbonic anhydrase in Micronesians from Guam and Saipan. *Science*. 140:53 (1963).

33. P.M. Hoffman, D.S. Robbins, C.J. Gibbs Jr., D.C. Gajdusek, R.M. Garruto, and P.I. Terasaki, Histocompatibility antigens in amyotrophic lateral sclerosis and parkinsonism–dementia on Guam. *Lancet* . 2:717 (1977).

34. C.C. Plato, M.T. Cruz, and L.T. Kurland, Frequency of Glucose–6–phosphate dehydrogenase deficiency, red–green colour blindness and Xg^a blood–group among Chamorros. *Nature*. 202:728 (1964).

35. C.C. Plato, R.M. Garruto, and D.C. Gajdusek, Further studies on the Genetics of the Chamorros of Guam: Dermatoglyphics. *Hum Hered* . 33:329 (1983).

36. R.M. Garruto, Pacific paradigms of environmentally induced neurological disease: Clinical, epidemiological and molecular perspectives. *Neurotoxicology* . 12:347 (1991).

PANEL DISCUSSION ON
GENES, ENVIRONMENT AND DISEASE [1]

Amitabha Basu, Ranjan Gupta
(Indian Statistical Institute, Calcutta, India)
Robert E. Ferrell
(University of Pittsburgh, Pittsburgh, U.S.A.)
Robert C. Elston
(Louisiana State University Medical Center, New Orleans, U.S.A.)
D.F. Roberts
(University of Newcastle-upon-Tyne, Newcastle-upon-Tyne, U.K.)
D.P. Mukherjee
(University of Calcutta, Calcutta, India)

It is now generally understood that there is a wide spectrum of manifestations of ill-being, in general, and diseases, in particular. The spectrum varies from those which are familial to the others caused by infection, malnutrition and pollution. Most of the common and chronic systemic or degenerative diseases are located in the middle of the spectrum. Many of these diseases have variable ages of onset, variable types and magnitudes of expression, and show results of teratogenic or accidental causes. Deciphering the mechanisms and processes of manifestation of these diseases provide a challenging frontier, and coordinated research involving specialists with expertise in different disciplines would be required for substantial progress in this field.

In course of the Panel Discussion, an attempt was made to probe into these mechanisms and processes, utilizing the varied expertises and experiences of the Panelists, in particular, and the participants of this conference, in general.

In the first part of the discussion, specific questions were raised with reference to certain empirical findings regarding the possible roles of genes and environments in the determination of health-related traits, and answers sought; in the second, the canvas was broadened to cover much ground in addition to the specific questions discussed in Part I, particularly the situation prevailing in the Indian scenario and the social structures determining health and well-being ; and in the third, the manifestations of ill-being /diseases were sought to be interpreted in relation to the social structures in which they occur.

Part I

J.B.S. Haldane (1949) in his article entitled 'Disease and Evolution' suggested the pos-

---

[1]D.P. Mukherjee and Amitabha Basu acted as the Chairman and the Coordinator, respectively, of the Panel Discussion.
This report has also been prepared by them. — *Editor*

sibility of "a particular race of bacteria or virus being adapted to individuals of a certain range of biochemical constitutions, while, those of other constitutions are relatively resistant". One may re-word this suggestion as follows: different groups of individuals (Haldane specifically mentioned mammals and birds) possessing different biochemical constitutions may respond differently to the same environmental stress as represented by, for instance, the attack of a certain microbial race (by being vulnerable or resistant to the attack). One may substitute 'human populations inhabiting different regions of the world' for 'different groups of individuals' and 'physical environmental stress generally referred to as high altitude' for 'microbial race', without losing the thrust of Haldane's argument.

A. Basu and R.Gupta in their presentations to the Panel Discussion briefly referred to some data relating to certain health-related traits (but not diseases *per se*) prevailing among some human populations, inhabiting Himalayan and Andean mountains, who are spatially located far apart, in order to ascertain whether their responses to exposure to the environmental stresses associated with high altitude, namely hypoxia, cold, terrain, etc., were similar or different. They framed certain questions as to the possible genetic and/or environmental determination of the differences with respect to these traits between the groups.

What are the effects of high altitude on human populations? Are these effects universal? If not, what could be the possible reason for the non-universality? Could the non-universality of environmental effects operate at lower altitudes, within the same region but between population, and if so, what could be the possible reasons?

The 'high altitude', defined as altitude above 3000 m, is generally associated with

"   -   a decline in atmospheric pressure,
    -   a decline in temperature, and
    -   a rise in short-wave radiation,

including ultraviolet and gamma wave lengths" (Baker, 1989). A decline in atmospheric pressure is associated with hypoxia, the most serious altitudinal stress unmitigated by cultural factors. Other altitudinal stresses are low humidity, high wind flow and rapid soil erosion (Little, 1981).

Baker (1989) suggests the following characteristics, among others, of high altitude residents in comparison with low altitude residents:

"1. lower fecundity and fertility as a consequence of poor gonadal function and reduced oxygen profusion of the placenta;

2. a higher neonatal death rate as a consequence of low birth weights and higher frequencies of *patent ductus arteriosus* ;

3. infants and adolescents grow more slowly in general body size (particularly weight) but develop large chests and lung volumes at all ages;" etc.

Baker (1989), however, has categorically mentioned that these findings, and the possible causes thereof, are based primarily on Andean populations. The question is, could these findings be generalized for other, particularly, Himalayan populations ?

Basu and Gupta's data on fertility, mortality, physical growth of children, adult body dimensions and hemoglobin level were collected from the Sherpas of high altitude (3500 — 4500 m.) in northeastern Nepal, and the Sherpas of moderate (1000 — 1500 m.) altitude in Kalimpong subdivision, Darjeeling district, West Bengal, India . The high altitude Sherpa sample has been subdivided into two sociocultural subgroups namely 'more urbanized' and 'less urbanized', and the moderate altitude sample into several occupational subgroups.

Table 1. Fertility and mortality among some populations and subpopulations

Population/ Subpopulation	Area	Altitude (m.)	Completed fertility	Infant mortality (%)
Popupulation				
Sherpa	Upper Khumbu, Nepal	High	4.5	7.2
Sherpa	Kalimpong, India	Moderate	7.4	6.6
Lepcha	– do –	- do -	5.7	11.7
Aymara	Belen, Chile	High	5.8	14.2
Aymara	Hullataire, Chile	- do -	7.3	33.3
Aymara	Chapiquina, Chile	- do -	8.5	13.0
Aymara	La Paz, Bolivia	- do -	5.7	16.9
Quechua	Nunoa, Peru	- do -	6.7	16.4
Quechua	Cerro da Pasco, Peru	- do -	7.7	10.0
Subpopulations of Sherpa				
'More urbanized'	Upper Khumbu, Nepal	High	4.4	5.0
'Less urbanized'	- do -	- do -	5.0	11.6
Rango (Plantation lab.)	Kalimpong, India	Moderate	8.3	13.8
Echhay (Agricultural)	- do -	- do -	5.6	6.1
Munsong (Plantation lab.)	- do -	- do -	6.9	6.8
Lava (Forest lab.)	- do -	- do -	7.8	6.6
Labdah (Agricultural)	Darjeeling, India	- do -	6.8	4.5
Mungpoo (Plantation lab.)	- do -	- do -	4.6	6.1

Source : Gupta, 1991; Gupta et al., 1989.

The data show that (1) completed fertility is lower in the high than in the moderate altitude as is generally found in the Andes, but a similar magnitude of difference exists between the 'more urbanized' and 'less urbanized' subgroups of the high altitude Sherpas as well as among the occupational and/or spatial subgroups of the moderate altitude Sherpas; (2) infant mortality hardly differs between the high and moderate altitude Sherpas, unlike in the Andes, but differs substantially between the 'more urbanized' and 'less urbanized' subgroups of the high altitude Sherpas as well as among some occupational and/or spatial subgroups of the moderate altitude Sherpas; (3) completed fertility is substantially higher and infant mortality lower in the Sherpas than in the Lepchas, both groups inhabiting the same moderate altitude habitat; and (4) completed fertility as well as infant mortality in the high altitude Sherpas are both generally lower than those in the Aymaras and Quechuas of the Andean high altitudes (Table 1).

As to the other traits, (5) physical growth of children is slower in the high than in the moderate altitude Sherpas, as is generally found in some Andean studies, but is more protracted, so that (6) the adult body dimensions are larger in the high altitude Sherpas, unlike in the Andeans; (7) the hemoglobin level is higher in the moderate altitude Sherpas, like in the Andeans; (8) the values of haematological parameters (hemoglobin level, packed cell volume and red cell number) and frequency of hookworm infestation, are higher, and eosinophil count is lower, in the Sherpas than in the Lepchas of the moderate altitude (Table 2); (9) the high altitude Sherpas' physical growth curves are uniformly lower than those of Andean ones; and (10) the mean hemoglobin level of the high altitude Sherpas is lower than that of the high altitude Andeans.

Table 2 . Hematological and other clinical traits

Population	Hb(g./dl.)	PCV(%)	RCN ×10⁻⁶	Eosinophil count(%)	Hookworm infestation(%)
Sherpas	12.3	43.4	4.4	12.3	73.8
Lepcha	9.2	35.5	4.2	18.7	75.8

The results suggest that (1) sociocultural factors may make as much or more impact on the basic demographic traits than the physical environmental factors; and (2) different ethnic groups sharing the same microenvironmental stresses (e.g. the Sherpas and Lepchas of Kalimpong), or the same macroenvironmental stresses like hypoxia, etc. (e.g. Himalayan and Andean high altitude populations) may respond differently to similar environments.

The point that emerges from these results, given in much further details by Gupta et al. (1989), is that there is no evidence of any universal effects of altitude-related factors on health-related traits.

What could such a lack of universality in the responses of human populations to the stresses associated with high altitude, or to similar environments, in general, be due to ?

One could visualize two possible causal mechanisms : (1) different high altitude populations respond to the high altitude stresses differently because of their different genetic compositions; and (2) although all high altitude environments possess some common stresses, e.g. hypoxia, cold, terrain, etc., they may differ with respect to others, which may cause differential responses despite similar genetic compositions.

The latter possibility will be examined first. Indeed, the high altitude environments the world over are not identical. The altitudes at which human populations reside permanently differ between Himalayan and Andean regions, particularly with respect to the concomitant stresses. Obviously, in the human case one can not conduct rigorous experiments to control the effects of concomitant factors like food availability and habit, health facilities and awareness, and the life-ways, in general, and must rest content with studying natural experimental situations, with their relative limitations. However, do these limitations of studies on humans adequately explain the situation on the ground ? Could the differential responses of different high altitude populations to the high altitude stresses be explained away in terms of our inability to conduct controlled experiments ? Or, do different high altitude human populations respond differently to similar altitude stresses due to their different genetic compositions ?

The possibility that "the Sherpas are *genetically*[2] better adapted to high altitude than are Amerindians living on Peruvian highlands" with respect to hematological traits, "possibly as a consequence of much more prolonged exposure to such an ecological factor of selection as high altitude", was suggested by Morpurgo et al. as early as 1976 but this possibility has hardly been investigated since then. Analogously, the Sherpas may be genetically better adapted than the Lepchas are to the Kalimpong microenvironment.

Now, getting back to Haldane's suggestion quoted in the beginning of this Part of the report on the Panel Discussion, and the implications of the suggestion which follows it, as well as the data presented above which may corroborate the suggestion and its implications, it appears quite plausible that some groups of individuals may be better adapted to given environmental stresses, while others are not and this phenomenon may occur at both micro and macrolevels. The question that we wish to pose is, why this is so ? specifically,

(1) if the reason is genetic, how do the genes actually function to produce different responses to similar environmental stresses ?

(2) if the reason is environmental, how do the environmental factors other than the ones the effects of which are being studied, mediate to produce different responses to similar environmental stresses ?

(3) if the reason is gene–environment interaction, which intuitively is most likely, how exactly does the interaction operate in case of specific traits in specific environments ? and

(4) what would be the best strategy to look for answers to these questions ?

R.E. Ferrell mentioned a few points which were presumably illustrated, to a certain extent, by what was presented by the first two speakers. One often wishes to understand gene-environment interaction, i.e. how the environment affects the expression of genes, an example being that of two populations which appear to be genetically similar but have very different biological descriptions. A problem of trying to understand gene-environment interaction is that one can neither define the genes, nor the environment. It is estimated that there are 100,000 genes in the human genome, of which about 2400 have been enumerated (i.e. established the existence of), but the genes for which the details of their operation are understood is less than 1% of the total number of genes in the genome. So, attempts are made to understand very complex biological problems, without even being able to enumerate the genes that underlie the phenotypes concerned. So, on the geneticist's side, one is virtually ignorant, at least at the biochemical level, of the genetics of most phenotypes. There are, of course, statistical methods of finding evidence that genes are acting. But, when it comes to actually understanding what genes are acting in a given biological system

---

[2]emphasis added

determining a particular phenotype, one is relatively ignorant. However, one wants to look at gene-environment interaction. Yet, in most studies, when one tries to specify the environment, it can only be specified in very vague and general terms, because it is not understood which of the virtually infinite number of environmental variables that could be measured are important. So, the current position is somewhat awkward: trying to understand the interaction between two systems neither of which can be defined in any detail, except in a very general descriptive sense. The questions that Basu and Gupta listed, and Haldane asked, are legitimate questions but they are questions that still can not be answered to any satisfactory extent at all.

As far as the last question regarding how one goes about trying to understand gene–environment interaction is concerned, everyone's answer would be dependent on what her or his area of expertise is. Given Ferrell's own laboratory bias, he thinks if one is going to understand gene-environment interaction, one has to define the genes and the environment. As an experimentalist, he despairs of ever being able to define the environment, but he is optimistic that genes can be defined at least in a general way. The technology available now may not allow identification of individual genes, but it already allows classification of individuals into groups based on knowledge about certain relevant segments of the genome. One can then follow those segments of the genome in a very specific way in families or in populations living under defined and possibly different environments. One may be forced to being with an experimental system, or a quasi-experimental system, at least since true experimental systems are hard to establish in human populations. If it is possible to enumerate all segments of the genome, it might be possible to determine which of those contribute to the difference in response that are seen. Once those segments of the genome that contribute to the response to an environment can be identified, then using molecular genetic methods one can go into a detailed and a finer level, and eventually identify those genes that are involved in that response. In a sense, the ability to look at gene-environment interaction demands that,first of all, can be evolved a set of tools by which can be defined the genes and, to a certain extent, the environment. Geneticists are only reaching the point just now that the tools are in hand. Now comes the application of these tools.

R.C. Elston commented specifically on the determinants of diseases. He had some reservations on N.E.Morton's assertions[3]. He presented what he thought the future of determining the genetic and environmental determinants of common diseases — those that are not simple Mendelian — were. Ferrell had mentioned that neither the genes nor the environment could be defined, which perhaps pose difficulties. Elston believed that the tools are now available and suggested how to go about it. He pointed out that geneticists now have the tools to perform a global search of the entire human genome by linkage analysis. They have been working on designs of doing this to make it cheaper. First of all, one can span the genome perhaps at 20 or 30 centiMorgan level with markers and then use appropriate statistical methods to maintain power. One may not look for lod scores of 3 — that wastes money — rather one may look for much larger lod scores, and then look at further genes on either side. It has been shown theoretically how one could maintain power, keep the signifcance level low and look for genes. How is one going to do this ? Elston remarked that Morton still maintains that linkage analysis is well-motivated only under two conditions: either one has to have a good candidate gene or one has to have demonstrated by segregation analysis that there is a single major gene segregating. But that does not make sense, he said. If one

---

[3]During the course of this Conference, Professor Newton Morton mentioned several times that affected sib-pair methods are of little use in the detection of linkage. He also mentioned that combined segregation and linkage analysis is the ideal method for genetic analysis. He pointed out linkage analysis is useful only when there is a good candidate locus, or when the trait is known to be primarily controlled by a single major locus. — Editor

has a disease that is caused by the interaction of three loci, one can never demonstrate that by segregation analysis. Classical linkage analysis by likelihood methods requires that one knows the mode of inheritance. If one does not know the mode of inheritance, one can not get very far. Initially, simple, robust and non-parametric methods of analysis that is going to point the way to how many genes are involved in a disease may be used. One of the simplest methods to do that is either to use the affected sib-pair methodology or something very similar, which Morton believed was useless and wrong. It may not necessarily be affected sib-pairs, but something like affected relative pairs, possibly triplets. Certainly, affected relatives are extremely powerful and provide more information per unit of money. In fact, if it costs as much to investigate affected grandparent-grandchild pairs as affected sib-pairs — which will not be true for diseases with late onset — then, affected grandparent-grandchild pairs will be the cheapest way to go. After that, possibly affected half-sib pairs, which are more informative than affected sib-pairs, may be investigated. Affected half-sib pairs are rarer in the population, and it is because they are genetically more distant that when they are both affected, there is something stronger between them. And, then, perhaps affected cousin pairs, or possibly affected uncle/aunt – niece/ nephew pairs, may be investigated.

By doing a global search, one could do various things. One could find out, for example, that there is a strong possiblity of linkage with a disease between, perhaps, three regions on the chromosome. That will tell one that probably there are three loci involved — may be two regions, may be four regions. These are within the bounds of possibility these days, with the technology that exists. Such methods are relatively cheap when compared with linkage studies performed on large pedigrees in the United States with various mental diseases, which, Elston thought, have led to nowhere. He was not against large pedigree methods; he thought they were very good. But, he thought one might find it cheaper to initially look for affected relative pairs. Then, when one knows that one has some tight linkages, it is far easier to perform some kind of segregation analysis, or perhaps combined segregation and linkage analysis, because one has a handle on the genes — one knows that one has a gene which is co-segregating with the trait or disease of interest. At that point, one could also bring in environmental variables. But, it is also important to allow for unmeasured environmental variability, he emphasized. This, Elston thought, was the way of the future for the analysis of those diseases that do not segregate in a simple Mendelian fashion.

About gene-environment interaction, Elston pointed out that there was a big difference between the joint action of genes and environment and statistical gene-environment interaction. Whenever analyses are performed either in plants or in animals to look at this, the first thing that is done is to get rid of any statistical interaction. Interaction is very often — not always — an assumption of the scale of measurement, just like dominance and epistasis. To the extent possible, it is better to make suitable change of scale and to use models that do not have interaction. They would, of course, have the joint action of genes and environment, and as much as possible, analyses would be done on an additive scale, because this again would give the most efficiency. If, of course, there is some interaction that can not be got rid of by a change of scale of measurement, then there would be huge sample size requirements. But, one should try the simple step first.

Part II

D.F. Roberts spoke on a canvas which covered much ground in addition to the narrow precincts of the specific questions raised by the first two speakers. He entirely agreed with Elston, but in a different context. He believed that Elston had sketched the way forward for improvement of the understanding of the genetic basis of disease in the Western context. He wished to consider how the Indian context could help us forward. Several times during this symposium, the great wealth of variation that there is in India has been mentioned. Speaking in terms of the peoples, has been mentioned the great wealth of environmental variation that there is. Basu has taken one particular extreme form of environment — the high altitude. There is scope for similar investigations of the other environmental extremes that exist here. The rigours of the deserts, the problems of tropical forest communities — each of those poses a string of challenges to the survival of the people who live in them. Then, there are many problems which can no longer be solved in the West. There appears a tendency to bypass the unsolved problems when a new technique becomes available. This has to be done because of the curious way of financing research in the West. Today, it is almost impossible to obtain a grant for doing any piece of research from the Medical Research Council in Britain unless it involves some molecular study. So, there is a bandwagon (who have jumped into it ?) in the West in order to survive. To do so, they have to leave unsolved so many problems. To revert to some questions to which Haldane drew attention: what about the selective significance — the advantages and disadvantages — of each of the polymorphisms that have been discussed during this meeting ? Perhaps, the selective advantage of many of those polymorphisms comes in early life. They will be detected by differential mortality and survival in infancy and childhood. One can not find those in the West because no longer there is sufficiently high levels of mortality in infancy and childhood. But in India, unfortunately, those still exist. But, instead of regarding this as 'unfortunately' from a human point of view, one may regard this as 'fortunately' from a scientific point of view. It is only a matter of time before those problems are solved, and that is why it is important that one gets on to them now and takes advantage of the opportunity that exists.

One may think of another feature of India. There are many many communities living in rural surroundings practicing the same way of life as their parents before them, and their parents before them, and their parents before them. There is a steady drift of individuals from those communities into the towns and cities. The effect can be seen in Calcutta. There are many problems of why do people move from stable situations to new ones about which they have no experience. But, the sort of biological problem that might deserve attention is the physiological-psychological stress that such a move engenders. It is possible to measure such stress by very simple biological variables — things like the excretion rate of catecholamines. There have been some interesting studies of this. For instance, in Oxford, in the Oxfordshire region, these levels have been assayed in inhabitants of a particular village everyday for a given period. This village is what is called a 'dormitory-type' village; people live there, go to work in the city in the daytime, come back at night and stay there for the weekends. The difference in the catecholamine levels between the unstressful weekends at home and the worrying workdays is interesting. There are similar opportunities here. What is the cost—physiologically, psychologically—to these individuals who move. Obviously, the most interesting way of doing this would be to do this prospectively. One would have to go into the villages and find out who is going to move. But, that is not necessary at the first step. Let us look at a comparison of these levels in Calcutta's street-dwellers and in the communities from which they came.

Another theme that is of interest would be the simple mapping of genetic diseases in this country. Such maps do not exist. It is not easy to do. But, with the developing of the services that are now available, it would not require a great deal of ingenuity. The doctors have to spend some time with a rural community after they qualify. It would not be a too difficult thing to do to organize as a part of that requirement an all-out attack on the number of cases of Huntington's chorea or of Duchenne dystrophy in each area. One may not consider the difficult cases that require high-powered techniques to diagnose, but take up the very visible disorders. The populations are sufficiently large for the incidence of those diseases to show up. It may be suggested that this might be an interesting thing to do.

Several times during this meeting a number of different mating systems were mentioned. Elston had mentioned the use of half-sib analysis. Roberts pointed out that another thing that India could do, which could not be done in many other parts of the world, would be to utilize that variation in mating systems to explore the extent of genetic and environmental contributions, not in terms of normal metrical measures, but in terms of those quantitative variables that are associated with particular diseases.

Roberts mentioned that one of the projects that was brought to his attention in a personal discussion involved the question of the etiology of cancer in women. Here, again, there is still scope for useful contributions to be made. It is known, for example, that about 5% of cases of ovarian cancer in English women are due to a single dominant gene; the remainder are not. What is the proportion of such cases in Indian women ? Simple collection of family material using the close association that the lady practitioners, particularly gynecologists, have with their patients will take care of the validity of the diagnoses, the correctness of the reports of affected members of the family, and so on.

The final thing that Roberts emphasized was that he often got the impression that the -scientists were collecting data merely for the sake of collecting data. This question was also discussed during this conference. However, to make the best use of the advantages that are available, one should think seriously before collecting the data as to why the data are being collected, what hypotheses are being tested. In other words, collecting data for the sake of collecting data is not really very efficient. One should set up a hypothesis and then collect data specifically to test it.

Part III

Taking the cue from Roberts' emphasis on the unique opportunity provided by the Indian Scenario, in general, and his reference to the variation in mating systems in India as a potential source of obtaining an insight into the relative contributions of genes and environment to the etiology of disease-related quantitative variables, in particular, D.P. Mukherjee suggested that an approach to understanding the combined roles of genes and environment, and their interactions, in the causation and manifestation of a wide range of diseases, involving studying them in well defined populations, along with data on their social structures, could be rewarding. Social structure, for this purpose, can be defined as the definite pattern in which the parts of a social system harmoniously work with the whole. Anthropologists have generally considered kinship and marriage rules as the main determinants of social structure. Radcliffe-Brown (1926), whose name is associated with the expression 'social structure', agreed with Malinwoski (1922) that biological ties are the origin of kinship ties. Levi-strauss (1962) described social structure as a three-tier structure of communication, in which kinship and marriage rules regulate the communication of 'genes between phenotypes'. He also found the genetical and demographic concept of 'isolate' helpful in clarifying 'social structure'. Thus, through the doorway of social structure we may enter the domain of genetical structures of human populations, in which are recorded the degrees of isolation of subpopulations, the way the genes combine, mating patterns, and size and spread of populations. Haldane referred to this central aspect of social structure when, in 1955, he stated "plenty of populations are wholly isolated for many generations" and "any such population has a structure". He almost depicted a characteristic feature of traditional endogamous populations of India. "Most populations have a structure somewhere between those found when mating is random, on the one hand, and inbreedig is complete, on the other", he stated.

A characteristic feature of the Indian society has been, for a long period of time, the continuous division and subdivision of populations — hierarchies of marriage isolation. At the first level of social grouping are the linguistic and local cultural divisions corresponding broadly to ecological regions. The bulk of the population of each region constitutes the traditional Hindu society which represents a system of maritally exclusive but otherwise interacting groups, called *jatis* or castes. Some of the isolated groups living near forests or hills have gradually entered into this caste system, while others have retained their identity as specific tribes. The other religious communities and sects like the Shia and Sunni Muslims, Sikhs, Christians of numerous denominations, sects of Jains and Buddhists have also grouped themselves into hierarchical categories, often into those that they belonged to before conversion or migration. There are also local groups like Parsees, Jews, Ahmediahs and others. Within each of these large populations, there are smaller and still smaller maritally isolated populations. However, endogamous groups are sometimes large and spread over wide areas in a region, which leads to marriage restrictions by geographical barriers, distance, socioeconomic and other cultural considerations. Thus, the actual breeding population or deme in genetical terms may not be identical with the notional endogamous groups. Besides, endogamous groups are sometimes socially split up into endogamous subgroups. For example, the endogamous Kaibarta caste of southern West Bengal, whose traditional occupation was fishing, has split up into the fishing (Jele) and cultivating (Chasi) Kaibartas; some Chasi Kaibartas have further transformed themselves into the Mahisya caste, which is spread over almost the whole of southwestern West Bengal. But, intensive studies revealed marital isolation of geographically separated subgroups of Mahisyas. Similar examples are also available from other regions of India (see, e.g. Basu, 1969). Such breeding isolations, as expected, lead to increased prevalence of disorders — especially monogenic recessive ones — resulting from increased homozygosity due to random genetic drift and/or inbreeding. For example, among the Pattusali, a very small weaving community residing in Tirupati (Andhra Pradesh), high incidence of genetic disorders — e.g. club feet, prelingual deafness, dwarfism, etc. — have been observed.

There are also many other social rules governing marriage. The principle of hypergamy is one such, which permits a woman to marry a man from the same endogamous subgroup or from a subgroup of a higher status, but not from a group of a lower status. For example, within an endogamous group called Rarhi Brahmin in West Bengal, members of certain lines of descent belong to the Kulin category, who claim themselves to be of a higher status than the non-Kulin categories. A Kulin male can marry a girl from a non-Kulin (Shrotriya) family, but a Kulin female is not permitted to marry a Shrotriya male.

The rule of lineage exogamy prohibits marriage between members of a descent group, which may be patrilineal or more rarely matrilineal, e.g. Nairs of Kerala, Garos or Khasis of North-East India, thereby preventing the population from splitting up into very small fragments. As noted by Haldane, such practice of lineage exogamy in India and Australia guarantees cross-cousin marriage. In fact, it permits, and is compatible with, a preference for marriage between offspring of a brother and a sister in southern India, in general, and also some tribal and low caste (e.g. the Mahar) populations in other parts of India. Lineage exogamy is generally absent among endogamous Muslim populations of India which explains a preference for parallel-cousin marriage, that is marriage between offspring of two brothers, although such marriages have been found to be extremely rare among Muslims of Andhra Pradesh. Among the local Hindus, cross-cousin marriages are more frequent than parallel-cousin marriages, and the highest frequency of maternal uncle-and-niece marriages is found in Andhra Pradesh and neighbouring regions, especially among the artisan castes and fishermen. Consanguineous marriages, however, are usually very rare, if at all, in the traditional caste populations in northern India and more particularly in populations of eastern India comprising West Bengal and eastern parts of Orissa and Bihar, in spite of the prevailing rule of *gotra* (clan) exogamy among some Hindu and tribal populations.

This is, however, not universal among Hindu caste populations. Analysis of inter-surname or inter-*gotra* marriages suggests not only absence of lineage-endogamy over that expected in random mating in some other populations, e.g. among the patrilineal Mahisya caste of West Bengal and the matrilineal Boro-Kachari tribe of Assam.

There is another rule of exogamy which indeed prevents closely consanguineous marriages among most Hindu caste populations throughout northern India, with a few exceptions. This rule is called *Sapinda* exogamy, which prohibits marriage between descendants of ancestors of several upward generations, usually seven in the paternal line and five in the maternal line. This was most strictly followed by the people of eastern India, though it is getting relaxed to some extent due to breakdown of kinship structure.

There is also a practice of village exogamy in parts of northern India. In southern India, however, village endogamy is generally preferred. These rules are often connected with various socioeconomic factors, such as inheritance of land by daughters or incapacity to pay bride price or dowry to the bridegroom, and so on. These are some loop-holes of the strict rules of endogomy and subdivision of endogamous groups through hypergamy, for example.

The social structures of the populations in different parts of India, in fact, form a part of the total environment which influences the mating systems, which in their turn affects health-related genetical traits. The impact of mating systems can be most directly illustrated by the results of consanguineous marriages, especially in small endogamous populations. Compared to a randomly drawn individual, the average inbreeding coefficient is higher among patients of various specific diseases, including even infectious diseases, like pulmonary tuberculosis, and systemic diseases, like bronchial asthma, apart from congenital malformations, as has been noted in hospital studies conducted at Waltair (Dromamraju and Meera Khan, 1963), Tirupati (Mukherjee, 1972) and Hyderabad (Murthy and Jamil, 1972).

The occurrence of specific congenital malformations or disorders like club feet, neurological disorders like prelingual deafness, anencephaly often exceed in neonates in areas like Tirupati around which closely consanguineous marriages are very common compared to that in Calcutta where such marriages, especially among the Hindus, are extremely rare.

Sanghvi, in 1966, formulated the hypothesis of a considerable decline of lethal recessive genes as a result of high inbreeding through hundreds of generations, assuming their maintenance by constant mutation rates or heterozygous advantage. However, the data collected from numerous small populations have confirmed an increase of congenital malformations, genetic disorders like thalassaemia (as in a sample of Muslims of North 24 Parganas of West Bengal), and infant and child mortality, in the inbred sample (Kumar et al., 1967; Rao and Mukherjee, 1975; Mukherjee et al., 1977; and others). This increase was, however, not detected in large surveys, e.g. in Tamil Nadu, in which the social structures of the populations involved were not taken into account. In fact, Sanghvi and Master (1974) also have reported some excess in the prevalence of malformations in the highly inbred sample of neonates in Bombay. It is true that lethal recessive genes may often tend to be lost in the highly inbred section of the population than in the less inbred, as Haldane had earlier pointed out. But, for non-lethal diseases and disorders, inbreeding effects can certainly be used as a tool for the study of non-dominant genes in India as in other parts of the world. The increase in morbidity in inbred subjects is not always statistically significant, however.

The small increase of homozygosity of deleterious genes that are expected in the range of inbreeding observed in human populations would require large samples to display statistically significant effects. But large samples for study of inbreeding effects on diseases are unrealistic due to aggregation of consanguineous marriages in families resulting in small effective size of the breeding population. Wide variation in consanguinity rates among neighbouring villages have been observed in a number of populations (Mukherjee, 1992). In some cases, the progress of autozygosity with inbreeding may be interrupted by increasing selection pressure (A. Mukherjee, 1990). For this and other possible reasons, the increase of homozyosity with inbreeding have also been observed to be non-linear in certain instances. Special care has to be taken to obtain adequate samples of individuals with the same degrees of inbreeding, in view of recurrence of consanguineous marriages in successive generations. Sometimes, the homozygosity in the background population, i.e., the offspring of unrelated parents, may have become already high for evolutionary reasons, and the increase of homogygosity at specific loci in the inbred families, may, therefore, not be detectable. Such problems can be minimized only through well-designed studies which would require specific investigations of the social structure influencing mating patterns in the population studied. For the study of genetics of diseases it is not enough to look for homozygosity of recessive genes with positive or negative effects. Homozygosity of different alleles might interfere with the well-being of individuals. Theoretically, both hypertension and hypotension may be reflective of disease conditions due to increased homozygosity which can be detected by comparison of frequency distributions of quantitative traits related to diseases which are often not recognised. There is consistent evidence of the absence of increased incidence of early abortions of inbred foetuses in several Indian populations (Mukherjee, 1992), as in Israel. In these studies, the reverse trend was, in fact, detected. It is necessary to be careful in controlling variation in physical environment while studying inbreeding effects on disease manifestations. In the available studies from Indian populations, increase of morbidity with inbreeding appears to vary considerably between populations, only a part of which could be attributed to the factors mentioned here.

The unique genetic background of certain genetically complex diseases, whether due to a single locus or multiple loci, whether recessive or additive, or a combination of major genes acting in polygenic or oligogenic background , can be better understood if the variation in their disease phenotypes are studied along with social structures, and mating patterns of the parents of affected and unaffected individuals in a deme inhabiting one small econiche.

The studies conducted so far indicate that the deleterious effects of high inbreeding is much more marked in specific populations in which environmental factors like high rate of natural radiation causing mutations or, as Pai (1972) suggested, extreme poverty in both inbred and non-inbred groups operate, than in others. There are other instances where the morbidity effects of homozygosity or inbreeding was more marked in the better-off section of the population within which inbred and non-inbred children were compared, than in the worse-off one. Such preliminary findings point to the prospect of further clarification of gene-environment interactions in specific instances with controlled experimental designs.

The rule of exogamous marriages in many populations of southern and central India and a few upper castes of northern India provide a unique opportunity for locating X-linked genes and also gene-environment interaction on specific diseases. This is because lineage-exogamy in the patrilineal societies lead to various degrees of marital migration and dispersion of X-linked genes along with the brides, of course, along with autosomal genes also. However, X-linked influence can be easily traced in such a situation. For example, a study of red-green colour blindness in successive generations among the Mahisyas of Uluberia in Howrah district, West Bengal, suggest a change in the frequency of the concerned genes with change in the marriage area and direction of receiving brides. Certain disorders may again have a higher frequency among inbred females than among inbred males which would indicate some influence of X-linked genes.

Since consanguineous marriages lead to increase of homozygosity at several loci, there would be a reduced probability of detectable recombination in the offspring of such marriages. Increased association between diseases and certain marker genes in the highly inbred individuals may sometimes indicate linkage relationship and shed further light on the genetic background of disorders. Linkage disequilibrium between disorders and markers may be expected to increase in the inbred sections of a demic subdivision of a population. Mapping of genes for diabetes or hypertension will be easier if these diseases are studied within one gene pool — where a single locus and perhaps a single allele is more likely to be involved in the disease. It will be even more convenient if the levels of inbreeding of individuals affected by a genetic disease in various degrees at various ages are systematically recorded. Allopatric branches of a population when exposed to diverse cultural or physical environmental stresses may often show differences in disease susceptibility. Again, in newly formed small populations, there is a greater probability of observing linkage disequilibrium between certain diseases and marker loci. This outline of the possible impact of social structures of the endogamous populations of India on genetic diseases is neither exhaustive nor static. It is basically meant to suggest some lines of investigation that can be pursued by using the knowledge of variation of social structure to shed further light on gene-environment interactions in the etiology of diseases. It is relevant to mention that Haldane's contribution in the field of human genetics had included his early contribution in exploring the possibilities of utilizing information about kinship and social structures of the endogamous populations of India in understanding complex genetical characters which are often greatly influenced by environmental variation. Haldane used to emphasize the need for taking stock of the deleterious genes in different endogamous groups living side by side in different parts of India before a breakdown of isolation takes place. As mentioned above, he was instrumental in collection of pedigree data on consanguineous and polygynous marriages along with the Census of India in 1961. Subsequent research along the lines suggested by Haldane has not only provided further data, but also suggestions and testable hypotheses, which require to be verified by collaborative studies between anthropologists and researchers in other disciplines. It is hoped that results of well-designed studies in this direction will supplement the findings of genetic epidemiological studies conducted throughout the world using the sophisticated statistical and molecular genetic techniques.

COMMENTS FROM THE FLOOR

A.R. Banerjee *(University of Calcutta, Calcutta, India)* narrated his impression, obtained from his long experience in analyzing data relating to childbirths in a Calcutta hospital, of an increasing prevalence of Down's syndrome. He wondered whether this increase was related to certain environmental factors.

K.C. Malhotra *(Indian Statistical Institute, Calcutta, India)* stressed the need for conducting research to assess the impact of modern medical and other interventions on genetic diseases. He cited three examples. G6PD deficiency is widespread among both tribal and non-tribal populations of India. Until anti-malarial drugs became easily available, deficient individuals were apparently well-adapted to the environment they lived in. However, after the introduction of such drugs, G6PD deficient individuals who consume these anti-malarial drugs, suffer from severe haemolytic crises, often resulting in death. Such individuals, of course, have no *a priori* knowledge of their G6PD deficient status, nor is it easy to assess their status in the rural Indian context. Similarly, with the changing of the traditional pattern of resource utilization in rural India, Hb AS heterozygotes are facing a lot of clinical problems. The program of distributing breeds of cattle with high milk-yield among Indian populations with a high prevalence of lactose intolerance, is causing a tremendous burden to the meagre medical resources currently available in rural India.

C.Y. Valenzuela *(Universidad de Chile, Santiago, Chile)* stressed two points. First, he stated that natural selection should be studied as a time-dependent phenomenon, because the most important period of selection is intra-uterine life — only about 20% of concep- tuses are actually born, and the infant mortality rate decreases with increasing age of the newborn. Second, he stressed that in biology it is important to set up sharp, meaningful biological hypotheses and then test them statistically. For example, to test the reasons for the biological differences observed between Himalayan populations (Sherpas) and Andean populations (Aymaras) by Basu and Gupta, sharp biological hypotheses should be formu- lated and tested. Valenzuela stressed that statistical significance is not necessarily biological significance; statistical interaction is not necessarily biological interaction.

J. Ott *(Columbia University, New York, U.S.A.)* pointed out that Elston's claim that affected grandparent-grandchild pairs are the most informative for detection of linkage is not necessarily true. He cited studies by Neil Risch and Timothy Bishop who had earlier shown that the informativeness of affected relative pairs varied with the nature of the disease under study — under certain circumstances affected sib pairs may be the most informative, while under a different set of circumstances affected first-cousin pairs may be the most informative. He also differed with Elston's contention that the mode of inheritance of a disease can be inferred from linkage analysis. He pointed out that this is feasible only if there is a prior established linkage of a marker with the disease under study. In the absence of such established linkage, inferring mode of inheritance from linkage analysis may not be a valid approach. One may change disease definitions and obtain a significant lod-score with a particular disease definition and a particular model of disease inheritance, but by this process one is neither sure whether there is linkage nor whether the assumed inheritance model is the true mode of inheritance of the disease.

R.C. Elston made the following clarificatory remarks. He stated that his claim is not that affected grandparent-grandchild pairs are the most informative, but that if such pairs are of the same cost to get as other types of affected relative pairs, then they will provide more information per dollar. He added that the question asked by him 'How much will it cost to find linkage by global search ?' is a slightly different question form that Risch and Bishop asked. He then stressed that if one defines a disease phenotype by linkage analysis, then it must be replicated or cross-validated.

A. Vetta *(Oxford Polytechnic, Oxford, U.K.)* stated that the only way to determine genotype-environment interaction is by conducting designed experiments, which is not pos- sible in humans. Inferences regarding such interactions obtained from observational studies should only be considered as suggestive.

The Panel Discussion ended with a vote of thanks to the Chair.

REFERENCES

Baker, P.T., 1989, Introduction, *in:* "Human Biology of Asian Highland Populations in the Global Context," Basu, A., and Gupta, R., eds., Indian Anthropological Society, Calcutta. p. ix.

Basu, A., 1969, The Pahira: a population genetical study, *Am. J. Phys. Anthrop.* 31: 399.

Dronamraju, K.R., and Meera Khan, P., 1963, The frequency and effects of consanguineous marriages in Andhra Pradesh, *J. Genet.* 58: 387.

Gupta, R., 1991, Culture and demography among the high altitude Sherpas, *in*: "Population Structure Among Tribes," Reddy, K.N., and Raghava Rao, D.V., eds., Tamil University, Thanjavur. p. 337.

Gupta, R., Basu, A., Pawson, I.G., Bharati, P., Mukhopadhyay, B., Mukhopadhyay, S., Roy, S.K., Majumder, P.P., Bhattacharya, S.K., Bhattacharya, K.K.,and Das, S.K., 1989, Altitude and human biology: A comparative study of Himalayan, Andean and Ethiopean data, *in*: "Human Biology of Asian Highland Populations in the Global Context," Basu, A., and Gupta, R., eds., Indian Anthropological Society, Calcutta. p. 1.

Haldane, J.B.S., 1949, Disease and evolution, *La Ricerca Scientifica* 19 (Suppl.): 68.

Haldane, J.B.S., 1955, Population genetics, *New Biol.* 18: 34.

Kumar, S., Pai, R.A., and Swaminathan, M.S., 1967, Consanguineous marriages and the genetic load due to lethal genes in Kerala, *Ann. Hum. Genet.* 31: 141.

Levi-Strauss, C., 1962, Social structure, *in*: "Anthropology Today: Selections," Tax, S., ed., Univ. of Chicago Press, Chicago. p. 321.

Little, M.A., 1981, Human populations of the Andes: The human science basis for research planning. *Mountain Res. Dev.* 1 (Spl. Issue): 145.

Malinowski, B., 1922, "Argonauts of the Western Pacific," George Routledge and Sons Ltd., London.

Morpurgo, G., Arese, P., Bosia, A., Pescarmona, G.P., Luzanna, M., Modiano, G., and Krishna Ranjit, S., 1976, Sherpas living permanently at high altitude: A new pattern of adaptation, *Proc. Natl. Acad. Sci. U.S.A.* 73: 747.

Mukherjee, A., 1990, Inbreeding effects on bilateral asymmetry of dermatoglyphic patterns, *Am. J. Phys. Anthrop.* 81: 77.

Mukherjee, D.P., 1972, I.C.M.R. Project Report, *cited by*: Mukherjee et al., 1977.

Mukherjee, D.P., 1992, Consanguineous marriages and their genetical consequences in some Indian populations, *in*: "Isolation, Migration and Health", Roberts, D.F., Fujiki, N., and Torizuka, K., eds., Cambridge Univ. Press, Cambridge. p. 63.

Mukherjee, D.P., Bhaskar, S., and Lakshmanudu, M., 1977, Studies on inbreeding and its effects in some endogamous populations of Chittor district, Andhra Pradesh, *Bull. Anthrop. Survey India* 26: 10.

Murthy, J.S., and Jamil, T., 1972, Inbreeding load in the newborn of Hyderabad, *Acta Genet. Med. Gemellol.* 21: 327.

Pai, R.A., 1972, Influence of environmental factors on consanguinity, *in*: "Genetics and Our Health," I.C.M.R. Tech. Rep., Ser. No. 20, p. 298.

Radcliffe-Brown, A.R., 1926, Father, mother and child, *Man* 26: 159.

Rao, A.P., and Mukherjee, D.P., 1975, Consanguinity and inbreeding effects on fertility, mortality and morbidity in a small population of Tirupati, *Proc. 2nd Annual Conf. Ind. Soc. Hum. Genet.*, Calcutta. *Mimeo.*

Sanghvi, L.D., 1966, Inbreeding in India, *Eugen. Quarterly* 13: 291.

Sanghvi, L.D., and Master, H.A., 1974, Effects of inbreeding on congenital malformations, (Abstract) *1st Annual Conf. Ind. Soc. Hum. Genet.*, Bombay, p. 17.

SPECIAL LECTURE

HOW WOULD HALDANE HAVE VIEWED THE SOCIETAL IMPLICATIONS OF TODAY'S GENETIC KNOWLEDGE?

James V. Neel

Department of Human Genetics
University of Michigan Medical School
Ann Arbor, MI 48109-0618 USA

All of us participating in this Symposium must be aware that J.B.S. Haldane was a man of strong political convictions, and indeed, it was those convictions that brought him here to India and ultimately set the stage for this meeting. Some of us, and I include myself, have, when he found human fallibility too much to bear quietly, on occasion followed him to Hyde Park. But his political interests were almost as broad as his scientific, and it would be most presumptuous of me to pretend on an occasion like this to guess in what intellectual direction Haldane would have loosed his barbs and wit. On the other hand, I suspect he would have felt that this program would not be complete without some consideration of the implications for society of the activities of contemporary human geneticists. If I fall short of his insights and eloquence, I trust you will bear with me.

In what follows, I propose to take a brief look at the sociopolitical interface of the subject of this Symposium, human population genetics. The issues to be discussed are common to the entire human gene pool; national boundaries are irrelevant to the following discussion. My thesis is very simple; I will argue that the principal issue geneticists have presented to the public since WWII, the genetic threat of increasing radiation and chemical pollutants in the environment, has assumed a magnitude inconsistent with the risks. In recent years, a similar amount of public attention has been focused on the promise of gene therapy, again attention probably far out of proportion to the societal significance of this development. Meanwhile, geneticists have been singularly uninvolved in what I now see as the development with the greatest potential for impact on the gene pool, the clash between the needs of a human gene pool expanding at a frightening rate and a rapidly shrinking resource base to support this gene pool.

1. THE GENETIC RISKS OF EXPOSURE TO IONIZING RADIATION AND CHEMICAL MUTAGENS

Driven initially by the eloquently expressed concerns of H.J. Muller, but then assuming a life of its own, the subject of the mutagenic impact on humans of a variety of the man-made exposures that come with civilization has been a major concern for almost five decades. Based largely on the selective use of the large body of data on the genetic effects of irradiating mice which accumulated during the 70s and 80s, a number of national and international committees estimated that the doubling dose of acute ionizing radiation was about 30-50 $\underline{r}$, and for chronic or low-level intermittent radiation, 50-250 $\underline{r}$ (UNSCEAR, 1972, 1977, 1986; Committee on Biological Effects of Ionizing Radiation, 1980; see also Lüning and Searle, 1971). Several years ago, from an analysis of data

Human Population Genetics, Edited by P.P. Majumder
Plenum Press, New York, 1993

accumulated over the past 40 years concerning children born to atomic bomb survivors in Hiroshima and Nagasaki, we suggested, from studies of some nine different end-points, that the genetic doubling dose for that experience was in the neighborhood of 200 rem, and, using the dose rate of factor of 2 which seemed appropriate to the distribution of gonadal doses that characterized this situation, that a doubling dose of 400 rem was a reasonable estimate for the effect of chronic ionizing radiation (Neel et al., 1990; Neel and Schull, 1991). The error to be attached to that estimate is admittedly large and somewhat indeterminate.

Stimulated by this outcome, Dr. Susan Lewis and I undertook a reanalysis of the murine data (Neel and Lewis, 1990). For a variety of reasons, a direct comparison of much of the mouse data with the human data was deemed inappropriate. The most nearly comparable data to the human data emanates from the mouse specific locus-specific phenotype test systems; when all the available studies of this type were considered, we derived a doubling dose estimate of 135 r for acute radiation and, using a dose rate factor of 3 because of the magnitude of the radiation doses employed in the mouse experiments, we suggested a murine doubling dose for chronic ionizing radiation of 405 r, in surprising agreement with the estimate for humans. I say surprising because there is no *a priori* expectation that two such different species as mice and humans should have the same genetic doubling doses. This estimate also carries a large but indeterminate error. So, however, did the earlier murine-based estimates.

A major new study of mutation at the DNA level in the children of a-bomb survivors and controls is now in progress (cf. Neel et al., in press). It is anticipated that the results of this study — which of course cannot supersede the earlier work — should substantially reduce the error term associated with the estimate of the doubling dose.

I should say that I am not so confident of the magnitude of the genetic threat posed by exposures to various chemical mutagens as I am concerning exposures to ionizing radiation. There is a need for a very major, very carefully designed study on the potentially most chemically mutagenized population that can be identified.

2. GENE THERAPY

More recently, especially in the U.S., the subject of gene therapy for genetic diseases has come to command a great deal of public attention. Given the spectacular developments in molecular genetics of recent years, it would be premature to set limits on what may be expected of this development. Thus far, however, the genetic diseases which seem most susceptible to this therapeutic approach are very rare. Viewed from the reference point of the issues I shall be discussing shortly, it is difficult to envision gene therapy as having more than a minor effect on the totality of gene pool problems. (The term gene therapy as used here does not include such developments as the use of genetically engineered leukocytes in the therapy of malignant disease.)

There is one aspect of research on human somatic cell gene therapy which must have high priority, namely, meticulous testing for unanticipated germ-line intrusions of the vectored DNA (cf. Neel, in press). The retroviral vectors employed in such therapy have been so enfeebled that there would appear to be no risk that they will carry their therapeutic passenger gene beyond the target tissue. But what of adventitial transmission into the germ line due to passing "helper" retroviruses or contamination of the vector strain by non-enfeebled virus? The human haploid genome contains of the order of one million sequences with greater or lesser homologies to retroviruses. Some of these must represent the footprints of viruses not specifically targeted to gonadal tissues, viruses which somehow strayed into the germ line. We have absolutely no idea of the price the species paid for these footprints. What we now see are the successful survivors — not those insertions resulting in genetic defect which were eliminated through selection. Studies on transgenic mice have shown that random insertion of vectored genes has at least a 10% probability of creating a "new" genetic effect (Palmiter and Brewster, 1986).

While the prospects of successful germ line intrusion in the course of retroviral-mediated somatic cell therapy seems slim (rev. in Cornetta et al., 1991), data to this effect are urgently needed. The use of DNA-liposome complexes for gene transfer would seem to lessen the likelihood of germ line intrusions, and, indeed, the first relevant studies reveal no evidence that DNA liposomal transduction was accompanied by uptake of plasmid DNA by gonadal tissue (Nabel et al., in press). I suggest that efforts to detect gonadal intrusion by foreign DNA should become a standard requirement for gene-therapy protocols; the gene therapy techniques with least evidence for germ line contamination should be the preferred techniques. It would be a thought-provoking therapeutic victory to enable a person who in consequence of genetic disease would die early, to live to reproduce, but with a significantly increased risk of transmitting mutational disease to any offspring.

3. THE MAJOR THREATS TO THE HUMAN GENE POOL

I suggest that there are impending developments with potentially vastly more important consequences for the gene pool of our species than current threats from ionizing radiation or the promise of gene therapy. Some of these developments are upon us and may be viewed in terms of consequences within the next several generations. Others of these developments have much more remote consequences, in terms of generations. In the remainder of the discussion, I shall occasionally make some rather didactic statements. The necessary documentation for these statements will be found in a book now in press (Neel, in press).

With respect to near-term developments with possible profound consequences for the gene pool, there are three that stand out:

3.1 A Burgeoning Population

Within my life time, spanning three generations, the world's population has gone from 1.8 to 5.3 billion persons, and in one more generation, by the year 2020, barring unforeseen developments, the population should reach approximately 8 billion if the United Nations projections are accurate. This is a rate of increase without precedent for our species. It results, on the one hand, from a continuation in many populations of traditional birth rates in the face of the improved infant and childhood survival resulting from modern sanitation, immunization, and antibiotics, plus, on the other hand, a blind faith, often reinforced by religious beliefs, that somehow technology will find a way to sustain these exploding numbers.

3.2 A Diminished Resource Basis to Support This Population

At the same time that the world's population is exploding, both the renewable and non-renewable resources to sustain this population are rapidly eroding. Given the abundant recent coverage of this matter, I need only mention a few salient facts. Globally, soil losses each year from erosion and other processes are now thought to be 24-26 billion tons in excess of new soil formation. By some accounts, more than half the top soil of the U.S. at the time of the discovery of the Americas has already been lost. On a worldwide basis, it appears that by the year 2000, at which time the world's population will be some 25% greater than at present, loss of top soil plus urbanization, desertification, and saline build-up will have claimed 25% of the arable land available 50 years earlier. In an effort to offset these losses and provide for population increases, an area of humid tropical forest equal to the size of Great Britain is every year being converted to other uses. Unfortunately, tropical soils are especially fragile and vulnerable. Much modern agriculture is highly dependent on irrigation, and all of these soil losses are further complicated by a growing shortage of the water available for such purposes.

It would be naive in the utmost to couch the resource-base problem in the generalities of an expanding population and diminishing resources. One has also to consider the pattern of the resource use which is responsible for the diminishing resources. The Ehrlichs have pointed out that a baby born in the United States currently represents 35 times the destructive potential on the Earth's ecosystems and resources as one born in India (Ehrlich and Ehrlich, 1990). Thus, although the population of India exceeds that of the U.S. by a factor of 3.6, the impact of the U.S. on the consumption of renewable and non-renewable resources is 10 times greater than that of India. India is as entitled to chide the U.S. concerning its profligate consumption of resources as the U.S. is entitled to preach population control to India or any other country. What makes the conspicuous consumption of energy in the U.S. especially grating is the extent to which it is achieved by deficit financing.

3.3 An Increased Susceptibility to Epidemic Disease

The projected population growth of the next generation will be disproportionately urban. The slums of Mexico City, Rio de Janeiro, New York, and — yes — of Calcutta can only become more crowded. At the same time, the nutritional status of slum dwellers, already bad, can only deteriorate. This is the stuff on which epidemics feed, and the historical impact of epidemics such as plague and influenza on selected segments of the population is only too well documented.

Given the available control measures for the known agents of epidemic disease, such as cholera and influenza, the greatest danger lies in the emergence of apparently new diseases, such as the acquired immunodeficiency syndrome (AIDS). As man presses the ecosystem more severely, other diseases, like AIDS, which apparently result from transfer of an infectious agent from an animal reservoir to humans, will emerge. Recent additional examples of such diseases, fortunately with less devastating consequences than AIDS, appear to be Lyme disease (Borreliaosis), due to a spirochaete transmitted by certain ticks whose life cycle is usually played out in deer or mice, or Venezuelan hemorrhagic fever, resulting from a previously unknown arenavirus, a member of a family of viruses chiefly encountered in rodents (Lederberg et al., 1992).

A companion danger is the resurgence of diseases thought to be under reasonable control, such as tuberculosis or malaria, because the agents of these diseases have become resistant to the chemotherapeutic agents which have held them in check for the past 40 years. Now, however, the greater population densities which cities have achieved will greatly facilitate the spread of these diseases.

Epidemic disease is of course a cruel but effective means of reducing population numbers. In the context of a concern for the total gene pool of humanity, it must be pointed out that epidemics, despite the large areas which they involve, still do not affect the world's populations uniformly. The decimation tends to be preferential with respect to ethnic group.

There are, in addition, longer range potential genetic threats to the species.

3.4 Departure from the Population Structure which Obtained during Human Evolution

A developments which has attracted genetic comment for the past century is the realization of the dramatic change in the genetic structure of human populations which was ushered in with the Agricultural Revolution and intensified with the Industrial Revolution. The isolates which were the competing units of human evolution are almost gone and, despite ethnic resurgence here and there, the human population is more and more one large interbreeding unit. Although the thesis is difficult to document rigorously, most of us share the intuition that the genetic component in differential survival and reproduction has been greatly diluted since tribal times. Furthermore, it is also clear that present human diets, especially in the affluent countries, are in many respects quite unlike those of our forebearers, with dietary excesses a problem in some countries, and dietary deficiencies an equal problem in others. We have in recent years, through studies of the Yanomama and other relatively unacculturated Amerindian tribes (rev. in Neel, in press), become greatly impressed by the extent to which the bases for survival and reproduction have altered since the tribal stage of human existence. But whereas the eugenicists of the first 40 years of this century could view genetic deterioration as proceeding very quickly, the principles of population genetics make it clear that the results of the relaxation of natural selection implied by those changes are slow to materialize. The immediate threat to humanity's genetic assets from this development is much less than from the preceding three developments.

3.5 The Emergence of a Gerontocracy

The final point I wish to make in this listing of developments which must be considered as we contemplate the world through genetic-tinged lenses is the emergence of a gerontocracy. Some representative facts: In the U.S., the percent of population over age 65, the nominal age of retirement, has reach 12.4%, and is predicted to reach 16.1% by the year 2020 and 19.2% by 2050. There are 19 nations in which the proportion of the population over age 65 is greater than the U.S. figure. By contrast, in a tribal population which we have studied in some detail, that proportion was estimated to be 0.3%. The significance of this, in relation to the genetic program we will be discussing shortly, is that the needs of the old are increasingly in conflict with the needs of the young, upon the realization of whose genetic potential so much of the progress in solving the problems before us depends.

4. THE GENETIC RESPONSE TO THESE DEVELOPMENTS

These genetically relevant developments are occurring against a background of profound challenge to our social and industrial structure. The Industrial Revolution was fueled by the availability of vast natural resources. The finiteness of these resources has now become very apparent. I suggest that the next several generations will witness a period of social readjustment as profound and difficult as any in the history of our species. The nature of the readjustment will vary form country to country, but no country will be unaffected.

In the remainder of this presentation, I would like to consider, with you, what contribution the genetic viewpoint can make to society's navigation through this difficult period. In this context, genetics should be viewed as just one of many disciplines whose viewpoint must be considered in the transition. I suspect that if geneticists fail to speak out, the world will not come to them for advice, and suggest that geneticists, who of all professional groups should be ombudsmen for future generations, promote discussion and action along the following five lines.

4.1 An Equitable Policy for Population Control

As a population geneticist, my first priority for the years immediately ahead is an egalitarian and equitable program of population control. I submit that not only do geneticists lack the knowledge to recommend a comprehensive eugenic program in the classical sense but such a program would be socially unacceptable. Rather, each couple the world over should be encouraged by all means available to limit itself to two children. Were this the stated objective, then, because of failures to marry and infertility, the realized number of children per couple might be 1.8. Because of the age composition of the world's populations, even if this program were initiated by the year 1995, population numbers will increase for at least four decades, peaking in the neighborhood of 6.7 billion in 2030.

This is not a eugenic program. It simply seeks for the foreseeable future to stabilize the human gene pool in all its wondrous diversity as it now is. Some will charge that this program legitimizes past social injustices, by virtue of which some ethnic groups have suffered whereas others have profited. Recognizing this, I can only suggest that even greater social injustice may overtake various ethnic groups without this effort at stabilization.

In a society where each couple is under strong social pressure to limit itself to only two children, the desire that these children be the best the couple can produce will be very strong. One can anticipate increased emphasis on the early prenatal diagnosis of such genetic diseases as thalassemia, sickle cell anemia, Huntington's chorea, and cystic fibrosis of the pancreas. I do not anticipate a society which mandates abortion for prenatally-diagnosed, severe genetic disease, but I do anticipate a society where the personal convictions of one group (in this case, an anti-abortion stance) cannot be imposed upon another group.

Some will argue that in seeking to stabilize the present human gene pool, this program thwarts continuing human evolution. It does. But the population structure that resulted in human evolution is already badly compromised. I do not suggest this reproductive schedule as necessarily continuing forever. Some day humankind may devise a better alternative with eugenic overtones, and it will be appropriate to reconsider this program. For now, however, the number one priority has to be simply to bring all human populations into better balance with their resources.

4.2 Control of Epidemic Disease

In any program which places emphasis on stabilizing the human gene pool, the control of epidemic disease looms large. Were the impact of epidemic disease (and famine) uniformly distributed over the world's populations, then epidemics would be a cruel but effective way of reducing the earth's populations to more sustainable numbers. Unfortunately, it is estimated that about 20% of the world's populations — say one billion persons -- are malnourished, with the malnourished more commonly concentrating in the large cities of the Tropical Belt. These are the persons most susceptible to the impact of an epidemic.

One does not have to be an epidemiologist to recognize that epidemics are best controlled early on, as soon as the causative agent has been identified and counter measures blocked out. The current epidemic of AIDS is the most recent appalling example of the result of relative neglect of a potential epidemic in its early stages. It is a reasonable surmise that had there been a more vigorous attack on this disease when it was largely confined to homosexuals, it would not now be the devastating disease of heterosexuals it has become throughout the world, but especially in parts of Africa and the Far East.

An aspect of the program of evenhandedness toward the preservation of the gene pool which I am espousing requires better 'early warning stations' for the occurrence of exotic diseases (see Lederberg et al. (eds.), 1992). Since many of these warning stations need to be located in poorer tropical areas, where the emergence of zoonoses is most likely, responsibility for coordinating these stations needs to rest with an organization like the World Health Organization, to whose budget the more advanced nations must contribute heavily, not only for humanitarian reasons but in their own self interest. After all, with modern transportation, epidemic disease can spread very rapidly.

4.3 Euphenics

A term not often encountered in the current genetic literature is euphenics, defined as the science of optimizing the phenotypic expression of the genotype. The term implies a conscious effort to structure all aspects of a group's culture so as to enhance the expression of its genetic attributes. This effort at enhancement in principle ranges from formulating an optimal diet to developing the intellectual milieu most appropriate to the long range survival and adaptation of humans.

The suggestion that euphenics emerge as a major genetic focus for the future sounds naively idealistic, but in fact on a small scale the practice of euphenics is already in place. In North America and Western Europe, the standard of living has encouraged a diet which appears to interact with genetic susceptibilities to promote hypertension, diabetes mellitus, and cancer, and government-sponsored efforts are under way designed, for instance, to reduce caloric intake, especially of fat, and increase the roughage of the diet. At the other extreme, there are parts of the world where the challenge of euphenics is to obtain the minimum calories for survival, but here, too, the kind of calories is an important issue. The greatest challenge is with respect to the enhanced functioning of the mind. We accept the worldwide need for the improve education that will help cope with the demands of modern technology, but the specifics of this generalization still elude us.

Euphenics may precede on a population level, with recommendations appropriate to all, as discussed in the preceding paragraph, or an on individualized level. Each of us is a bundle of inherited predispositions which vary widely from one person to the next. These range from inherited food intolerances to cancer susceptibility genes. Many of these genetic predispositions are still poorly understood, and represent a fertile field for the type of research best subsumed under the term, genetic epidemiology. Although measures directed at populations are the most obvious immediate aspect of euphenics, measures directed at individuals in the long run may be equally important.

The implementation of euphenic measures will require a degree of self discipline which our species is not now manifesting. The extensive measures which society has developed to ensure quantity of life must now be supplemented by measures to ensure quality of life. The pursuit of euphenics may require governmental reinforcement which some will equate with coercion and loss of individual freedom. I submit that the government which pays the health bills has the right to encourage healthy eating, and the government which enforces the law should be in a position to discourage the kind of TV programs which glorify anti-social violence and crowd out programs with educational content, to the detriment of the euphenics of the mind.

4.4 Improved Therapy for Genetic Diseases

That an important social aspect of genetics is a continuing effort to improve the therapy of genetic disease is a given. Much of such tomes as "The metabolic basis of inherited disease" (Scriver et al., 1989) is concerned with therapy for genetic disease, which because of its chronicity assumes an importance in Western medical practice out of proportion to its numbers. The continuing identification and detailed molecular analysis of alleles associated with disease will yield important therapeutic clues. For instance, the identification of the gene responsible for cystic fibrosis has brought with it the understanding that the protein encoded by the normal allele is an "ion-channel" protein primarily concerned with the transport of chloride ions. The defect in cystic fibrosis is a block in chloride efflux, accompanied by an increased uptake in sodium ions. This insight suggests clinical approaches as diverse as blocking sodium uptake with amiloride or the delivery of aerosolized adenine or uridine triphosphate to respiratory epithelial cells to encourage chloride efflux (rev. in Collins, 1992). As the molecular basis of additional hereditary diseases is elucidated, similar therapeutic insights will emerge.

This is the place to return to the subject of somatic cell gene therapy. Viewed against the totality of the impingement of genetically-oriented measures on human affairs, I see gene therapy *sensu strictu* as a very minor player that has received publicity all out of proportion to its impact for the foreseeable future. What humans can, humans do, and this subject will be actively investigated in the future, perhaps with a much more major return that I can visualize. It has the beguiling quality that, once its safety is assured, its practice does not involve the touchy issue of abortion. It will be, however, a very high tech field for the foreseeable future, one not easily transplanted from a relatively few research centers.

Earlier I mentioned some of the implications of the aging populations of many nations. That the aging process is subject to a still poorly understood genetic control is clear from the differences in life spans between related species, and evidence for the existence of key genes in this control is beginning to emerge (e.g., Ning et al., 1991). The ability to modulate the expression of these genes so as to extend average life expectancy by a year would have a far greater societal impact than that of all the currently contemplated somatic cell gene therapy. It is strange, how little public discussion this possible development has evoked, as contrasted with the debates over the issues surrounding gene therapy for inherited disease. One has to wonder about the asynchrony of funding research which may lead to the manipulation of "aging switches," when most nations with aging populations are encountering such difficulties in caring for their present, let alone projected, senior citizens.

4.5 Protection Against Mutagens

Mutagenic exposures, whether from ionizing radiation or from the variety of naturally occurring chemicals in food (cf. Ames, 1989), are a fact of life. The challenge is to hold any increased exposures to a level commensurate with the societal gains accompanying these exposures. The extensive genetic follow-up studies on the children of the survivors of the Hiroshima and Nagasaki bombings have been paralleled by equally extensive studies on the parents of these children. These latter studies have revealed an increase in leukemia and a variety of solid tissue tumors, and these findings, together with earlier data on the development of cancer in individuals receiving ionizing radiation as therapy for ankylosing spondylitis, are setting the standards for permissible occupational exposures to radiation.

Although the calculation can only be approximate, it appears that with respect to radiation, the doubling dose for malignancies in a-bomb survivors (Committee, 1990) can be shown to be somewhat less than we have calculated for transmitted germ line genetic effects in the same population. Societal measures to protect against the carcinogenic effects of radiation should protect adequately against the genetic. Conversely, if a properly monitored cohort exposed occupationally or otherwise shows no increase in cancer, it is unlikely to have sustained significant transmitted genetic damage.

The corresponding bridges between carcinogenesis and mutagenesis in human cohorts exposed to chemical agents for the most part remain to be built, and this would seem a worthy line of investigation for the future.

SOME GENERALITIES

The genetic guidelines for the future which we have been discussing will be rather derisively branded as impractical by many. My response: look where practical leaders have taken us. Now should be a time of profound reassessment. In this process, one thing is very certain: humans will need to exhibit much more self discipline than in the past. This is true not only in the realm of population control but also in euphenics. For the euphenics of the body, discipline in diet is needed. For the euphenics of the mind, discipline in our intellectual fare is needed. I doubt that the content of American television, 90% pandering to the lowest common denominator in human entertainment, and commanding an average of four hours of daily viewing time on the part of American youth, is quite the correct preparation for the future.

Some will find that what I am espousing smacks of coercion, an affront to 'free enterprise' and 'human individuality'. I remind them that in times of perceived crisis in the past — which, unfortunately, for the most part translates into times of military action — humans have accepted some mobilization and loss of individual freedoms. The transition we face will require no less a concerted action than is associated with military stress.

And what of the precise role of human geneticists in all of this? There is no professional group better qualified, or more oriented, to be ombudsman for the future, but yet we haven't been doing much of that recently. In the matter of population control, obviously we are not the people to pass out condoms or be responsible for the implantation or injection of subcutaneous contraceptive chemicals. But we are the people expected to voice concern over the impact on the future human gene pool of unbridled human population expansion. I can't find that any genetics organization has in recent years taken an official public position on this matter. Of the role of geneticists in elucidating the genotype-phenotype interactions upon which euphenics is founded, there is no doubt, nor is there any doubt of what we can expect from genetic research aimed at improving prenatal diagnosis and the understanding of the basis for inherited disease, which is the foundation for improved therapies. This brings us, finally, to gene therapy. Although this is now seen in some quarters as the crowning achievement of human genetics, the perspective I have attempted to develop cannot assign to this development a very high priority among the genetic issues of the immediate future.

BACK TO HALDANE

Let us now attempt to consider how Haldane might have viewed some of the foregoing suggestions. In a major statement in 1963 entitled, "The implication of genetics for human society" (Haldane, 1963), he wrote:

"...it is no doubt desirable to prevent, if we can, the production of the most socially inefficient five per cent or so of most populations. With our present knowledge we could not do much more than halve it. There are two quite distinct methods of negatives eugenics available. In the first place persons with well-marked dominant and sex-linked recessive defects should not have children. Nor should the daughters of haemophilics and others with similar sub-lethal sex-linked conditions.... Equally important, perhaps more so, is the screening of whole populations for rare and harmful autosomal recessive genes. Almost every one of us is heterozygous for a lethal or sub-lethal, and the average load may be as high as four. There is no hope of eliminating them for many generations to come. But it should be possible to discourage marriages between persons heterozygous at the same locus. If, say, 2 per cent of all babies are homozygous for unwanted recessives, (and this is an overestimate on our present knowledge) this would only mean that 8 per cent of intended marriages would be contra-indicated. At present we can only detect heterozygotes at a few loci. But a start could be made.... By the end of the century we may hope that in advanced countries everyone will undergo routine tests for heterozygosity at various loci on reaching the age of puberty, so that they may know whether marriage with a given person will be genetically dangerous." (pp. *xcv-xcvi*).

This statement on what he labelled negative eugenics of course was written before the advent of prenatal diagnosis; I suspect he would have accepted the latter coupled with optional abortion where indicated as a more practical approach to reducing the impact of genetic disease than the marriage counselling he advocated. It can also be presumed that he would have been fascinated by the improving therapeutic approach to genetic disease that result from biochemical and molecular studies.

It is clear Haldane espoused euphenics. For instance, one reads in the same publication referenced above (pp. *xcvii-xcviii*):

"I believe that any satisfactory political and economic system must be based on the recognition of human inequality.... We do not at present know how to recognize any but the crudest forms of innate human inequality. Our school examinations, and to a large extent our university examinations, are largely tests for precocity and memorization rather than for more important qualities. Some psychological tests may be slightly better. At present we find our adult positions in society by a crude compromise between the effects of personal choice, examinations and other tests, and economic inequality. Many of us are frustrated in consequence. It is common to speak of my country, India, as undeveloped. Here economic inequality is still the predominant factor. But for a geneticist, all countries are undeveloped, for they have not learned to use their human resources as some of them have learned to use the resources of their soil and minerals. This state of affairs is due to ignorance, which will only be overcome by an immense joint effort of psychologists and geneticists lasting through several centuries."

I have not in my perusal of Haldane's writings encountered opinions on reconciling population numbers with resources, perhaps because he believed at that time that inequities in resource allocation were more important to perceived shortages than actual resource limits. However, I would hazard the guess that given the strong emphasis I have placed on prenatal diagnosis and abortion, a form of negative eugenics, he would accept as an accompanying temporary measure an egalitarian program of human reproduction, designed to perpetuate the genetic status quo until such time — if ever — that humankind can devise a sound eugenic program.

— — —

As the last formal lecturer on this program, may I — since I have already been so presumptuous as to speak in Haldane's name — express the further opinion, that Haldane would have been delighted in his intellectual legacy to India, as witnessed by the scope and vigor of the presentations of his Indian students and associates at this memorial program. We from abroad feel privileged to have joined you of India in this commemorative event.

REFERENCES

Ames, B.N., 1989, Mutagenesis and carcinogenesis: Endogenous and exogenous factors. *Environ. Mol. Mutagen.* 14, Suppl. 16:66.

Collins, F.S., 1992, Cystic fibrosis: Molecular biology and therapeutic implications. *Science* 256:774.

Committee on the Biological Effects of Ionizing Radiation, 1990, Health effects of exposures to low levels of ionizing radiation (BEIR V), Washington, DC: National Academy Press. pp. *xiii* and 421.

Committee on the Biological Effects of Ionizing Radiation, National Research Council, 1980, The effects on populations of exposure to low levels of ionizing radiation: 1980. Washington: National Academy Press. pp. *xv* and 524.

Cornetta, K., Morgan, R.A., and Anderson, W.F., 1991, Safety issues related to retroviral-mediated gene transfer in humans. *Hum. Genet. Ther.* 2:5.

Ehrlich, P.R., and Ehrlich, A., 1990, "The Population Explosion," Simon & Schuster, New York.

Haldane, J.B.S., 1955, Genetical effects of radiation from products of nuclear explosions. *Nature* 176:115.

Haldane, J.B.S., 1965, The implications of genetics for human society, *in:* "Genetics Today (Proc. XI Intern. Congr. Genetics)" S.J. Geerts, ed., Pergamon Press, Oxford.

Lederberg, J., Shope, R.E., and Oaks, S.C. (eds.), 1992, "Emerging Infections: Microbial Threats to Health in the United States," National Academy Press, Washington, DC.

Lüning, K.G., and Searle, A.G., 1971, Estimates of the genetic risks from ionizing radiation, *Mut. Res.* 12:291.

Nabel, E.G., Gordon, D., Yang, Z.-Y., Ling, X., San, H., Plautz, G.E., Wu, B.-Y., Gao, X., Huang, L., and Nabel, G.J. Gene transfer *in vivo* with DNA-liposome complexes: Lack of autoimmunity and gonadal localization., *Hum. Gene Therapy* , in press.

Neel, J.V., Germ line gene therapy: Another view, *Hum. Gene Therapy*, in press.

Neel, J.V., "Physician to the Gene Pool," Wiley-Liss, New York, in press.

Neel, J. V., and Lewis, S. E., 1990, The comparative radiation genetics of humans and mice, *Ann. Rev. Genet.* 24: 327.

Neel, J.V., Satoh, C., and Myers, R., Report of a workshop on the application of molecular genetics to the study of mutation in the children of atomic bomb survivors, *Mut. Res.*, in press.

Neel, J.V., and Schull, W.J. (eds.), 1991, "The Children of Atomic Bomb Survivors," National Academy Press, Washington, DC.

Neel, J.V., Schull, W.J., Awa, A.A., Satoh, C., Kato, H., Otake, M., and Yoshimoto, Y., 1990, The children of parents exposed to atomic bombs: Estimates of the genetic doubling dose of radiation for humans, *Am. J. Hum. Genet.* 46:1053.

Ning, Y., Weber, J.L., Killary, A.M., Ledbetter, D.H., Smith J.R., and Pereira-Smith, O.M., 1991, Genetic analysis of indefinite division in human cells: Evidence for a cell senescence-related gene(s) on human chromosome 4, *Proc. Natl. Acad. Sci., USA* 88:5635.

Palmiter, R.D., and Brewster, R.L., 1986, Germ-line transformation of mice, *Annu. Rev. Genet.* 20:465.

Scriver, C.R., Beaudet, A.L., Sly, W.S., and Valle, D., 1984, "The Metabolic Basis of Inherited Disease," McGraw-Hill, New York.

United Nations, 1972, Report or the United National Scientific Committee on the Effects of Atomic Radiation. General Assembly Official Records: Twenty-seventh Session. Supplement No. 25 (A/8725).

United Nations Scientific Committee on the Effects of Atomic Radiation, 1977, "Sources and Effects of Ionizing Radiation," United Nations, New York.

United Nations Scientific Committee on the Effects of Atomic Radiation, 1986, "Genetic and Somatic Effects of Ionizing Radiation." United Radiation, New York.

INDEX

Acquired immunodeficiency syndrome
  (AIDS), 334
Adaptation, 182
Adaptive evolution, 182
Adaptive gene combination, 183
Adaptive landscape, 32
Adaptive surface, 43, 182
Adaptive threshold, 43
Adaptive zone, 43
Additive genetic covariance, 39
Additive genetic effect, 33, 42,
  239
Additive genetic variance, 18, 39,
  44
Additive polygenic effect, 173, 292
Additive polygenic genotype, 293
Admixture, 82, 202, 207, 255, 259,
  265, 267
  genetic, 255
  proportion, 134
Adopted relative, 298
Adoptee, 298
Adoptive family, 298
Affected relative pair, 321
Affective disorder, 176
Affinity
  genetic, 255, 265, 267, 269, 272,
    273
  socio-cultural, 265
African mother hypothesis, 79
Age of onset, 307
Albinism, 156
  tyrosinase-positive, 156
Allelic genealogy, 59, 60
Allelic independence, 158
Allelic lineage
  persistence time of, 54
Aluminium, 305, 311
Alzheimer's disease, 281, 305, 309,
  311
Amyotrophic lateral sclerosis, 305,
  306, 309-311
Ancestral heredity
  law of, 280
Anencephaly, 325
Ankylosing spondylitis, 280, 337

Antigen diversity, 51
Antigen recognition site, 50
Apolipoprotein, 286
Archaic lineage, 78, 83
Artificial selection, 43
Ascertained sample, 291
Ascertainment, 132, 280, 297, 298
  complete, 115
  complex, 297
  incomplete, 115
  multiple, 134, 297
  nonrandom, 295
  probability of, 115, 282
  scheme, 297
Association, 280
Assortative mating, 183, 186, 299
Ataxia telangiectasia, 161
Australoid, 257, 258, 270
Austro-Asiatic, 212
Autoimmune disease, 52
Automated ordering, 162
Autozygosity, 325
Average waiting time, 107

Balanced polymorphism, 13
Balanced translocation, 282
Baseline risk, 130
Beta amyloid, 307, 308
BETA model, 293-296, 301, 302
Biased estimator, 197
Bilineal pedigree, 174
Biochemical diversity, 50
Biodiversity
  conservation of, 191
Blastomere, 89
Blastula, 90
Body dimension, 317, 318
Body odor, 97
Bootstrap, 148
Breast cancer, 175
Breeding isolation, 218, 324
Breeding size, 83, 190, 200
Breeding unit, 190
Breeding value, 35, 42
  moments of, 39
Bronchial asthma, 325

Genotypic selection, 36
Germ line intrusion, 332
Germinal mosaicism, 159
Gerontocracy, 334
Gibbs-sampler algorithm, 145, 146, 147
Gibbs sampling, 172, 173
Graded susceptibility, 280
Graft rejection, 182
Group selection, 23

Half-sib analysis, 323
Haplophase, 282
Haplotype, 280
  frequency, 236, 237, 239, 242, 250, 252
Hearing impairment, 277, 286
Hemoglobin level, 316, 318
Hemophilia, 16, 159, 166
  A, 159
  B, 159
Heritability, 44, 45, 121, 129, 134, 134, 154, 302
  cultural, 294, 302
  genetic, 294, 302
  overestimation of, 300
Heterogeneity
  allelic/genetic, 139
  genetic, 279
  phenotypic, 139
Heterostyly, 4
Heterozygosity, 60, 76, 186, 229, 259, 270, 339
Heterozygote advantage, 57, 325
High altitude
  effects, 316
  stress, 318, 319
Histocompatibility antigen, 49, 89
Histocompatibility genes, 95, 97
Homologous loci, 278
Homoscedasticity, 129
Homozygosity, 187, 324-326
Hookworm infestation
  frequency, 318
Household effect, 311
Human diversity, 75
Human genome
  mapping, 162
Human leucocyte antigen (HLA), 67-69, 95, 182, 280
Human phylogeny, 76, 77
Huntington's chorea, 158, 159, 322, 325
Hypergamy, 324, 325
Hyperlipidemia, 286
Hyperparathyroidism, 311
Hypertension, 326, 327
  genetic susceptibility, 336
Hypertrichosis of pinnae, 3
Hypervariable DNA loci, 286
Hypervariable DNA polymorphism, 202

Hypotension, 326
Hypoxia, 316, 318

Identifiable
  statistically, 141
Identity by descent, 184, 278, 279
Importance sampling, 169
Inbred load, 283, 284
Inbreeding, 89, 95, 96, 158, 183, 184, 219, 221, 224-226, 283, 324, 327
  coefficient, 186, 224, 284, 308, 325
  effect, 283, 325, 326
Incidence, 307
Inclusive fitness, 25
Index marker, 281
Indo-European, 212
Industrial melanism, 183
Infinitely many alleles model, 25
Influenza, 334
Inherited predisposition, 336
Interaction, 3
  gene-environment, 299, 311, 319-321, 326
  gene-gene, 299, 300
  genotype-by-environment, 300, 328
  genotype-by-genotype, 300
  metal-enzyme, 308
  metal-gene, 308
Interference, 154, 158, 167, 278
  positive, 167, 170
Intermarriage, 306, 310
Intraclass correlation, 195
Invariance, 192
In situ hybridization, 282
Ionizing radiation, 331-333, 337
Isolate, 186, 323
  genetic differentiation of, 186
Isolated case, 140
Isolation
  cultural, 235
  genetic, 189
  geographical, 76, 235, 239, 310
  marital, 324
  social, 235
Isolation by distance, 200

Kinship, 183, 323, 327
  bioassay, 200
  local, 186
  system, 218

Lactose intolerance, 327
Language family, 236
  Indo-European, 236
  Uralic-Altaic, 237
Lethal equivalent, 226
Leukemia, 337
Liability, 280
Liability class, 122

Skeletal evidence, 209
Skewness, 123
Social organization, 213
Social structure, 323, 325, 326
Sodium ion, 337
Solid tissue tumor, 337
Somatic cell therapy
  retroviral-mediated, 332
Sperm cell receptor, 98
Sperm chemotaxis, 98
Spinal muscular atrophy, 161
Spirality
  floral, 4
  foliar, 4
Sporadic case, 142, 282
Sporadic proportion, 141
Spouse correlation, 124, 127, 299
Stabilizing selection, 39, 41, 42,
  44, 45
Stationary state, 16
Strong selection, 44
Structural equation, 292
Structural relationship, 291
Substitutional load, 9-11
Susceptible genotype, 140
Synonymous substitution, 50
  rate, 55

T cell, 50-52, 56-58, 62
T lymphocyte, 50
TAU model, 292, 294-296, 298, 301,
  302
  pseudopolygenic, 301
Temporal trend, 118
Territoriality, 93, 94
Tetrad, 157
Thalassemia, 325, 335
Thymocyte, 50
Tibeto-Burman, 212
Topology, 73
Transformation, 117, 123,124
Transgenic mice, 332
Transition probability, 118,120,
  123,130
Transmissible influence
  sibling, 292
  spousal, 292

Transmissibility
  polygenic model of, 292
Transmission probability, 119,120,
  123, 133
  model, 117-119, 123, 131, 134
Tribe, 215, 217, 221, 229, 255,
  256, 259
Triglyceride, 301
Truncation, 297
  direct, 297
  indirect, 297
  latent, 297
Truncation selection, 34,38, 60
Tuberculosis, 334
Tuberose sclerosis, 158
Tumor resistance factor, 49
Tumor suppresor, 286
Tumor transplantation, 49
Twins, 298, 299
  dizygotic, 298
  monozygotic, 298
Two-locus disease model, 174 ,175

Unified model, 123, 129-131
UPGMA method, 69, 70, 73

Variable age of ouset, 139-141
Variance component, 129, 195, 201,
  280,
  estimator, 197
Vectored genes
  random insertion of, 332
Venezuelan hemorrhagic fever, 334
Vitiligo, 141
Vomeronasal organ, 89, 91

Weak selection, 42, 44
Weighted estimator, 161
Weighting algorithm, 281

X-chromosome, 159, 282

YAC, 282
Y-linked gene, 3

Z-score,162